Palgrave Studies in the History of Science and Technology

Designed to bridge the gap between the history of science and the history of technology, this series publishes the best new work by promising and accomplished authors in both areas. In particular, it offers historical perspectives on issues of current and ongoing concern, provides international and global perspectives on scientific issues, and encourages productive communication between historians and practicing scientists.

More information about this series at
http://www.palgrave.com/gp/series/14581

Ariane Dröscher

Plants and Politics in Padua During the Age of Revolution, 1820–1848

palgrave
macmillan

Ariane Dröscher
Bologna, Italy

ISSN 2730-972X ISSN 2730-9738 (electronic)
Palgrave Studies in the History of Science and Technology
ISBN 978-3-030-85342-6 ISBN 978-3-030-85343-3 (eBook)
https://doi.org/10.1007/978-3-030-85343-3

Cover illustration: Carlo Matschegg, Veduta dell'Orto botanico di Padova con il platano (Padua, c. 1862), by permission of the Biblioteca dell'Orto Botanico dell'Università degli Studi di Padova, identification no. 249027.

This Palgrave Macmillan imprint is published by the registered company Springer Nature Switzerland AG.
The registered company address is: Gewerbestrasse 11, 6330 Cham, Switzerland

To my master Renato G. Mazzolini, constant source of inspiration

Preface

Do plants have an influence on historical events? My curiosity in the question at the heart of this book was aroused when I noticed the many parallels between botany and politics in the public life of Padua during the age of revolutions. In the early nineteenth century, both fields underwent a period of profound change that, in the long run, brought the former to revolutionize the field of the life sciences, and the latter, notwithstanding the failure of the revolutions of 1848, to transform the socio-political landscape of Europe. The parallelism of both these revolutions is a well-known Europe-wide phenomenon, yet hitherto no study has investigated the interplay between them in a specific locale.

I first came to know one of my protagonists, the Paduan botanist Giuseppe Meneghini, in the 1990s during research into the history of cell biology in Italy. I returned to him about twenty years later when I participated in the project *Politics of the Living: A Project on the Emergence and Reception of the Cell Theory in France and Germany, ca. 1800–1900*, which was directed by Florence Vienne and Marion Thomas. On this occasion, I realized that Giuseppe's elder brother Andrea was in his days prominent, yet today poorly known politician. My subsequent studies drew me deeper and deeper into the fascinating Paduan world of the pre-1848 period. Each step uncovered new facets of the intimate relationship between botany, politics, and public life, and forced me to embark on the risky journey of interdisciplinary research and to venture into areas distant from history of science. Yet, the dimensions of garden culture, political economy, and agricultural philosophy turned out to be indispensable for piecing together the overall picture and transformed my original plan of a short essay into a book project.

Bologna, Italy Ariane Dröscher

Acknowledgments

This research was kindly supported by a one-year post-doctoral fellowship granted by the Department of Cultures and Civilizations (University of Verona) for the academic year 2018–2019.

I am grateful for the bibliographical help from Giovanna Bergantino (Biblioteca Antica del Seminario Vescovile di Padova), Alessandro Bison (Villa Contarini—Fondazione G.E. Ghirardi), Nicola Boaretto (Archivio di Stato di Padova), Giuseppe Bonafé (Centro per la ricerca e la documentazione sulla storia locale—Battaglia Terme), Loredana Capone and Ilario Ruocco (Biblioteca Universitaria di Padova), Marco Favretto (Biblioteca Civica di Padova), Marina Francini and Fulvia Lora (Biblioteca Civica Bertoliana di Vicenza), Barbara Lapucci (Biblioteca di Scienze Naturali e Ambientali dell'Università di Pisa), Manola Ramon (Fondazione di Storia Onlus di Vicenza), Concetta Rociola (Biblioteca del Seminario Vescovile di Padova), Maria Sacilot (Sistema Bibliotecario Urbano di Padova), Mirco Travaglini (Biblioteca BES di Bologna), and the librarians of the Biblioteca dell'Accademia Olimpica di Vicenza.

I am much obliged to Matteo Ceriani for the permission to reproduce his canvas and to my brother Till for his graphic arts. I give my sincere thanks to Stefano Dal Santo (Biblioteca Capitolare di Padova), Sebastiano Miccoli (Centro di Ateneo per le Biblioteche), Francesco Leone, Bernardo Falconi, Paola Mario (Biblioteca dell'Orto botanico dell'Università di Padova), Lucia Baroni (Biblioteca universitaria di Pisa), Luciana Battagin (Biblioteca Nazionale Marciana), Benedetta Basevi (Genus Bononiae), Marina Gentilini (Biblioteca Statale di Cremona), and Riccardo Ghidotti for their assistance in finding and obtaining the images for this book.

I am extremely grateful for the numerous suggestions, comments, and advice I received from David Cahan, Renato Camurri, Moreno Clementi, Luca Ciancio, Pietro Corsi, Pietro Del Negro, Carmelo Donà, Enrico Francia, Riccardo Ghidotti, Matthew Herron, Christiane Liermann, Laurent Loison, Paolo Marangon, Martina Massaro, Giuliana Mazzi, Sabrina Minuzzi, Valeria Mogavero, Antonella Pietrogrande, Paolo Pombeni, Marc Ratcliff, Andrew Reynolds, Gianfranco Tusset, and Agenese Visconti. A special thanks goes to the participants of the project *Politics of the Living: A Project on the Emergence and Reception of the Cell Theory in France and Germany, ca. 1800–1900*, in particular to Florence Vienne, Marion Thomas, and especially Lynn K. Nyhart, who has been tremendously helpful. Maura Flannery, Marianne Klemun, Renato G. Mazzolini, and Sandro Minelli kindly read preliminary versions of the manuscript and made precious comments. I thank Luca Ciancio for his moral support and Adam Bostanci who has made his best to improve my English. Naturally, the responsibility for any statement and error in this book lie on me alone.

Contents

About the Author

Ariane Dröscher studied history and biology at the universities of Hamburg and Bologna and received her PhD with a dissertation on the history of cell biology. She worked as researcher and Lecturer of History of Biology, History of Science, Philosophy of Science, Science Policies, and Communication of Science at several Italian universities. She has published four monographs, two edited volumes, one translation, and over 120 essays and papers. She is vice-president of the *Deutsche Gesellschaft für Geschichte und Theorie der Biologie* and member of numerous editorial boards and scientific societies and networks.

Abbreviations

BCP	Biblioteca Civica di Padova
BOBP	Biblioteca dell'Orto Botanico di Padova
BSNAP	Biblioteca di Scienze naturali e ambientali di Pisa
HSHÖK	Hof- und Staatshandbuch des österreichischen Kaiserthums. Wien: K.k. Hof- und Staats-Aerarial-Druckery
HSSÖK	Hof- und Staatsschematismus des österreichischen Kaiserthums. Wien: K.k. Hof- und Staats-Aerarial-Druckery
NSIRAS	Nuovi Saggi della Imperiale Regia Accademia di Scienze, Lettere ed Arti in Padova

List of Figures

CHAPTER 1

Introduction

During the night of June 12, 1848, the lawyer Andrea Meneghini (1806–1870) (Fig. 6.1 in Chap. 6), his younger brother, the botanist Giuseppe Meneghini (1811–1889) (Fig. 7.1 in Chap. 7), their families, and closest friends hastily fled Padua. About 6000 left Padua under similar circumstances during the following months. This dramatic event and the experiences of the years of military defeat, exile, and persecution deeply marked all their lives. But one by one, the revolutions of 1848 ended in all Italian and European cities, leaving behind a disillusioned generation of young intellectuals and their ideals, among them a noticeable number of scientists and naturalists. The thoughts and actions of the Meneghini brothers during this 'springtime of the peoples', the prominent role of plants in Padua's emerging civil society, and parallels between scientific-naturalistic and socio-political views are the principal themes of this monograph.

Hence, to put it starkly, this book tells a loser's story set in the periphery of early nineteenth-century history and history of science. Padua's 1848 revolt was only of secondary importance, even in the Italian context. Historians agree that the revolutions which spread over the European continent arose from multiple causes. The movements show several common features—a remarkable synchronism, similar demands for a constitution and more individual freedom, involvement of wider parts of the

A. Dröscher, *Plants and Politics in Padua During the Age of Revolution, 1820–1848*, Palgrave Studies in the History of Science and Technology, https://doi.org/10.1007/978-3-030-85343-3_1

population in policy-making, the role of technological progress and the popular press and other—but also a great variety of national and local peculiarities (e.g., Sperber 2005). Likewise, there is much debate as to when the age of revolution started. Some scholars see the years between the French Revolution of 1789 and the revolutions of 1848 as a historically coherent period (e.g., Hobsbawm 1962), others start with the turmoils of the 1810s and 1820s (Church 1983), and still others concentrate only on the biennium 1848–1849. The main supporters of the insurrections were liberal aristocrats and ascending bourgeois, yet all belonged to a very broad spectrum of currents and movements. This inherent diversity, which often degenerated into open conflicts about the means and the goals of the revolution, is considered to have been the main reason for its quick suppression. Nevertheless, in many European countries, mainly Austria, France, Germany, Denmark, and the Netherlands, the revolutionary movements left their mark and the restored governments maintained or enacted a series of liberal reforms. Not so in Italy.

The revolution of 1848 was a seminal episode in Italy's history. Insurrections broke out all over the peninsula. Denis Mack Smith (2002), among others, emphasizes that not the upheavals of Paris, but those happening in January in Sicily led to the first 1848 constitution. More important, the Italian age of revolutions coincided with the First Independence War of the Risorgimento, the national movement that resulted in the unification of the Italian states. For centuries, most of the peninsula had been under Spanish, French, and Austrian rule. In the 1840s, aside from the minuscule states of San Marino and Monaco, Italy was divided into seven independent states: the Bourbonic Kingdom of the Two Sicilies in the south, the Papal states, the Grand Duchy of Tuscany, the duchies of Parma and Modena, the Piedmont with Nice, Savoy, and Sardinia, and finally Lombardy-Veneto, which was subject to the Habsburg empire. In Lombardy-Veneto, the main revolts broke out in March in Venice and Milan. Both cities became places of fierce battles and sieges. Other cities, like Padua and Vicenza, seized the opportunity of the Austrian retreat to join the insurrection. Nevertheless, General Radetzky (1766–1858) soon reconquered the whole territory, and a long period of repression followed.

Most historical accounts of 1848 do not even mention Padua. The main scenes were other cities. During the so-called (pacific) reformist period of 1846–1847, Florence, Rome, Genoa, Livorno, Naples, and Turin were important centers of debate and diplomacy, but also places where political activity poured out of private cabinets into the streets, cafés, and bourgeois social circles. Palermo and then Milan and Venice gained their renown as the strongholds of the military revolutions of the 1848–1849 period, and, finally, Venice, Bologna, Livorno, Rome, Brescia, and Genoa became symbols of ultimate resistance and sacrifice (Francia 2013, p. 12). According to standard narratives, the fact that the Habsburg army did not need to fire a single bullet to retake Padua has led to the assumption that Padua had no real 1848 revolution. In the chapters that follow, I show that the relationship between the Paduan notables and the Austrian regime was indeed ambiguous, yet this made the events no less dramatic and disruptive for people like the Meneghini brothers.

Historiography, in particular history of science, tends to focus on a few exceptional sites—Renaissance Italy, eighteenth-century Paris, late-nineteenth-century Germany, twentieth-century USA—and takes an interest in other venues primarily to see how these received or assimilated the dominant theories and approaches. However, recent scholarship stresses the European character of the 1848 revolutions but likewise pays heed to local peculiarities. A narrow local focus bears the risk of producing provincialist or nationalist views, yet may also provide important insights. Emma Spary, for instance, has compellingly shown the local and historical contingencies of the foundation of the *Muséum d'Histoire Naturelle* in 1793 in the Parisian Botanical garden (Spary 2000). The *Muséum* was to become a leading scientific institution of its time. Padua was much less successful. In the 1830s and 1840s, its position as leading center of European science and education was a relic of a bygone era. Nevertheless, the city still enjoyed regional cultural and scientific prestige and liveliness. In particular, the naturalistic research of Veneto scholars was of a notable standard and attracted international recognition. Looking at minor sites, like Padua, can therefore help to avoid excessive simplification in evaluating the causes of major events as well as do justice to historical diversity. Historians of biology, for instance, have rarely considered the variety and the local particularities of the early nineteenth-century conceptions of the organization of organic forms. As I will argue in the following chapters, the specific local socio-political situation heavily influenced the development of these conceptions. As David Livingston (2003, p. 7) frames the challenge, "the task is to make particular sense of particular rules in particular places".

The Meneghini brothers epitomize many of the typical, but also some untypical features of the place and time. They were cultured, learned, and farsighted, pragmatic, yet enterprising and ambitious, and soon attained a position of cultural leadership in their native city. The changing financial situation of their father made them at the same time members of the wealthy emerging Paduan elite and part of the young generation whose place in society was contingent on the hoped-for new order. Even more interesting for the purpose of this book, they exemplify various forms of entanglement between politics and botany, similar to the better-known example of the German plant and early cell scientist Hugo von Mohl (1805–1872) and his elder brother, the political scientist and activist Robert von Mohl (1799–1875). In this period of close affinities between socio-economic ideas and concepts of the natural sciences, the political and scientific dimensions often merged in one and the same person. The numerous scientist-politicians of these years include the physiologist Emil Du Bois–Reymond (1818–1896), zoologist Carl Vogt (1817–1895), pathologist Rudolf Virchow (1821–1902) and the naturalist and democratic politician Emil Adolf Rossmässler (1806–1867) in Germany (Lenoir 1992, p. 18–52; Daum 2002), Stephan Endlicher (1804–1849) and Franz Unger (1800–1870) in Austria (Klemun 2016), Émile Küss (1815–1871) and François-Vincent Raspail (1794–1878) in France (although both were somewhat isolated from the countries' mainstream bio-medical circles) (Thomas forthcoming; Loison 2017; Vienne 2017), and Barthélemy Charles Joseph Dumortier (1797–1878) in Belgium.

Andrea and Giuseppe Meneghini were anything but secondary figures. Andrea was the author of several treatises on economics and political economy and became the first mayor of Padua after its annexation to the Italian Kingdom. Similarly, German-speaking botanists praised Giuseppe in the influential *Botanische Zeitung* as "the most famous Italian phycologist of modern time" (W. 1843, p. 370), one of his monographs was translated by the Ray Society (G. Meneghini 1846; G. Meneghini 1853), and he later became head of an important school of geology in Pisa. Both were central figures of Padua's reformist period and its 1848 revolution. Andrea was engaged in the renewal of political and economic thought and in the foundation and promotion of new forms of civil togetherness, expression, and collaboration. Like his older brother, Giuseppe was involved in all main Paduan associations, newspapers, and networks (Chap. 2 and Sect. 9.3 in Chap. 9). The thinking of both brothers touched topics that were central in the European and Paduan debate, like the relationship between the whole and its parts, the dignity of the lower classes, developmental and social progress, hierarchical constitution, and the role of a superior divine principle. Both held views that were in tune with, but also significantly different from, the view of many other colleagues.

A focus on only two historical figures, in this case brothers, cannot provide a complete picture of the complex events, nor of the multilayered intellectual currents of their milieu—not even within the spatial and temporal constraints chosen for this inquiry. However, it allows us to concentrate on individual experiences and on specific circles and ideas, which I consider highly significant. For instance, the tropes found in Giuseppe's scientific works (Chap. 7) were characteristic of Paduan liberal circles in general. Both brothers' claim to a leading role in Padua's society made them particularly receptive to the intellectual currents and the emerging local, regional, and international trends of their time. Their case is therefore suited to investigating the broader Paduan milieu and to analyze the impact of this specific setting on their thought and *vice versa*. Their case thus sheds light on the broader intellectual and cultural milieu of their time and place and, at the same time, epitomizes an interesting local variety of early nineteenth-century conceptions of organic and civil organization.

Both brothers held important political positions also after Italy's unification. Still, theirs is a story of juvenile failure, and this may be the reason why it is almost untold. Andrea's major political project, the establishment of a new civil society in Padua, as well as Giuseppe's main scientific project, the establishment of a new botany based on cell theory, fell apart in June 1848. Andrea is today almost completely unknown. His treatise *Elementi di economia sociale*, or Elements of Social Economy (A. Meneghini 1851), is not even mentioned in Massimo Augello and Marco Guidi's comprehensive inquiry into the history of the popularization of economics in Italy (Augello and Guidi 2007). For Giuseppe, historical scholarship has produced several remarkable essays on his life and work as a celebrated geologist in Pisa (Ciancio 2013; Corsi 2001, 2008), but nothing on his time in Padua.

Alongside the Meneghini brothers, plants are a guiding thread of this book. On the scientific level, botany played a key role in studies of the organization of living forms, in particular in cell theory, during the period under consideration. Vegetal metaphors and analogies will therefore receive particular attention. Theresa M. Kelley reminds us that historically the debate about plants and their manifold manifestations was particularly distinctive for the Romantic age:

> Botany is the cultural imaginary of romantic nature and, as such, is at issue there wherever nature matters, including nature as matter. [...] As the most popular and to a significant degree the most visible of romantic natural histories, botany was the site romantic writers used to stage practical, figurative, and philosophical claims about nature. (Kelley 2012, p. 11)

On the socio-cultural level, plants connected the members of the Paduan economic and social elite in a number of ways. In the guise of poetry, gardens, agriculture, floriculture, and pharmacy, they acted as objects of communication and interrelation that transcended social and geographical affiliations and led to new types of cultural and political sociality and institutions (Chaps. 2, 4, 8, and 9). Moreover, Paul Ginsborg has recently pointed out that Italian nationalists displayed a peculiar romantic sensibility to nature and landscape (Ginsborg 2012).

In the 1830s and 1840s, 'botany' still connoted a broad range of activities. René Siegrist has shown that in the seventeenth century for Joseph Pitton de Tournefort (1656–1708) agriculture and gardening were integral parts of botany, while leading scholars of the following century, like Carl von Linné (1707–1778), increasingly dismissed these fields as not truly part of botany (Siegrist 2013, p. 206–214). Giuseppe Meneghini's diverse activities indicate that in Northern Italy the variegated social characteristics and demarcation criteria of early eighteenth-century plant studies still held for a good part of the early nineteenth century. Notwithstanding its long and rich institutional and conceptual history, botany as a profession was still poorly developed. Beyond that, the pool of interested and engaged people was exceptionally broad. More than any other branch of the natural sciences, botany depended (and still depends) on wide-ranging networks and on the involvement of a huge number of skilled amateurs. This is especially true for Veneto, where plant knowledge and commerce traditionally enjoyed a high status. Professionalized plant scientists, mostly working in the botanical gardens, strove to become the intellectual and institutional leaders of these heterogeneous groups. As the following chapters show, in the late 1830s and 1840s Giuseppe Meneghini did his best to distinguish himself as such a leader. In his *Lectures on Popular Botany* (1844–1846), he drew a neat line between popular science and what he called 'philosophy of science':

> According to our view, popular science does not have other aims, because the people can leave the scabrous path of analysis to the real professionals,

> and know about, savour and use those applications which are the results of long, continuous and tedious studies. (G. Meneghini 1844–1846, 3(2), p. 318–319)[1]

Meneghini thus exemplifies Europe-wide efforts by his peers to promote the status of professional botany without isolating it from nonprofessional amateurs. Matthias Schleiden's (1804–1881) battles of the early 1840s to render botany institutionally and scientifically autonomous from medicine and pharmacy are well known (Jahn and Schmidt 2005, p. 86–95). In the first half of the nineteenth century, botany indeed ceased to be a purely descriptive and classificatory discipline in support of pharmacy. Naturalistic and taxonomic studies were still important, yet a growing number of young botanists took up the microscopic and experimental studies that stood at the forefront of general morphological, physiological, and anatomical theories and innovative mathematical and chemical approaches. Cryptogams in particular played a seminal role in the foundation of biology as a science. Cryptogams are a group of seemingly 'lower' plants like ferns, algae, mosses, lichens, fungi, and some bacteria that reproduce by spores. The apparently simple organization of these organisms made them ideal objects for basic chemical, physiological, anatomical, and developmental inquiries. Unfortunately, their incredible variety and diversity of forms and vital manifestations also confused researchers. Today, many of these species are no longer considered 'plants' but put into separate kingdoms. Be that as it may, in the nineteenth century, cryptogam research was fundamental for the advancement of botany and biology in general. Without cryptogam research, cell theory as we know it today would not exist. North Italian scholars were among the most expert cryptogamists of that period, and Giuseppe Meneghini was one of them.

Sections 9.3 and 9.4 in Chap. 9 will illustrate Giuseppe's deep commitment to the promotion of applied botany and to the popularization of plant science, and his endeavors to take a prominent role through his expertise in general theoretical knowledge. Expertise was indeed an important currency in career making and in the self-image of the emerging civil meritocratic society in general. Moreover, I will argue in Sect. 2.4 in Chap. 2 that the Paduan elite used science in general and plant knowledge in particular as a means to gain more and more widespread authority over Veneto's provinces and greater independence from Venice, which continued to exert political and social dominance after the fall of the Republic of Venice in 1797. This book will therefore investigate the social and economic dimension of the broader field of plant knowledge, considered as the vitally important 'undergrowth' of scientific botany. This includes the gentlemen botanists and those engaged in the cultivation of useful (Chap. 9) and aesthetically appealing plants (Chap. 8), vegetal representations in gardens (Chap. 4) as well as the imagery of plants in social and economic thought (Chap. 6).

Science and society, and, more specifically, biology and politics have never been neatly separate. Nonetheless, the early nineteenth century was a period of exceptional conceptual cross-fertilization. The book is primarily concerned with theories of the organic state and constitution of the body, ideas of progress, perfection and (pre-Darwinian) evolution, and the emerging views of associationism and cooperation between parts. Debates about the ontological status and the relationship between parts and wholes were not limited to that age. Rather, they date back to the seventeenth century, in particular to the proponents of atomistic ideas. Nor were these debates confined to organismal anatomy (e.g., Bouchard and Huneman 2003; Wolfe and Kleiman-Lafon 2021). Yet, from the late eighteenth century onwards, questions about organization, self-organization, cohesion, and individuality were central to biological and socio-political agendas.

The most innovative element in nineteenth-century discussions on organization was cell theory. The path from late eighteenth-century fine anatomy to nineteenth-century cell theory was not as linear and uniform as often described. Schleiden and Schwann's *Zellenlehre* was certainly central, yet its importance needs to be qualified. Several alternative conceptions were put forward in the 1830s and 1840s and continued to be discussed during the following decades (Dröscher 2002). The same happened in Padua: different ideas about cells and their role in the organization of life coexisted (Sect. 7.4 in Chap. 7). Conceiving and establishing the cellular level of vital organization depended on more than just the observation of cells through high-resolution microscopes. The conceptualization of the relationship between parts and wholes required new concepts of individuality and constitution (Sheehan and Wahrman 2015). The possible solutions were multiple, leaving much space for diverging views.

Similar questions also exercised the political thought of that time, especially among those who sought a new kind of civil society that would offer greater individual freedom, democratic participation, and economic liberalism. These utopias required a new relationship between the state and its citizens. New ways of conceiving the interaction between the people and the state went hand in hand with new ideas about the organization of living bodies. In the first decades of the nineteenth century, the concepts of the organization of the state and the body were both still poorly defined, multilayered, and had considerable differences. Yet, this does not preclude the possibility that they had a powerful impact on the thinkers of these years. In particular, the fuzziness of the early organismal ideas made them particularly amenable and thus a rich source of inspiration for a variety of different political concepts. For historians, the major challenge is to avoid oversimplification. Another one lies in the fact that during that period organismic ideas were often not explicitly formulated, even less codified.

The definition of politics that I adopt is as broad as that of botany. This book will not concern itself with the history or the administrative aspects of the Austrian rule of Padua. Only the final chapters will describe the impact of the political events of 1848 and the post-revolutionary period. Rather, priority will be given to the debates and activities of the members of the emerging civil society. Special attention will be devoted to the development of organic theories of the state, progressive theories of social organization (Chap. 6), the 'spirit of associationism' (Chap. 2), and the role of science in the societal projects of pre-revolutionary Padua (Chap. 9). The emergence of industrial floriculture illustrates the transformation of aristocratic conventions into bourgeoise economically profitable sectors (Chap. 8). Conceptually, economics and political economy were just emerging as autonomous disciplines, and their discourses were often still part of general reflections about natural processes. During this foundation phase many scholars of the political and economic sciences drew inspiration from apparently perfectly organized and functioning living bodies, adopting—or discarding—them as models of harmonious interaction in the hoped-for administrative and societal order. These ideas will be the principal subject of Chap. 6 (from the socio-economic point of view) and Chap. 7 (from the biological-anatomical point of view).

Analogies and metaphors are often found in theories of political as well as biological organization because they both help to deal with an imperceptible phenomenon (Fox Keller 1995; Maasen et al. 1995; Reynolds 2018). One can visualize the participating entities, that is, citizens and cells, institutions and organs or bodies, but not how or why they cooperate. Social and biological theories of organization thus faced the same problems of conceptualization, evidentiality, and communicability. In such a situation, metaphors are useful, albeit not easy to handle. Nineteenth-century cell researchers were aware of what philosophers of science today discuss under the rubric of nomadic concepts (Minelli 2020), meaning concepts, terms, and metaphors that wander between different disciplines. In his seminal *Beiträge zur Phytogenesis* (1838), Matthias Schleiden, himself a creator of many metaphors, advised his readers, referring to the use of the term 'growth':

> We must here be on our guard against two dangerous rocks: first, when we transfer words from one science to another, without first accurately testing whether they fit their new situation as respects all their accompanying significations also; and, secondly, when we voluntarily lose sight of the signification of a word consecrated by the spirit of the language and its historical development, and employ it without further ceremony in compound words, where perhaps, at the most, only some unessential part of its signification suits. (Schleiden 1847, p. 249–250)

Analogies exert great power in persuasion and scientific legitimization. However, identifying the exact form and level of congruence is far from easy. The spectrum of possible conceptualizations of political and biological organization, for instance, is manifold. For this reason, case studies are helpful, and Giuseppe Meneghini provides an important one. He was a pioneer of cell theory, and the cooperation of parts was a fundamental aspect of his conception of organization (Chap. 7). However, as I will argue, analogies drawn by researchers like Meneghini in the 1830s and 1840s were far more fuzzy and subtle than those that had been studied by historians of science in the period post-1848. The fact that the language of both Meneghini brothers was unremarkable may be another reason why they have received little attention. Going beyond an analysis of a shared vocabulary, the investigation of the Paduan context therefore requires research that reads between the lines and verges into the somewhat woolly field of thought styles.

Greater linguistic conformity can be found in the field of Italian progressivist thinking. The emerging concepts of social organization and anatomical constitution were both deeply entangled with those of development and evolution. Since the second half of the eighteenth century, political as well as biological thinking had increasingly paid heed to dynamism, and this produced very different ideas about the causes of past, present, and future events. Chapters 6, 7, and 9 illustrate the influence of historicist and progressivist currents on the new generation of Paduan intellectuals. The reception of the philosophies of Giambattista Vico (1668–1744) and Gian Domenico Romagnosi (1761–1853) demonstrates that understanding oneself as a member of a historically contingent society also focused the mind on possible future scenarios and improvements. Analogously, the case of Giuseppe Meneghini illustrates the role of historicist thinking in understanding living beings as products of time and linked to one another by genealogical bonds. The tragedy of most political activists of the reformist period was in fact that their profound belief in unstoppable and ineluctable progress toward democracy and civil emancipation suffered a severe setback, politically and psychologically. The last chapter briefly deals with this, namely, the 1848 defeat. Andrea and Giuseppe Meneghini's dramatic professional and personal experiences are typical of many Italian scientists of the period. Moreover, I will argue that the repression had a particularly negative effect on the development of the natural sciences in Italy in general and of cell biology in particular.

The institutional, economic, and socio-cultural dimensions of the interaction between the plant sciences and politics are another main topic of this book. The 1840s distinguish themselves through the creation of new forms of civil organization and the foundation of newspapers, societies, and associations. Chapters 2, 3, 4, 7, 8, and 9 illustrate Andrea and Giuseppe Meneghini's active involvement in the scientific and political events of their time, and their skill in establishing local, Italian, and international networks. Networks indeed played a crucial role in their lives. Special attention will be drawn to the Paduan circle of young botanists, plant lovers, and agronomists, many of whom became central figures in the political history of Padua. They belonged to the same generation and, in spite of differences in social status and degrees of involvement in political and naturalistic debates, shared a number of basic convictions.

Gardens and gardening occupied center stage in the new institutional, economic, and socio-cultural order—in more ways than one. The way wealthy families designed their gardens reveals much about Padua's social elite, their self-representation, and their concept of nature (Fig. 1.1). A comparison of the Meneghini garden and other Paduan parks, for instance, reveals the level of involvement of Agostino Meneghini (1775–1844), Andrea and Giuseppe's father, in Padua's notable society (Sect. 4.3 in Chap. 4), an aspect that has hitherto been completely ignored. Arguably, the preference of certain plant species and certain garden styles expressed much more than just a personal predilection (Chaps. 4 and 8). Yet political convictions were rarely displayed explicitly. Padua's Festival of Flowers, for instance, was careful not to become a political event (Sect. 8.2 in Chap. 8). For this very reason it represented a moment of geographic, political,

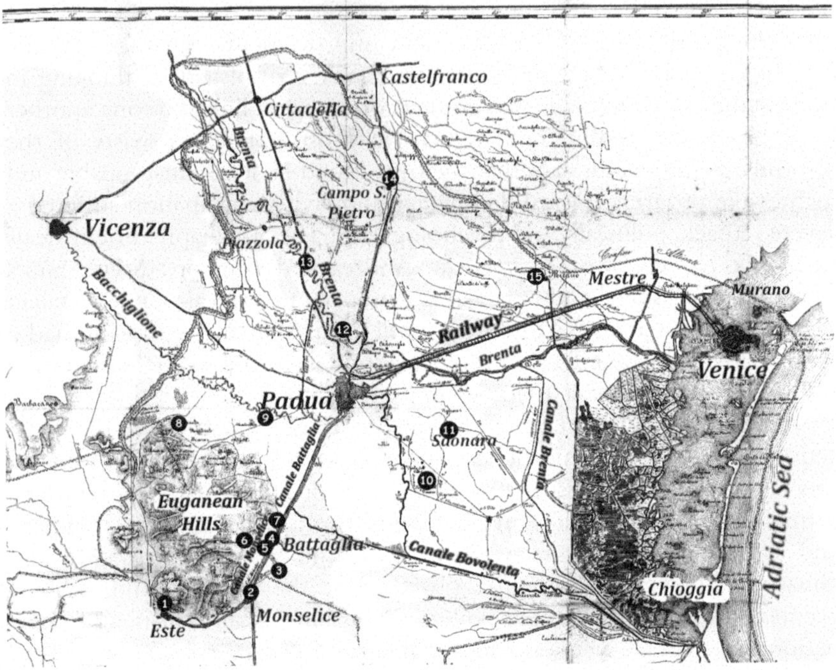

Fig. 1.1 Famous parks and gardens in the province of Padua. Graph art by Till Dröscher. By courtesy of Till Dröscher Graphic & Design. (1) Garden of Villa Benvenuti (at Este); (2) Cromer garden (at Monselice); (3) garden of Villa Emo; (4) garden of Villa Meneghini (previously Villa Selvatico, then Villa Wimpffen); (5) garden of Villa Corinaldi (then Villa Sgaravatti and Villa Italia di Lispida); (6) garden of Villa Barbarigo (at Valsanzibio); (7) Renaissance garden of Catajo Castle; (8) garden of Villa Papafava (at Frassanelle); (9) garden of Villa Cesarotti (at Selvazzano); (10) town of Legnaro; (11) garden of Cittadella Vigodarzere (at Saonara); (12) garden of Villa De Zigno (at Vigodarzere); (13) garden of Villa Trieste (near Piazzola sul Brenta); (14) garden of Villa Polcastro (then Villa Wollemborg); (15) garden of Villa Belvedere (at Mirano)

social, and gender transgression. It provided women an opportunity for public participation and brought together people from different social backgrounds. Moreover, an interest in horticulture was a means for Paduan plant lovers to get in contact with like-minded high-ranking Austrian officials, a relationship that paid off for Giuseppe Meneghini in the months of imminent danger after the 1848 defeat (Sect. 10.3 in Chap. 10). The state of Lombardy-Veneto had been a stable part of the Austrian empire since 1814. Yet, despite the anti-Austrian sentiment of the Italian Risorgimento, the ties between Paduan naturalists and notables and their Viennese counterparts continued to be strong even during and after the Independence Wars that broke out in 1848 and ultimately led to Italian unity (1860–1861) and the annexation of Veneto to the Italian kingdom (1866). Such 'botanical bonds' thus shed some light on the complex relationship between Padua's elite and the Austrian government in general. Still another crucial aspect of the Festival of Flowers is its function as a catalyst of industrial floriculture and of the growing social status of the gardeners. Finally, I will demonstrate in Sect. 9.2 in Chap. 9 that influential landowners like Andrea Cittadella Vigodarzere (1804–1870) (Fig. 9.1 in Chap. 9) cherished agrarian and social utopias that were inspired by gardening. Their concept of cultivation encompassed men as well as plants.

The analysis of the interplay between plants and politics thus opens up an unforeseen expanse of interacting areas. In the Veneto, the first half of the nineteenth century probably represents the heyday of the convergence of these spheres. During the second half of the century, we observe the beginning of a slow decline in the design and care of great private gardens. In parallel, the festival of flowers, highly successful in the 1840s, stopped for good in the 1860s. Many traditional and successful gardener families shifted toward industrial floriculture (Sect. 8.2 in Chap. 8). At the same time, agriculture became progressively less important for the economy of Veneto and was superseded by manufactory, industry, and the tertiary sector. In the late 1840s already, political interest in botany became increasingly instrumental and utilitarian (Chap. 9). Even though a close relationship to the vegetal world has survived to our days, the special bond and close interweaving between botany and politics came to an end.

Note

1. G. Meneghini 1844–1846, 3(2), p. 318–319: "E la scienza popolare non ha, a nostro parere, altro scopo, perché può per essa il popolo lasciare a chi ne fa davvero professione la scabrosa via dell'analisi, ed intendere, assaporare ed usare quelle applicazioni che sono il frutto di lunghi, incessanti e faticosi studii."

References

Augello, Massimo M., and Marco E.L. Guidi, eds. 2007. *L'economia divulgata: Stili e percorsi italiani (1840–1922)*. Milano: Franco Angeli.

Bouchard, Frédéric, and Philippe Huneman. 2003. *From groups to individuals: evolution and emerging individuality*. Cambridge, MA: MIT Press.

Church, Clive H. 1983. *Europe in 1830: revolution and political change*. London: George Allen & Unwin.

Ciancio, Luca. 2013. I segni del tempo: Teorie e storie della Terra. In *Il contributo italiano alla storia del pensiero*, ed. Antonio Clericuzio and Saverio Ricci, 332–343. Roma: Istituto dell'Enciclopedia Italiana.

Corsi, Pietro. 2001. La geologia. In *Storia dell'Università di Pisa*, 2 (1737–1861), 889–927. Pisa: Pisa University Press.

———. 2008. *Fossils and reputations: a scientific correspondence: Pisa, Paris, London, 1853–1857*. Pisa: Edizioni Plus/Pisa University Press.

Daum, Andreas W. 2002. Science, politics, and religion: Humboldtian thinking and the transformations of civil society in Germany, 1830–1870. *Osiris* 17: 107–140.

Dröscher, Ariane. 2002. Edmund B. Wilson's *The Cell* and cell-theory between 1896 and 1925. *History and Philosophy of the Life Sciences* 24: 357–389.

Fox Keller, Evelyn. 1995. *Refiguring life: metaphors of twentieth-century biology*. New York: Columbia University Press.

Francia, Enrico. 2013. *1848: La rivoluzione del Risorgimento*. Bologna: Il Mulino.

Ginsborg, Paul. 2012. European romanticism and the Italian Risorgimento. In *The Risorgimento revisited: Nationalism and culture in Nineteenth-Century Italy*, ed. Silvana Patriarca and Lucy Riall, 18–36. London: Palgrave Macmillan.

Hobsbawm, Eric. 1962. *The age of revolution: Europe 1789–1848*. London: Abacus.

Jahn, Ilse, and Isolde Schmidt. 2005. *Matthias Jacob Schleiden (1804–1881): Sein Leben in Selbstzeugnissen*. Halle (Saale): Deutsche Akademie der Naturforscher Leopoldina.

Kelley, Theresa M. 2012. *Clandestine marriage: Botany and romantic culture*. Baltimore: Johns Hopkins University Press.

Klemun, Marianne, ed. 2016. *Einheit und Vielfalt: Franz Ungers (1800–1870) Konzepte der Naturforschung im internationalen Kontext*. Wien: Vienna University Press.

Lenoir, Timothy. 1992. *Politik im Tempel der Wissenschaft: Forschung und Machtausübung im deutschen Kaiserreich*. Frankfurt a.M.: Campus Verlag.

Livingston, David N. 2003. *Putting science in its place: Geographies of scientific knowledge*. Chicago: The University of Chicago Press.

Loison, Laurent. 2017. Cellule cancéreuse ou théorie cellulaire? Sur la réception de la théorie cellulaire à Strasbourg au XIXe siècle. *Bulletin d'Histoire et d'Épistémologie des Sciences de la Vie* 24 (1): 27–42.

Maasen, Sabine, Everett Mendelsohn, and Peter Weingart, eds. 1995. *Biology as society, society as biology: metaphors*. Dordrecht: Springer.

Mack Smith, Denis. 2002. The revolutions of 1848–1849 in Italy. In *The revolutions in Europe 1848–1849: from reform to reaction*, ed. Robert J.W. Evans and Hartmut Pogge von Strandmann, 55–82. Oxford: Oxford University Press.

Meneghini, Andrea. 1851. *Elementi di economia sociale ad uso del popolo*. Torino: Cugini Pompa e comp.

Meneghini, Giuseppe. 1844–1846. Lezioni di Botanica Popolare. Giornale Euganeo 1: 314–320 and 394–402; 2(1): 197–207 and 407–412; 2(2): 31–39, 266–273, and 301–311; 3(1): 251–259; 3(2): 194–208, 432–450, and 497–519.

———. 1846. *Sulla animalità delle Diatomee e revisione organografica dei generi di Diatomee stabiliti da Kützing*. Venezia: Tipografia Pietro Naratovich.

———. 1853. On the animal nature of the Diatomeae, with an organographical revision of the genera established by Kützing. In *Botanical and physiological memoirs*, ed. Arthur Henfrey, 343–513. London: Ray Society.

Minelli, Alessandro. 2020. Disciplinary fields in the life sciences: evolving divides and anchor concepts. *Philosophies* 5: 34. https://doi.org/10.3390/philosophies5040034.

Reynolds, Andrew S. 2018. *The third lens: metaphor and the creation of modern cell biology*. Chicago: University of Chicago Press.

Schleiden, Matthias Jakob. 1847. *Contributions to phytogenesis*. London: The Seydenham Society.

Sheehan, Jonathan, and Dror Wahrman. 2015. *Invisible hands: self-organization and the eighteenth century*. Chicago: University of Chicago Press.

Siegrist, René. 2013. On some social characteristics of the eighteenth-century botanists. In *Scholars in action: the practice of knowledge and the figure of the savant in the 18th century*, ed. André Holenstein, Hubert Steinke, and Martin Stuber, 205–234. Leiden and Boston: Brill.

Spary, Emma C. 2000. *Utopia's garden: French natural history from Old Regime to revolution*. Chicago: The University of Chicago Press.

Sperber, Jonathan. 2005. *The European revolutions, 1848–1851*. 2nd ed. Cambridge: Cambridge University Press.

Thomas, Marion. forthcoming. De l'individualité biologique à l'individualité sociale selon l'agenda bio–républicain d'Émile Küss (1815–1871).

Vienne, Florence. 2017. Worlds conflicting: the cell theories of François-Vincent Raspail and Theodor Schwann. *Historical Studies in the Natural Sciences* 47 (5): 629–652.

W. 1843. Memorie della reale accademia delle scienze di Torino. Serie seconda, tomo IV, Torino 1842. *Botanische Zeitung* 1: 368–376.
Wolfe, Charles T., and Sylvie Kleiman-Lafon. 2021. Unsystematic vitality: From early modern bee swarms to contemporary swarm intelligence. In *Active materials*, ed. Peter Fratzl, Michael Friedman, Karin Krauthausen, and Wolfgang Schäffner. Berlin: De Gruyter.

CHAPTER 2

Padua's Networks

Before we embark on the story of the Meneghini brothers and their agency between plants and politics, this chapter introduces the Paduan world of sociality in the first half of the nineteenth century. Always and everywhere, personal relationships and social networking have been and still are of fundamental importance. Even more so in urban contexts, such as Padua, where sociality expresses itself in particularly numerous and variegated ways. Today, social network analysis is a major field of sociological research and classics such as Linton Freeman's *The Development of Social Network Analysis* (Freeman 2004) and Bruno Latour's *Reassembling the Social* (Latour 2005). Historians of science instead have often paid attention primarily to the interaction between individual scientists (Browne 1995–2002; Moon 2016; Parker et al. 2010). It is important to bear in mind that in the period with which this book is concerned, the natural sciences were still not neatly separated from other realms of social and intellectual life. It is therefore worth devoting some discussion to the subtle and multilayered world of Padua's social networks. While offering a complete analysis is beyond the purpose of this book, an understanding of its social groups and the way they interacted is essential.

The most important features of Padua's social networks were continuity in group membership over time, the relatively small number of active members, and effectiveness in addressing civic claims. An example

A. Dröscher, *Plants and Politics in Padua During the Age of Revolution, 1820–1848*, Palgrave Studies in the History of Science and Technology, https://doi.org/10.1007/978-3-030-85343-3_2

illustrates the value and performance of networks during crucial events. In Padua, a culturally significant event was the international public fundraising for a monument to honor the sculptor Antonio Canova (1757–1822) in 1827. The initiative had started out in Venice, but many Paduans contributed and appeared on the public list of donors. Among them, we find Agostino Meneghini, Antonio Vigodarzere, Giuseppe Jappelli (1783–1852), Giovanni de Lazara (1744–1833), the counts Papafava, bishop Modesto Farina, Stefano Gallini (1756–1836), and many others (Anon. 1827, pp. 27–28). If we compare these names with the 755 names—among them most important Paduan families and many university professors, high and low clergymen, and Jews—donating in 1848 a total of 151,984.97 Lire to the revolutionary *Comitato provvisorio dipartimentale* (Provisional Committee of the Department of Padua), headed by Andrea Meneghini, we find a high degree of overlap (Anon. 1848a). If one then examines these donor lists together with the registers of Padua's most important formal associations, one quickly detects a small group of socially highly engaged people who were active over several decades and in profoundly different events (Table 2.1).

During the first decades of the nineteenth century, the principal meeting places of Padua's élite were still the classical sites of late eighteenth-century social life, that is, salons (regular gatherings in patrician houses), casinos (quarters rented for conversational gatherings), and cafés (coffee-houses), but also shops (tailors, butchers, and especially pharmacies), which were much more than just premises for selling goods. In contrast to the exclusiveness of salons and casinos, cafés and shops were informal meeting places where boundaries associated with social rank and gender were transgressed. In the 1830s and 1840s, networking in Padua assumed a new quality when, alongside the more traditional forms, new types of sociality emerged, namely, associations and journals. In this context, it is well-nigh impossible to draw a distinction between scientific, cultural, and political circles. As in other places throughout the centuries, members of Padua's élite—or those who wanted to belong to it—were often committed along all three dimensions. Likewise, power, representation, entertainment, exchange, knowledge, and action were central to all of these spheres.

The Meneghini brothers were constitutive and representative of the local cultural and intellectual milieu. Both were active at all aforementioned levels, in socially closed circles (salons), scholarly closed circles (academies), socially semi-open circles (associations), scholarly semi-open circles (societies, congresses), and sites of free encounter (theaters,

Table 2.1 The most active members of Padua's associations and the Academy of Science, their roles in the University of Padua, the *Istituto Veneto*, and the *Comitato Provvisorio Dipartimentale/National Guard*, and their attendance at the Congresses of Italian scientists[a]

Name		*Reading Cabinet (1836–1843)*	*Gardening Society (1846–1847)*	*Society of Encouragement (1841)*	*Academy (1838–1847)*	*Istituto Veneto (1839–1849)*	*University (1840–1848)*	*Congresses (1838–1847)*	*Comp. provv. Dipart./Nat. Guard (1848)*
Agostini, Stefano	(1797–1877)	o			o		p/d	P	C
Bazzini, Carlo Augusto		+	+				p/r	PiPLMV	
Beggio, Francesco			+	+				V	
Benvenisti, Moisé	(1818–1888)	o			+			PV	
Bernardi, Giuseppe	(1788–1851)	+			o			TPMV	
Bonomi, Bortolo			+	+					
Catullo, Tommaso	(1782–1869)	+			+	+	p/r	FPMV	
Cavalli, Ferdinando	(1810–1888)			o				PV	C
Cicogna, Giovanni		+	+				p	PV	
Cittadella Vigodarzere, Andrea	(1804–1879)	o	+	o	o	o		PMV	C
Cittadella, Giovanni	(1806–1884)	+	+	o	+	o		PV	g
Configliachi, Luigi	(1787–1864)	+	o	+	em		p	PMV	
Conti, Carlo	(1802–1849)	+	+		+	+	p/d	PV	
Cortese, Francesco	(1802–1883)	+			+	+	p	PV	C
Cristina, Giuseppe			o	+					g
Dondi Dall'Orologio, Francesco	(b. 1806)?	+		+					
Dondi Dall'Orologio, Galeazzo	(b. 1787)?	+	+						
Fabeni, Vincenzo	(1799–1861)	+					p	PMV	
Faccanoni, Antonio	(1797–1848)?		+	+					
Fannio, Gio. Francesco	(1798–1849)	o					p/r	PV	
Festler, Francesco Saverio	(b.1801)	+			+		p/d	PMV	
Fioravanti Onesti, Gaetano		+	+	+					
Foscarini, Giacomo (Jacopo)	(1810–1880)	+						PMNV	
Galvani, Antonio	(1797–1869)	+	+			(+)		PV	(C?)
Giacomini, Giacomo Andrea	(1796–1849)	+			o	+	p	PiTPMNV	
Girardi, Antonio			+	+					

(*continued*)

Table 2.1 (continued)

Name		*Reading Cabinet (1836–1843)*	*Gardening Society (1846–1847)*	*Society of Encoura-gement (1841)*	*Academy (1838–1847)*	*Istituto Veneto (1839–1849)*	*University (1840–1848)*	*Congresses (1838–1847)*	*Comp. provv. Dipart./Nat. Guard (1848)*
Giustinian Cavalli, Nicolò Ant.	(b. 1781)	+	+	+					
Gritti, Giovanni			+	+					
Gusella, Antonio		+	+						
Jappelli, Giuseppe	(1783–1852)	+	o		o	+		PMV	
Lazara (de), Nicolò	(1790–1860)	+	+	+					
Mainardi, Gio. Antonio		+		+					
Manfrin, Domenico			+	+					g
Marchi, Alessandro			+	+					
Maritani Sartori, Domenico	(d. 1868)	o	+					NV	
Meneghelli, Antonio	(1765–1844)	+			em		p	P	
Meneghini, Andrea	(1806–1870)	o	+	o				MV	C
Meneghini, Giuseppe	(1811–1889)	o	o		o	+	p	PiFPMNGV	C
Menin, Lodovico	(1783–1868)	+			o	o	p/r	PiPLMV	
Minich, Serafino Raffaello	(1808–1883)	o			+	+	p	PMV	
Montalti, Arnaldo			+	+					g
Morpurgo, Isacco Vita	(1805–1896)	+	+						g?
Moschini, Giacomo	(d. 1895?)	+	+						
Mugna, Giambattista	(1799–1866)	+			o			PMV	C
Nardi, Francesco	(1808–1877)	+	+		+		p	PV	
Orsato, Fabrizio	(1779–1848)	o	+	+					
Papafava, Alessandro	(1784–1861)	+	+	+					
Papafava, Francesco	(1782–1848)	+	+	+					
Pettenello, Giovanni		+	+						
Pezzini, Giuseppe	(d. 1851?)	o		+					C
Piccini, Daniele			+	+					
Pivetta, Gio. Batt.	(d. 1867)	+		+					
Poli, Baldassarre	(1795–1883)	o			+	+	p/d	PV	
Racchetti, Alessandro	(1789–1854)	+	+			o	p	PV	
Rio (Da), Nicolò	(1765–1845)	o			+	+	d	PiTFPM	

Ronconi, Giambattista	(1812–1886)		o				a	V	g?
Sanfermo, Marc'Antonio	(1783–1849)	+		o				PV	C
Santini, Giovanni	(1787–1877)	+			o	o	p/d	PMV	
Selvatico Estense, Giovanni			+	+					
Selvatico Estense, Pietro	(1803–1880)	+		o	o			PMV	
Sinigaglia, Pietro	(d. 1876)		+	+					
Sorgato, Gaetano	(b. 1802)	+	+					V	
Spongia, Filippo	(1798–1880)		+		o		d	PNGV	
Steer, Martino Francesco	(1800–1881)	+	+				p	FPNV	
Treves de Bonfili, Giacomo (Jacopo)	(1788–1885)	+	+	+				V	
Treves dei Bonfili, Isacco	(1790–1855)	+	o	+					
Trevisan, Vittore	(1818–1897)	+	o					PMV	
Trieste, Gabriel del fu Jacob	(1775–1847)	+	+	+					
Trieste, Gabriel del fu Maso	(1784–1860)	+		+					
Trieste, Giacomo			+	+					
Trieste, Leon	(b. 1827)	+		+					
Turazza, Domenico	(1813–1892)	o			o	+	p/d	PV	g
Valsecchi, Antonio	(1799–1882)	+	+				p/r	PMV	
Valvasori, Gio. Batt.			+	+					
Vecchia (Dalla), Francesco			+	+					
Vecchio (Dal), Benedetto	(b. 1800)?	o	+	+				V	C
Visiani (de), Roberto	(1800–1878)	+	o		o	+	p	PiTPMNGV	
Zacco, Teodoro	(1806–1869)	+	+	+				PV	g?
Zambon, Giovanni Battista			+	+					g
Zigno (de), Achille	(1813–1892)	+	+	+	o			PLNGV	

Sources: *Members of the Istituto Veneto*: Gullino 1996, pp. 242–251; *of the university*: HSSÖK 1840, pp. 238–243; HSSÖK 1841, pp. 241–245; HSSÖK 1842, pp. 194–197; HSSÖK 1843, pp. 194–197; HSHÖK 1844, pp. 199–202; HSHÖK 1845, pp. 202–206; HSHÖK 1846, pp. 209–213; HSHÖK 1847, pp. 216–220; HSHÖK 1848, pp. 229–233; *of the Academy*: NSIRAS 1838, pp. xiii–xix; NSIRAS 1840, pp. iii–viii; NSIRAS 1847, pp. iii–xi; *of the Gardening society*: Anon. 1846c; Anon. 1847a; *of the Reading cabinet*: Solitro 1930, appendix 8; *of the Society for encouragement*: Solitro 1930, appendix 5; *of the Comitato*: Solitro 1930, p. 51; *of the congresses*: the attendance lists published in the *Atti* of the nine congresses

[a]Key: + = associate; o = assumed official function; em. = emeritus; p = professor; d = faculty director; r = university dean; a = assistant; Pi = Pisa; T = Turin; F = Florence; P = Padua; L = Lucca; M = Milan; N = Naples; G = Genoa; V = Venice; C = Comitato provvisorio dipartimentale; g = national guard

festivals, meeting places, newspapers). Both played crucial roles in bringing together members, in laying down the formal statutes of collaboration, and in trying to instigate general reform impulses through these associations. Hence, an investigation of Andrea and Giuseppe's social engagement provides a useful introduction to Paduan life and culture and some of its protagonists. This also sheds light on the 'spirit of association' as a distinctive feature of the social philosophy of this period. A closer look at the Gardening Society will demonstrate that plants acted as a common scaffold or trellis on which several forms of political and intellectual activity grew, while finally a brief discussion of the *Istituto Veneto* and of the science congresses in Padua and Venice illustrates that science was an important currency by means of which Padua's elite gained greater regional influence.

2.1 The Spirit of Association

Sociality was a distinguishing trait of the Veneto society up until the outbreak of the First World War (Fincardi 2004). Giuseppe and Andrea Meneghini participated in various types of networks, achieving different levels of publicity, confidentiality, and reach of action. Their father Agostino had successfully established close ties with important notables in the first decades of the nineteenth century (Chaps. 3 and 4). His sons followed his lead and became involved in formal and informal, scientific, cultural, and economic circles on the local, regional, national, and international level. During the most critical periods of both their lives, when death sentence was hanging over Andrea[1] and Giuseppe was searching a safe refuge to build up a new life in exile, many of these relationships turned out to be of fundamental, even vital importance (Chap. 10). In the following, I will introduce the most eminent forms of sociality in Padua.

Salons were typical institutions of the Enlightenment. They were private, informal, and exclusive gatherings of the cities' social elites and those who wanted to stay in close contact with them, that is, of aristocratic and rich bourgeois circles and their hosts, foreign guests, poets, artists, and scientists. Liberal aristocratic circles consisted of local notables, who performed a relevant influence on European politics and socio-cultural development. In their heyday, these circles played an intermediary role between state, nation, and city (Mogavero 2014). For Max Weber,

> Notables (*honoratiores*) are persons (1) whose economic position permits them to hold continuous policy-making and administrative positions in an

> organization without (more than nominal) remuneration; (2) who enjoy social prestige of whatever derivation in such a manner that they are likely to hold office by virtue of the member's confidence, which at first is freely given and traditionally accorded. Most of all, the notable's position presupposes that the individual is able to live *for* politics without living *from* politics. (Weber 1978, p. 290)

Padua until the first decades of the nineteenth century conforms to this definition. Local power was almost completely a monopoly of a handful of families often interconnected by marriages. The most prominent were Vigodarzere, the various Cittadella families, Maldura, Lazara, Polcastro, and Degli Oddi. They succeeded in holding onto their power regardless of whether the government was French or Austrian. From the 1810s onwards, a gradually increasing number of emerging families and of less wealthy but learned citizens participated in these circles, a trend encouraged by Napoleon. These newcomers—Andrea and Giuseppe's father was one of them—were quickly assimilated into new alliances to cement or expand the influence of the traditional elite (Dal Cin 2019, pp. 204–218). In the 1830s and 1840s, in contrast, ever more bourgeois acquired public influence in their own right, thus reducing the importance of the aristocratic circles.

Compared to Venice's exceptional social life (D'Ezio 2012; Plebani 2004), Padua's salon culture was much less opulent, prestigious, and cosmopolitan. Yet, along with literature, art, and politics, the natural sciences played a prominent role. In the late eighteenth century, the salons of Arpalice Papafava and Francesca Maria Bragnis, mother of the naturalist Alberto Fortis (1741–1803), were places of literary, political, and often scientific conversation. Some salons were notorious for their botanical orientation, like that of Enrichetta Treves (1758–1832), aunt of Giacomo (1788–1885), and Isacco Treves dei Bonfili (1789–1855), who in the late 1820s created a famous romantic garden (Sect. 4.2 in Chap. 4 and Sect. 8.1 in Chap. 8) and were founding members of Padua's Gardening Society (Sect. 2.3). Enrichetta's salon was in close contact with the poet Melchiorre Cesarotti (1730–1808) and famed for discussions of politics, literature, and botany, and her house, in particular her rich library, hosted young botanists such as Alberto Parolini (1788–1867) and Roberto De Visiani (1800–1878), the future director of the Botanical Garden (Massaro 2014–2015, pp. 60–61; Massaro 2019).

Giuseppe Solitro (1927, pp. liv–lv) lists as among the most important salons in Padua those of Cittadella Vigodarzere, Sartori, Pivetta, Giustinian Cavalli, Maldura, Gaudio, Ferri, Wollemborg, and Guerrieri Gonzaga. Due to his wealth and social activism, Count Andrea Cittadella Vigodarzere (Fig. 9.1) arguably became the central figure in the network of aristocratic and wealthy patrician Paduan families (Mogavero 2014). He was an offspring of two well-known patrician families, and the owner of one of Padua's most famous English gardens (Sect. 4.2 in Chap. 4). In 1835, he acquired great wealth by inheriting the possessions of his uncle Antonio, which enabled him to express his cultural, scientific, and economic philanthropy (Dal Cin 2019, p. 134). As the following chapters demonstrate, his social circle had a decisive influence on cultural and political events and shaped Padua's 1848 revolution, which was characterized by deep Catholicism, philanthropism, moderate reformism, and moderate patriotism, despite the many bonds with the Austrian nobility.

Important weddings give useful insights into the 'who's who' of the exclusive spheres of Padua's social elite, too. In the absence of official guest lists, newspaper gossip and such like, the then widespread custom to gift publications—poems and other literary forms, but also short scientific treatises—that had been produced *ad hoc* to mark an occasion, provide a valuable, if limited, source that reveals the membership of certain circles and the topics of choice to entertain the bridal couple and their families. In Sect. 3.2 in Chap. 3, I draw on this source to illustrate Agostino Meneghini's successful marriage strategies. One of the numerous poems and pieces of prose composed for the *fastosissime nozze* (sumptuous wedding) between his daughter Anna and Francesco, member of the prominent Gaudio family in 1823, was from his "most affectionate friend", Antonio Vigodarzere (Vigodarzere 1823, p. 6). The dedication points to Agostino Meneghini as one of the newly rich who were assimilated into the aristocratic circles.

Less exclusive settings than salons and aristocratic weddings were provided by public celebrations, popular and ecclesiastical festivals, theatres, casinos, and cafés. Many of these had a long tradition. Padua's *Scuola di Cavallerizza*, or Equestrian School, had Renaissance origins,[2] and trotting races, the *Padovanelle*, took place as early as 1808 at Prato della Valle. From the 1840s, Padua's most famous public café was arguably the Caffé Pedrocchi (Fig. 2.1, I; Fig. 10.1). Created by Giuseppe Jappelli, the architect who had also designed the gardens of Antonio Vigodarzere, Agostino Meneghini, and many others, it opened its doors, day and night, in 1842,

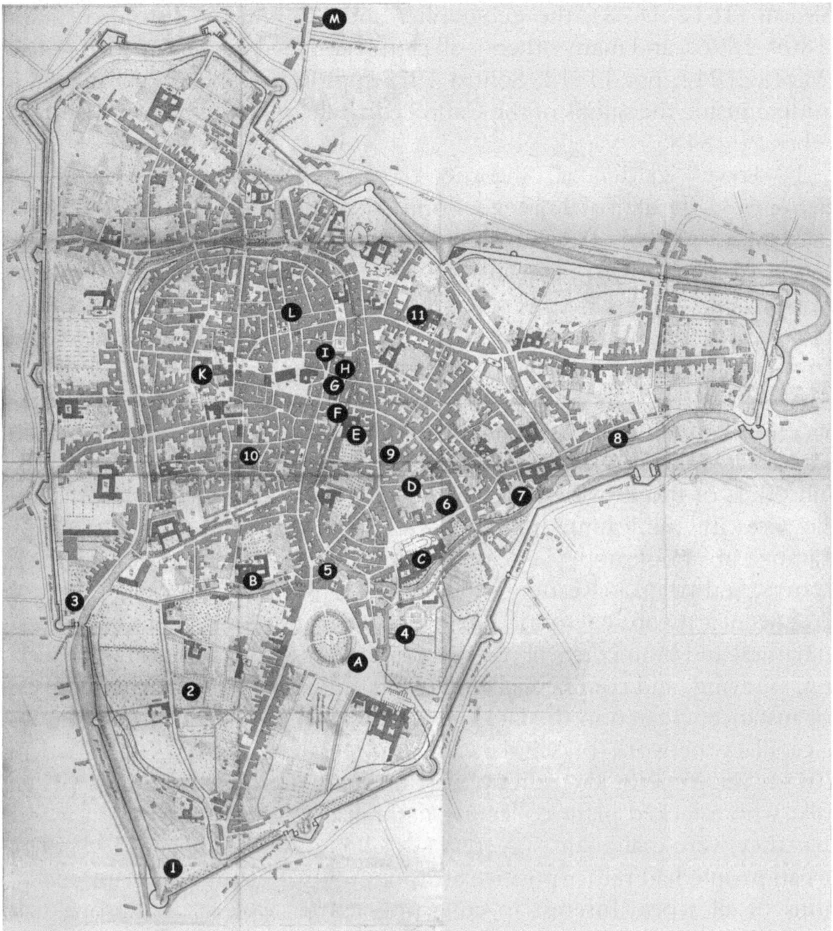

Fig. 2.1 City-map of Padua in 1842, with some of its famous sites and gardens. Graph art by Luigi Patella. Detail from Anon. (1842, foldout). The numbers and letters were added by me. By courtesy of Biblioteca Capitolare di Padova, Coll. 700.AA5.7

and soon became the favorite meeting place of nobles, intellectuals, and students. Patrons included the Meneghini brothers and their friends—Jappelli, the physician Giovanni Battista Mugna (1799–1866), the romantic poets-cum-politicians Giovanni Prati (1815–1884), and Aleardo

Aleardi (1812–1878), the geographer and economist Cristoforo Negri (1809–1896), and many others—all closely watched by the Austrian police (Miotto 1942, pp. 11–12; Solitro 1927, pp. liv–lv). The wounding of a student inside the salons of the Caffé Pedrocchi unleashed the upheaval of February 1848.

1—Treves garden all'Alicorno (created 1865); 2—Agrobotanical Garden; 3—garden of Lorenzo Priuli (sixteenth century); 4—Botanical Garden; 5—garden of Gian Antonio Cortuso (sixteenth century); 6—garden of Alvise Cornaro (sixteenth century); 7—Treves garden; 8—garden of Francesco Morosini (seventeenth–eighteenth century); 9—Giacomini garden; 10—Papafava dei Carraresi garden; 11—garden of Pietro and Torquato Bembo (sixteenth–seventeenth century);

A—Prato della Valle square; B—Seminario vescovile (Episcopal seminary) and office of the Seminario publishing house; C—Sant'Antonio Church; D—Borgo Vignali; E—I.R. Lyceum-Gymnasium Santo Stefano and office of the Provincial Congregation; F—Triest palace and office of the Crescini publishing house; G—University of Padua 'Il Bo'; H—Zambeccari bookshop; I—Caffè Pedrocchi; K—Academy of Science, Letters, and Art; L—Reading cabinet; M—railway station.

A peculiarity of Veneto sociality was a longstanding and marked interest in natural and biomedical objects as well as a venerable tradition of collecting, studying, and commercializing plants and fossils. Venetian geologists, for instance, created in the late eighteenth and early nineteenth centuries a capillary network of scholars and 'masters' (Ciancio 1994; Ciancio 2014). Yet, Veneto, like other places, had its share of superstitious village folk, who attacked plant collecting naturalists in the countryside for fear that they were malignant sorcerers (Dal Prete 2008, pp. 197–198). The urban people had rather positive attitudes toward plants and plant collections of all types. Interest in collecting natural objects in Europe had steadily grown during the eighteenth century, particularly among educated people (Pomian 1987, pp. 142–143). Many gardens also fulfilled museal functions. To attract interest and visitors, their proud owners treasured, listed, and advertised special specimens or categories of plants. This is true especially for Veneto. According to Roberto De Visiani (1854, pp. 21–23), between the sixteenth and eighteenth century, Venice possessed more private gardens than anywhere else in Italy. The reasons for this were manifold, but predominantly of an economic, ornamental, representational, and medical nature. The rich literature on early modern Venetian pharmacy (Palmer 1985; Pugliano 2017; Pugliano 2018;

Minuzzi 2016) and the close relationship between Venice's apothecaries and Padua's scholars attest to the exceptional role of the port of Venice in worldwide commerce and in the provision of exotic plants for gardens and pharmacies (Parrish 2015). Both cities had powerful colleges of pharmacists: the *Collegio dei Speziali* in Venice (founded in 1565) and the *Collegio dei farmacisti* of Padua (founded in 1260). In the early years of the Botanical Garden, robberies of herbs and other plants were so frequent that the government was compelled to engage Luigi Anguillara (Squalerno) (1512–1570) as keeper and built him a house within the garden (Paganelli 1988, p. 43). As De Visiani (1840, pp. 3–7) complained, the decline of the Republic of Venice decisively diminished the import from overseas.

Yet, not only commercial aims lay behind the spread of gardens and botany. Often, the practice of *scientia amabilis* kindled profound passion for the plant world. More importantly for this book, the study and use of plants—in all its professional and amateur, scientific and magic-hermeneutic, commercial, artistic, agronomical, and medical guises—also played a remarkable social role. Several historians have emphasized that Venetian pharmacies and—for socially lower stations—barber shops were traditional meeting places for people from different social classes. Along with providing medical remedies and the latest scientific information, they were also social venues for planning editorial enterprises, exchanging news and gossip, making new acquaintances and connections, and to generally combine business and pleasure (De Vivo 2008; Franceschi 1999; Stössl 1983). Daniel Jütte (2012), for instance, has shown that in early modern Venice trading in (secret) alchemical-medicinal-pharmaceutical knowledge could open doors that were otherwise closed to Jews and pave the path to the centers of power. In the 1790s, Vincenzo Dandolo (1758–1819), who had been molded scientifically as well as politically by studying in Padua, ran the pharmacy *Spezieria di Adamo ed Eva* in Venice that was notorious for spreading of French chemical (Lavoisieran) as well as political (revolutionary) ideals (Giormani 1988; Pederzani 2014, pp. 32–35; Preto 1982a).

In the first half of the nineteenth century, botanical interests in Padua continued to give rise to broad networks of persons of different social and political rank, yet bonded together by the long tradition of a shared passion for private and public gardens, herbaria, healing powers, joint field excursions, and the study of exotic species (Bussadori 1988; Egmont 2010, pp. 78–89). The main center of these activities was the Botanical Garden of the university (Fig. 2.1, no. 4; Fig. 2.2; Fig. 8.1), founded in

Fig. 2.2 The Botanical Garden of Padua in 1842. Lithograph by G. B. Cecchini. From Anon. (1842, p. 335). By courtesy of Biblioteca Capitolare di Padova, Coll. 700.AA5.7

1545 (see e.g., Minelli 1995). In the sixteenth and seventeenth centuries, many of its Prefects came from north of the Alps (Minelli 2017), but by the early nineteenth century, it had lost most of its former splendor and its international ties. Already during the eighteenth century, the geographical reach of Italian botanical epistolary networks had shrunk, a symptom of international marginalization (Siegrist 2013, pp. 223–225). Nevertheless, plant studies in Padua still enjoyed prestige and were remarkably lively. Around eighty Veneto amateur botanists produced significant scientific studies during the first decades of the nineteenth century, apothecaries, physicians, noblemen, parish priests, landowners, lawyers, and schoolteachers among them (Busnardo 1998). A list of local and international botanists, who made important contributions to the knowledge of Padua's flora, is also provided by Augusto Béguinot (1909, pp. 52–74).

Hence, the botanical-agronomical network was larger, geographically broader, socially more varied, and less exclusive than the aristocratic-political one. This facilitated the diffusion of ideas and practices into the less wealthy classes, but, as we will see in Sect. 2.3, Sect. 8.2 in Chap. 8 and Sect. 9.3 in Chap. 9, it also provided an opportunity for forging relationships between persons of different ranks. Furthermore, the botanical-agronomical circle provided politically active notables with scientific knowledge and authority with which to expand their sphere of influence (Sect. 2.4). Giuseppe Meneghini was particularly active in these networks. He first was De Visiani's assistant, then his colleague, and collaborated with him in running the Gardening Society (*Società di Giardinaggio*) (Sect. 2.3) and the Festival of Flowers (Sect. 8.2 in Chap. 8). Among the numerous botanophiles that gravitated around the Botanical Garden were socially influential figures like the bankers Giuseppe and Giacomo Treves, the *podestà* (mayor of Padua) Achille De Zigno (1813–1892) and his future counselor, the lichenologist and politician Vittore Trevisan Earl of San Leon (1818–1897), scientific botanists like Giovanni Zanardini (1804–1878), Giuseppe Clementi (1812–1873), Abramo Massalongo (1824–1860), and many others.

The most innovative element of sociality in Padua in the 1830s and 1840s was the formation of the first associations. Andrea and Giuseppe Meneghini soon understood the potential of these new forms of congregation and became their active promoters. Depending on the criteria adopted, the decisive steps toward the establishment of North-Italian associationism took place before 1848 or after 1866. Marco Meriggi considers the 1830s and 1840s as a turning point, because during those years the nature of associationism evolved from intimate and closed clubs to societies that were open to subscribers, and from meeting places where private affairs were negotiated to venues dedicated to public utility (Meriggi 1992a). Steven Soper (2013, p. 11 and 25) and Renato Camurri (2004), among others, recognize pre-Unitarian efforts, but argue that only after 1866 the number of associations surpassed a critical level and they became truly public and open institutions. In Italy, the first associations were founded somewhat later than in other countries (Banti and Meriggi 1991; Malatesta 1988). Nonetheless, these associations played a fundamental role in building of nineteenth-century Italian society. They represented the first instances of extra-familiar organization, and defined new aristocratic-bourgeois classes. According to some scholars, the first associations acted as laboratories of democracy and platforms for the construction of public opinion. They thus anticipated the creation of political parties, worked as centers for the diffusion of news and staying up to date with cultural matters, and, most of all, as places for networking and political arrangements. Besides that, Meriggi and Banti consider them as the places where the notables decided on the career of local candidates (Banti and Meriggi, 1991, p. 361).

It is misleading to regard the early associations as open institutions, however. The level of the membership fees determined the level of exclusivity. Another effective means of selection and control was the admission of new members by cooptation. According to Steven Soper, associationism in Veneto bore some distinctive features. Here, "although in theory they embraced the principles of voluntary association and individual self-help, in practice they created the association as a means by which a small group of enlightened organizers might intervene to transform a backward society" (Soper 2013, p. 6). Hence, associations were participatory, but by no means republican or democratic institutions. Several scholars emphasize that the members of the Paduan economic and cultural elite feared the lower classes and the possibility of an uprising. My analysis broadly confirms this, yet not all active members fit this picture. Andrea Cittadella Vigodarzere did (Sect. 9.2 in Chap. 9), the Meneghini brothers, themselves upstarts, less so.

Padua's first associations faced a double challenge. On the one hand, they wanted to create lobbies and spaces of civil and public influence in harmony with the imperial-monarchal administration, on the other hand, they aimed to develop local forms of power. Consequently, the Austrian government as well as its officials in Venice watched these activities with some suspicion. The foundation of associations was hence difficult to manage and potentially dangerous, yet promised the creation of spaces of political, even if limited, influence. Founders were keen to keep the statutes as apolitical as possible. However, even after the initial obstacles had been overcome, there was no general outburst of liberty and congregation. The increasing popularity of associations in Padua did not mean that people now crowded into assemblies where they passionately debated the future or exploited new, proto-democratic opportunities. Andrea Meneghini repeatedly published calls reminding associates that they were obliged to participate in the yearly general meetings. These appeals went unheard. On July 28, 1847, the first general assembly of the Society for the Encouragement (*Società d'incoraggiamento per l'agricoltura e il commercio*) was deferred, owing to insufficient attendance, and the second, three weeks later, saw the participation of only seventeen members, representing 31 votes out of 200 (Anon. 1847k; Anon. 1847m; Anon. 1847n).

The actual running of the associations was in the hands of a very narrow circle of activists, surrounded by a slightly larger group of tacit sympathizers.

Two further forms of associationism deserve a brief introduction. The first are the mutual aid societies for doctors, surgeons, and pharmacists that emerged in great numbers throughout Italy from 1838 onwards (Soresina 1985). On February 21, 1847, Giovanni Berselli, who was not a member of any of the formal local associations, held the inauguration speech of the *Istituto medico-chirurgico-farmaceutico di mutuo soccorso in Padova* (Berselli 1847). The project was proposed to the Austrian government in 1844 and permission was granted in January 1847. Andrea Meneghini acted as notary (Anon. 1848b). Although many of the founding members were wealthy physicians or university professors, the principal aim of the society was to support the many poor medical officers and their widows and underage children. Yet, despite the initial 118 subscribers, its success was limited. It offered no platform for the exchange of information, and the economic support it provided was meagre. A similar mutual aid project was launched in 1847 for the landowners of Veneto (Sect. 9.3 in Chap. 9). An outline proposal by an anonymous Venetian was published in *Il Tornaconto* on 25 March 1847 (Anon. 1847j) and endorsed by Andrea Meneghini (1847), but never came into fruition.

Newspapers are the second further form of associationism worth mentioning. From the late 1830s, newspapers gradually (though never completely) supplanted the role of salons and cafés in the diffusion of news. In 1846, the associationist spirit also prompted the foundation of Andrea Meneghini's journal *Il Tornaconto* (The Profit). He treated his readers as 'associates', who subscribed for an annual fee of 24 Lire in Padua, and 26 Lire in the rest of the monarchy. It may sound odd to describe a newspaper as an association, but both were based on the same idea. Etymologically, in fact, Marco Meriggi (1992b, pp. 87–88) maintains that in Italy the term associationism had a double origin: on the one hand, in the joint stock companies and other forms of joint financing of traders and entrepreneurs; on the other hand, in the subscriptions to journals and expensive book editions. For Meneghini, creating a circle of associates was more than a way of raising money. He conceived of his *Il Tornaconto* as a means of instruction, but not dogmatic or school-like (Anon. 1847e, p. 9). Rather, he dreamt of a 'conspiracy of forces' (Anon. 1846d, p. 1). He used the term conspiracy not in the sense of plotting, but in the literal sense of *con-spiring*, breathing together. He saw his journal as a forum where everybody was invited to share their opinions on agricultural, technical, or commercial questions (Sect. 9.3 in Chap. 9). Similar to the *Society for the encouragement of agriculture*, he thought of his newspaper as creating a network among the landowners, agriculturists, and industrialists of all Veneto provinces. However, it also aimed to provide information specific to the interests of this circle: remedies for plant pests and parasites, crop yields, new laws and regulations, technical innovations, recent publications, news from agricultural congresses and societies, infrastructure. *Il Tornaconto* aspired to become the voice and organ of Veneto's economy.

Local administrative committees were a final important, yet poorly investigated site of socio-political encounters. They regarded only a very small group of actors, who were involved mostly in local administrative and technical questions. But for this reason, they represent an interesting source for identifying politically powerful groups and for investigating how these groups interacted with the citizenship. Renzo Derosas argues convincingly that, for example, real estate sales were social events, rather than mere economic transactions. Maintaining important relationships and participating in influential circles and salons were as indispensable for receiving positions and commissions as it was for negotiating the purchase of properties and the price to be paid (Derosas 2000, pp. 201–206). From this point of view, Agostino Meneghini's acquisition of the prestigious villa Selvatico (Figs. 3.1 and 4.2) was part of a complex societal play—especially in Veneto, the 'land of villas' (Sect. 4.3 in Chap. 4). Likewise, his son Andrea's appointment to three public offices—the Civic Deputation of Ornament, the Commission for Public Beneficence, and the Provincial Congregation of Padua—most probably had the blessing of the circle of notables.

2.2 Paduan Associations

In the 1830s and 1840s, a general institutional upturn revitalized life in Padua. Several institutions and events were revived and others were newly founded. Among the already existing institutions that enjoyed a revival in these years we can count the Paduan Academy of Sciences, Letters and Arts (Fig. 2.1, K) and the Botanical Garden (Sect. 2.3 and Sect. 8.2 in Chap. 8) (Fig. 2.1, no. 4). The new institutions included journals and three associations. The Gabinetto di Lettura or Reading Cabinet opened its doors in 1830 (Fig. 2.1, L). Its associates—aristocrats, patricians, bourgeois, and students—met in the bookshop of Antonio Carrari Zambeccari's

Tipografia della Minerva (Fig. 2.1, H), where they had Italian and foreign journals at their disposal, at least those permitted by Austrian censors (Callegari 2012, pp. 146–149). About forty years earlier, the physician Francesco Luigi Fanzago (1764–1836) had already founded a Reading club (*Società di lettura*) as a halfway house between the rigorous scientific debates of the Academy and the genteel conversations of the salons. Yet, this club was short-lived (Del Negro 1992).[3] The *Società promotrice del Giardinaggio*, or Society for the Promotion of Gardening (Sect. 2.3), and the *Società d'Incoraggiamento per l'Agricoltura e il Commercio*, or Society for the Encouragement of Agriculture and Commerce (Sect. 9.2 in Chap. 9), sprang into life in 1846.

The women and men who came together in these three associations did not represent all of Padua's society. Besides their social and economic distinction, the members shared at least a good part of a set of values, attitudes, and norms. The promoters of the Reading Cabinet and the Agricultural Society gravitated toward moderate liberal ideals yet mixed with paternalistic and protectionist elements. Like their European counterparts, most Paduan liberals agreed on the ideals of progress, individual freedom, respect for property, freedom of expression, equality before the law, equitable distribution of taxes, and access to public office based on personal merit (Freeden and Fernandez-Sebastián 2019, p. 11). The creation of associations was an evident sign that conceptions of individuality and well-working togetherness had undergone considerable change. I will deepen this argument in Chaps. 6 and 7. For now, I just want to say that ideas of the interdependence of free individuals presupposed, at least ideally, the worth of each and every person and a compromise between maximum personal freedom and maximum public wellbeing, rather than self-centered individualism. Associations were neither atomistic nor homogeneous, but collaborative forms of congregation. In contrast to late-nineteenth-century debates, when political elites began to interject that the masses curtail the liberties of single individuals (Meriggi 1992b, pp. 92–100), during the pre-1848 period, cooperation was considered not a denial of individual needs but rather as a means to increase the advantage of every single member of a community. All of this created a hotbed for organic theories of the state.

In a European context, Italian associationism emerged with some delay. In Great Britain and Ireland, at least 82 agricultural societies were active prior to 1810, about 20 societies with 29 local branches existed in pre-revolutionary France, in Spain there were some 70, in the German-speaking lands, depending on the definition, between 48 and 146, in Denmark 57, in Norway 26, and in Sweden 25 (Jones 2016, pp. 68–72; Stapelbroek and Marjanen 2012, pp. 7–9). If one counts associations other than that of the agricultural kind, around mid-century, there were some 14,000 in the German lands, 2234 in Austria, and 32,000 associations (counting four million members) in England in 1874 (Meriggi 1992b, p. 102). In Padua, only 327 people were officially members of the three pre-1848 associations. Meriggi (1992b, p. 91) estimates that associations in Milan represented about 0.66% of the local population (1000 associates out of 150,000 citizens). In Padua, the percentage was almost identical: 0.65% (327 associates out of 50,433 Paduan citizens), 0.1% if we consider the province of Padua with 312,450 inhabitants.[4] Although the statutes stipulated openness to all, the individuals who stepped out of their privacy and were willing to assume public visibility and join their forces in associations actually represented a very thin slice of the population.

In 1846–1847, the Gardening Society had a total of 159 Paduan associates (another 68 were from other Veneto provinces); in 1839–1843, the Reading cabinet had 178 associates, and the Society for the Encouragement of Agriculture had 91 founding members in 1841. Sixteen people (about 5% of the 327 Paduan members of the three associations) were members of all three, and forty-five (about 14%) were members of two associations. Thirty-three men assumed some kind of office. Eighty-one persons (25%) show multiple attachments if we count the university, the academies, and the congresses of Italian scientists (Table 2.1 and Sect. 2.4). There were surely others, many women among them, who were influential but stayed behind the scenes. However, we may identify these eighty names as Padua's cultural-intellectual élite. Unsurprisingly, in 1848, several of them felt the calling to assume political charge when the Austrians withdrew (Chap. 10). A comparison with the members of the *Comitato Provvisorio Dipartimentale*, the provisional revolutionary government of Padua, indeed confirms a significant overlap. Thirteen of the thirty-five officers of the *Comitato* were members of at least one association.

Fifteen of the luminaries were highly active: Andrea Cittadella Vigodarzere, the historians Giovanni Cittadella (1806–1884) and Abbot Lodovico Menin (1783–1868), the physician Giacomo Andrea Giacomini (1796–1849), the architects Giuseppe Jappelli and Pietro Selvatico Estense (1803–1880), the mathematicians Serafino Raffaello Minich (1808–1883) and Domenico Turazza (1813–1892), the astronomer Giovanni Santini (1787–1877), the agronomist Luigi Configliachi (1787–1864), the botanist Roberto de Visiani, the geologist Nicolò Da Rio (1765–1845), the naturalist Achille de Zigno, and—obviously—Andrea and Giuseppe Meneghini.

Eighty-five (26%) associates attended at least one of the Italian scientific congresses. Equally significant is the rate of university professors. Eight men (53%) of the 'inner circle' and forty-three (13%) of the associates were professors or deans at the University of Padua. Conversely, half of the university professors and deans of the 1840s were members of at least one of Padua's associations. Apart from the theologians, who made for only three associates, the others came from the other three faculties in fairly even proportions. However, if we consider the fifteen most active men, we notice that most of them cultivated medical-scientific-mathematical interests, and of these, three were professional botanists and two were eminent amateur naturalists. This means that, contrary to several scholars who assume that economic leadership dovetailed only with humanistic-literate interests (e.g., Casalena 2018; Fumian 1996, p. 10),[5] the natural sciences, in particular the plant sciences, had branched out into Padua's new civic institutions, whereas the fine arts were confined to their own institutions and thus developed looser ties to the new networks.

In this context, both Meneghini brothers stand out for their zeal. Not only did they, together, cover the whole spectrum of associations and institutions analyzed here, they also occupied important offices. Andrea was a member of the Gardening Society, member and (in 1846) commissioner of the Agricultural Society, and member, two times (1836 and 1838) treasurer and one time (1844) secretary of the Reading cabinet. He attended the congresses of Milan (1844) and Venice (1847). In addition, in 1838 he was member of the *Civica Deputazione all'Ornato* (which determined whether new buildings or architectural changes fitted into the townscape) and the *Commissione di Pubblica Beneficenza* (Commission for Public Beneficence), and from 1841 deputy to the Provincial Congregation of Padua (Sect. 5.3 in Chap. 5). Finally, he owned and managed the weekly newspaper *Il Tornaconto*. Giuseppe was vice-president of the Gardening Society; member and president (1845–1848) of the Reading cabinet; professor at the university; member and librarian of the Academy of Science; and member of the *Istituto Veneto*. He participated in seven of the nine national scientific congresses, as an attendee (Pisa), as deputy of the Paduan Academy (Milan, Naples, and Genoa), and/or as secretary or vice-president of the botanical section (Florence, Padua, Naples, Genoa, and Venice). Moreover Charles Lucien Bonaparte (1803–1857), nephew of

Emperor Napoleon and renowned ornithologist, appointed him to a commission charged to elaborate norms for a unified biological nomenclature (Minelli 2008, p. 37). He missed only two of the congresses, participating anyhow via letters and reports. Finally, he regularly contributed to *Il Tornaconto* and to the *Giornale Euganeo.*

2.3 Plants and Politics in the Gardening Society

In order to emphasize the nature of the new forms of civil congregation and the role of plants in it, it is worth putting the spotlight on the *Società promotrice del Giardinaggio*. This association was founded on 15 January 1846, as the first of its kind in Italy. Elsewhere, similar horticultural and floricultural societies appeared only from the 1850s, for example in Tuscany, Lombardy (together with the first specialized journal, *Il Giardiniere*, from 1854 called *I Giardini*), Piedmont, and Rome (Maniero and Macellari 2005, pp. 188–200). Not surprisingly, the Paduan Society collaborated closely with the Botanical Garden. Its president was Roberto de Visiani and its secretary the pharmacologist Giovanni Battista Ronconi (1812–1886). In December, Giuseppe Meneghini became vice-president, and Isacco Treves dei Bonfili became treasurer. Andrea Cittadella Vigodarzere did not assume any official position, but he was actively engaged in elaborating the statutes.

In contrast to the exclusiveness of the private salons and to the elective membership of the Academy of Science, the Gardening Society was open to anybody willing (or able) to pay an annual membership fee of 24 Austrian pounds (*lira austriaca*, corresponding to eight Lombardy-Venetia florins) for a minimum of three years (Anon. 1846b, p. 1). In the opening year of 1846, the society had 190 members. In 1847, there were 223. The membership list mentioned princes alongside simple citizens, a mixture of high and low aristocracy along with bourgeois. Moreover, this included persons of different, even opposing political orientations: 'Jacobins' like Girolamo Polcastro and Elisa Napoleona Camerata Baciocchi (1806–1869), nephew of Napoleon Bonaparte; future volunteer nurses in the 1848 upheavals like Teresa Papadopoli Mosconi (1807–1854), Catterina Polcastro Querini (c1796–1869), Angelina Sartori, and Carolina Zucchetta Steyer; but also high-ranking officers (or their wives) of the Austrian army in Padua. Among the latter were Wilhelm Karl von Thurn und Taxis (1801–1848), deadly wounded during the siege of Venice, Field-Marshal-Lieutenant Konstantin Freiherr d'Aspre von Hoobreuck (1789–1850), the military commander of the Austrian troops in Veneto, and Maria Anna von Wimpffen (1802–1862), wife of Austrian Feldmarschall Franz von Wimpffen (1797–1870). A third noteworthy aspect was the geographical reach of its membership. The Gardening Society was an urban Paduan association, yet not confined to it: 159 (70%) of its members indeed resided in the province of Padua, but thirty-three (14.5%) in the province of Venice, eighteen (8%) in the province of Vicenza, thirteen (5.7%) in that of Treviso, three (1.4%) in that of Udine, and one in Rovigo.

Finally, but importantly, women were also represented. The other Paduan associations did not explicitly exclude women, either. Yet Loredana Morosini Gatterburg (1806–1884), last descendant of a famous Venetian family, and the Austrian Maria Anna von Wimpffen were members of the Società d'incoraggiamento only in 1853 (Anon. 1854, p. 61). Cittadella Vigodarzere invited women to attend the Congresses of Italian scientists (Sect. 2.4), where many indeed participated in the botanical sections (Klemun 2021). Nevertheless, in the 1840s, only the Gardening Society actually had female members. Women were also prominent participants at the floral exhibitions in the botanical gardens of Modena and Padua (Maniero and Macellari 2005, pp. 75–78). In Padua, of the 227 members, 52 were women—almost 23%.[6] Most of them were noblewomen, some were bourgeois. The activities of women of all social ranks had steadily moved from the private to the public in Venetian culture from the late eighteenth century onwards (Filippini 2006; Filippini and Gazzetta 2011). Several scholars of the history of this period have illuminated the role of *salonnières*, female writers, and philanthropists (e.g., Betri and Brambilla 2004; Soldani 2007, pp. 198–202), few of them have touched on the role of women in early modern gardening, floriculture, and horticulture (e.g., Egmont 2010, pp. 45–72). Since the early modern age, Paduan female authors treated nature and flowers in their noteworthy pastoral-poetic works (Sect. 8.1 in Chap. 8). The Paduans Valeria Miani Negri (1563–1620) and Isabella Andreini Canali (1562–1604) respectively wrote *Amorosa Speranza* (Venezia, 1604) and the fable *Mirtilla* (Verona, 1588). Maddalena Campiglia (1553–1595) from nearby Vicenza received praise from Torquato Tasso (1544–1595), undisputed leader of the pastoral genre, for her fable *Flori: Favola boscareccia* (Vicenza, 1588) (Cox 2011; Rees 2008). Botany was hence a prominent feature of Veneto female culture, and in the 1840s, not only literature, but also flowers and the participation in 'unsuspicious' gardening societies gave women higher social visibility and the opportunity to overcome their marginality in official, academic, and institutional terms.

Even if most associates probably did not meet each other (although the statutes foresaw the participation in the general assemblies), the social, political, geographical, and gender diversity is striking. More than the other two Paduan associations, and much more than the academic institutions, the Gardening Society transgressed such boundaries. Contact between members of different social rank was facilitated by the fact that, compared to parks, which require land and considerable funds, the cultivation of plants and flowers did not necessarily demand large spaces. Historians have described many examples throughout the modern age of how naturalistic knowledge facilitated access to and even friendships with persons of much higher social status. A compelling example of the function of naturalistic expertise as a bridge between lovers of nature of different social ranks in the sixteenth century is that between Giuseppe Casabona (Joseph Goedenhuyze) (1500–1596) and the archdukes of Toscana (Olmi 1991, pp. 15–23). The close relationship of the 'Flower Emperor' Francis II of Habsburg (1768–1835), a learned gardener, and his botanist Franz Antoine (1768–1834) is but one example. Another is that of De Visiani and King Friedrich August II of Saxony (1797–1854), an amateur botanist who had travelled to Istria, Dalmatia, and Montenegro in spring 1838. As a reward for De Visiani's services, the king is even said to have offered to correct the manuscript of the three volumes of De Visiani's *Flora dalmatica* (De Visiani 1842–1852), published in Leipzig (Clementi 2017, p. 87; Curti and Menegalle 1996, p. 80). From the late eighteenth century onwards, this phenomenon extended to even lower ranks. The love for plants began to enter middle-class gardens and homes, which ceased to be mere sleeping and eating places. People from this background, too, entered the Gardening Society, and De Visiani gladly took any opportunity to promote the love of plants also in simple houses (Sect. 8.2 in Chap. 8).

The fact that the directory of the society encompassed such a heterogeneous group of people suggests that the majority of its members did not regard the associations in general, and the Gardening Society in particular, as places for proto-democratic political efforts, even less of conspiracy. Practical or reasons of prestige motivated the membership. Moreover, then and now, gardens, plants, and flowers were considered beautiful but sufficiently irrelevant. Even historical scholarship on associations neglects the gardening societies. Yet, precisely this seemingly apolitical character as well as the practical usefulness of membership paved the way for a tacit social revolution. The Gardening Society blurred the boundaries of gender and social rank more than any other institution of that time. It is worth mentioning that members of high nobility agreed to appear—not ordered by social rank, but alphabetically—in the same register as members from the lower ranks and subscribed to statutes that gave equal rights to all associates.

One should also bear in mind that, in these decades, a devotion to plants and flowers was everything but politically neutral. Some examples illustrate the multilayered and diverse nature of 'plant messages'. As we will see in Chap. 4, pre-1848 liberals preferred landscape gardens—the praise of the irregularity in their design was a response to the austerity of absolutistic symmetric parks. Likewise, the romantic emphasis on the unfamiliar and the monstrous—hence definition-eluding and class-transgressing specimens—for some was a countermove against the strictness of prevailing taxonomic as well as political order (Kelley 2012).

Particular plants could be vectors of political messages, too. Numerous recent studies emphasize the symbolic role of plants in literature, philosophy, and politics (e.g., Goody 1993; Jacobs and Kranz 2017; Kittelmann 2017). Not all connotations were positive. Jean Jacques Rousseau (1712–1778), for instance, detested horticulture and flowers as luxuriant but sick and aberrant products of society (Cook 2012; George 2005). A nice example of the political symbolism of plants are 'revolutionary trees'. Trees, due to their magnificence and direction toward the universe, occupy a special place in the mythology of many peoples (Le Quellec 2008). In recent European history, the tree metaphor has played a special role during Northern Italy's French period when the new governors put up 'liberty trees'—sometimes tall oak trees, but usually just simple wooden poles crowned by a red cap—in the middle of the village squares as symbols of revolutionary and Napoleonic achievements. During the anti-Jacobin raids of the Restoration, many of these trees were felled or renamed (Ambrosini 2014; Brofferio 1857–1861, I, pp. 21–28).[7] In Padua, the French conquerors erected and inaugurated such a propagandistic tree on 30 April 1797. The pole stood in the middle of the great square, the *Prato della Valle*. Songs celebrated the new liberty (Ronchi 1906, p. 29):

> *Sorgi felice pianta / Sorgi beato regno / Caro e beato pegno / Di Nostra Libertà / Ecc.*
>
> *Rise joyful plant / Rise blissful reign / Dear and blessed pledge / Of our Liberty / Etc.*

The population, however, showed general indifference. A popular jingle went:

> *Bareta senza testa / Albero senza vesta / Libertà che no resta / Quattro minchioni che fa festa.*[8]
>
> *Beret without pate / Tree without wear / Liberty that will not abide / Four idiots who celebrate.*

Upon the arrival of the Austrians on 19 January 1798, the 'tree' was cut down.

More commonly, flowers have played a prominent metaphoric and symbolic role in many of the revolutions since 1789, for example in the flower power movement of the 1960s and 1970s as well as during the Kyrgyz Tulip Revolution of 2005. Flowers exert an extraordinary emotional and stimulating power on humans that many have sought to exploit to communicate political and social messages and to create identities (Sittler 2016). A rather different kind of botanical metaphor, which will be object of Sect. 9.4 in Chap. 9, is the potato, a plant that assumed strong political connotations in the late 1840s. Also the subject of important agricultural and scientific debates during the famine of 1846–47, the potato stood for *il Tedesco*, that is, *the German*, which at the time meant *the Austrian*.

Hence, the preference for certain types of flowers was not only a means of aesthetical, but also of social and political differentiation and expression (Hyde 2002). Yet, the meaning that particular species carried for different people often varied greatly. This holds true even for the sentimental meanings of flowers, very common in the nineteenth century (Seaton 1995). The *Biedermeier* period was a period of flower mania, and preferences for specific plants often expressed political convictions. Around 1800, American gardeners proudly planted native species as symbol of patriotism (Wulf 2011, p. 2). Supporters of absolutist regimes preferred the durability of evergreens, conservatives revered the oak as a symbol of solemn composure and slow cyclical change, whereas liberal bourgeois and enlightened aristocrats preferred flowering annuals, because the image of the developing plant for them held a powerful metaphor of the unfolding of the nation and bourgeois freedom (Klemun 2016, pp. 21–22). During the 1810s, Lorenz Oken (1779–1851) transposed plant imagery onto political debate, drawing explicit parallels between the progressive development (*Entfaltung*) of organisms and the birth of the German nation

(Breidbach et al. 2001). For him, both processes were natural, teleological, and unavoidable. Later we recover this analogy in Giuseppe Meneghini's biological and political concept of progress (Chap. 7).

The Gardening Society, too, was a showcase of political plant metaphors, yet the metaphors were subtle and only promulgated by individuals. Niccolò Tommaseo's (1802–1874) interpretation of the meaning of flowers was similar to Oken's. Tommaseo was a famous linguist, one of the future leaders of the 1848 insurrection and the short-lived Venetian Republic of San Marco, and a close friend of De Visiani. Both were born in Sebenico, Dalmatia, and both graduated from Padua in 1822, the former in law, the latter in medicine. Tommaseo (1846, p. 28) advertised and praised De Visiani's Flower Festival and the Gardening Society in the journal *La Dalmazia* (1845–1847), a center of gravity for a group of Dalmatian intellectuals of romantic-Mazzinian creed. He complained that the society's name was 'disfigured' by the word *giardinaggio* (deriving from the French *jardinage*).[9] In fact, the Italian term had been introduced to Padua, and probably to the whole Veneto, by the *Società promotrice del giardinaggio* (Clementi 1845, p. 3094). Tommaseo would have preferred an Italian term. He preferred *giardineria* but admitted that *orticoltura* did not mean exactly the same and therefore included *giardinaggio* in his vocabulary, a work he had started around 1835 and that was to become the most important Italian dictionary of the Risorgimento period (Tommaseo and Bellini 1861–1879, II, p. 1068).

Tommaseo seized the opportunity of the short article on De Visiani's Society, also re-published in other North Italian journals like *Il Felsineo* and *L'Amico del Contadino*, to make some floral-political allusions. He emphasized the moralizing effect of flowers, because, in his view, through the very process of cultivating, flowers taught to also cultivate oneself (Tommaseo 1846, p. 29). Horticulture, according to him, was full of symbolism and taught us 'immortal things'. Paraphrasing Dante's famous metaphor "the blossom shall be followed by good fruit" (Dante 2007, p. 741),[10] Tommaseo saw Italy as the fruit and interpreted the flower as announcing resurging hope (*risorgente speranza*). A few lines later, his call for a Risorgimento became more explicit:

> I hope that man will finally realize that the great things have to be undertaken, and will be accomplished without waiting for the favour of chance or invoking the charity of the mighty; oh yes, the private forces, associated with persevering order, and with unanimous fervour directed towards an aim.

Only in this way, nations can expect renovation and greatness. (Tommaseo 1846, p. 29)[11]

During the Risorgimento, Dante was indeed hailed as the prophet of Italy's unification. His *Divine Comedy* was replete with vegetal metaphors (Di Santo 1993; Tomazzoli 2017). Yet, De Visiani did not swing behind Tommaseo's attempts to create a floral symbol of the nationalist movement. Nor did he take this as an opportunity to express his own political convictions in any other form. This was probably also due to his origin. Despite his Italian patriotism, having been born in Dalmatia, he risked remaining an Austrian citizen even after the Italian unification (Moreno Clementi, e-mail correspondence with author, 14 Aug 2016). The multiculturalism of people like De Visiani, who grew up in regions populated by patently diverse linguistic and religious groups, may have constituted another reason for his ambiguity (Kirchner Reill 2012). In 1840, his rhetoric in attempting to restore the magnificence of the Botanical Garden and in expressing his love for plants and gardens in Veneto showed indeed nationalistic elements by recalling Italy's past splendor and appealing to the national pride (De Visiani 1840).[12] However, in view of the many important Austrian members of his Gardening Society and the fact that the Austrian government was the main (if not only) financial sponsor of the Garden, he could hardly alienate powerful allies and transform the society into a site of political subversion.

Along with nationalistic-risorgimental sentiments, Padua's Gardening Society indeed provided room for Austrophile sentiments as well. One of its members was Baron Carl von Hügel (1796–1870), a high-ranked army officer, diplomat, explorer, and botanist, and a man who crossed Giuseppe Meneghini's path in several, decisive ways. He celebrated blossoms for their magical beauty and as symbols of self-sacrifice (Rotenberg 2002, p. 162). For him, flowers not only were full of symbolic meaning. He also realized the great economic potential of the expanding market around aristocratic, bourgeois, and middle-class gardens and even home installations. He became involved in floriculture as a hobby when he bought a villa with a vast park in Hietzing, a district of Vienna. Encouraged by the success of his own plants and flowers, he established a company that soon became the leading commercial garden establishment in Vienna. In 1837, he founded the *k.k. Gartenbau-Gesellschaft*, or Imperial Horticultural Society, of which he was president until 1848. The popularity of gardens and plant collections in Vienna and surroundings was also actively supported by Empress Maria Theresa (1717–1780) and subsequently, even more vigorously, by the 'Flower emperor', Francis I of Austria, and his brothers, who had spent a happy childhood in the Florentine Boboli Garden and had a real passion for gardening (Hlavac and Göttche 2016; Riedl-Dorn 1989). For them, especially the exotic plants gained symbolic value as a result and an expression of the Habsburgs's world-spanning relationships. The tropical plants were housed in several *Palmenhäuser*, built in the 1820s, and recorded on the over 1800 large watercolors of the *Florilegia* (Anon. 2019). More common and broadly accessible were flower bouquets composed of fritillaries, roses, auricles, narcissus (daffodils) and zinnia, whose initials formed the name 'Franz', reproduced also on greeting cards, china, and paintings (Lack 2006, pp. 16–17).

At first glance, it may seem paradoxical that a Paduan association developed latent political potential under the auspices of the Austrians. As a matter of fact, without the financial aid and promotion of the Habsburgs, there probably would not have been a renaissance of the Botanical Garden and its activities. By the 1830s, the garden had lost most of its former splendor. To make matters worse, on 26 August 1834, a severe hailstorm destroyed many of Padua's roofs and almost all of the plants of the garden. When De Visiani took over its direction in 1836, only 3000 species out of the over 5500 previously cultivated still existed (M. Clementi 2017, p. 62). Yet, De Visiani managed to find financial resources, in large part from the Viceroy of Lombardy-Venetia, archduke Rainer Joseph of Austria (1783–1852), to restore the walls, buildings, fences, pools, and gates, and to rebuild the water pumps and erect two greenhouses. During these crucial years, respectively from 1835–1839 and from 1839–1845, his assistants were Giuseppe Meneghini and Giuseppe Clementi. Under the management of these three men, the garden resumed its traditional role as a site of education and study, with new research facilities (albeit no laboratory), a botanical lecture theatre, and about 6000 tinplate tags for the names of the species being cultivated. Through travels, exchange, and purchase, the number of species quickly grew to a new peak in biodiversity (M. Clementi 2017, pp. 62–63; Curti, 1993; De Visiani 1840).

In order to bring together as many plant-loving people as possible, De Visiani avoided any political undertones and banked on the aesthetical and economic benefits that the Gardening Society would bring to each member. Indeed, the statutory purpose of the Paduan Society was "to promote the best culture of gardens in particular in the provinces of Veneto" (Anon. 1846, p. 1). And it was rather successful in facilitating the creation of networks of producers and consumers of plants and flowers, and backing the nascent floricultural economy in Veneto (De Visiani 1847a) (Chap. 8). Moreover, De Visiani's activities as president of the Society made the Botanical Garden a favored site of sociality in Padua—and beyond. In 1847, the attendees from the scientific congress of Venice came to Padua by train to visit the Flower Festival (Fig. 8.1). That year, a visit to the Botanical Garden—along with the Church S. Antonio, the Caffé Pedrocchi, the university, and a number of other sites—was on an official program prepared for illustrious visitors, like the entrepreneur and political leader Richard Cobden (1804–1865) (Anon. 1847l).

After the main structural alterations had been completed and the brilliance of the flowers and plants restored, the foundation of the Gardening Society and the launch of the spectacular Festival of Flowers, the Botanical Garden thus became an integral part of Padua's cultural life. Under De Visiani's direction (Fig. on cover), the Botanical Garden did not restrict itself to private study and contemplation, but transformed itself into a marketplace of social, political, aesthetical, and economic ideas. In 1847, De Visiani enthusiastically emphasized that: "this festival happily grafted the useful and the delightful, made science sensible, enjoyable, and popular, and fraternized and melted all social inequalities in one sole sentiment of natural beauty" (De Visiani 1847b, p. 419).[13] Giuseppe Olmi (2007) has pointed out that in the early modern age the passion for flowers and gardens went beyond political and religious divides. Similarly, in pre-1848 Padua, the apolitical nature of the Gardening Society was the basis of its tacit, yet remarkable political impact, that is, the meeting of science, culture, and politics beyond the boundaries of gender, social status, and geographical affiliation.

2.4 Science as an Instrument for Padua's Regional Influence

With the fall of the Republic of Venice in 1797, Veneto ceased to be an autonomous territorial unit and acquiesced to being incorporated into larger political organizations. Even worse, the French as well as the Austrians regarded Lombardy-Veneto, even though it was a proper state or kingdom, not as an equal among others, but as a peripheral province and military buffer. As humiliating as this was, in particular for Veneto's political and social elite, this constellation presented a new opportunity for

Padua. The new political organization had considerable repercussion also on the relationship between Venice and Veneto (see, e.g., Antonielli 2001; Ferrari 1958). For centuries, the region had been Venice's *hinterland*. Most decisions regarding Padua's economic, scientific, and educational destiny had been taken forty kilometers to the east in the lagoon city, which had a long history in mistaking its own interests for those of the whole state. Many decisions had been advantageous for Padua, like the foundation of the university and the Botanical Garden. Yet, many people were sore from continuous dictate and condescension and longed for greater independence. Andrea Meneghini's letter, in 1848, to the president of the re-created Republic of San Marco, Daniele Manin (1804–1857) is emblematic of this. Meneghini expressed a fear of many Paduans of an 'overly restrictive brotherhood' with Venice and a return of Venetian hegemony (Ginsborg 1979, p. 121). While Padua was behind in the arts and literature and economically backward, its biggest advantage was its intellectual and scientific superiority, and this was a card to lead.

In the 1840s, Padua, Venice, Vicenza, and Verona were tied connected closer together. The construction of the Ferdinandea railroad made journeys between these cities faster, easier, and more affordable (Fig. 1.1; Fig. 2.1, M; Fig. 8.1). In 1843, it linked Padua to Venice, in 1846 Padua to Vicenza, and in 1849 Vicenza to Verona. The whole Milan-Venice railway was completed in 1878. Against this background—even though travel to Venice or Padua continued to be burdensome for people living in Udine, Treviso, Trieste, and other cities in the eastern part of Veneto—it made sense and was reasonable to open the Paduan associations explicitly to all Veneto provinces, and to create overarching institutions. The *Istituto Veneto di Scienze, Lettere ed Arti* (not to be confused with the *Ateneo Veneto di Scienze, Lettere ed Arti*, whose members were inhabitants of Venice) was such a regional institution. Revitalized in 1838 by the Austrian government, it was located in Venice, yet the members from Padua soon dominated the scene. The Austrian rulers insisted that the *Istituto* should not become a leisurely meeting place for the Venetian aristocracy, but a union of scholars selected on technical-scientific merit (Gullino 1996, pp. 32–34). Paradoxically, therefore, it was the Habsburg monarchy that insisted that the *Istituto Veneto* lived up to bourgeois scholars and ideals.

All members received a uniform and many of them an annuity. A Cabinet of Technology was installed to promote the advancement of mechanics. To please the Venetian elite, the president was chosen from Venice's upper echelons, yet the writer Leonardo Manin (1771–1853), nephew of the last Doge Ludovico Manin, turned out to be inept. In 1845, when the second president, the astronomer Giovanni Santini, likewise showed little interest in advancing the *Istituto*'s affairs, the governmental commission nominated Andrea Cittadella Vigodarzere, even though the statutes prohibited the appointment of honorary members like him. This choice was fortunate. Cittadella Vigodarzere did not miss the opportunity to show his political-administrative skills. Under his presidency, the institution was finally profitable, sponsored initiatives and encouraged the advancement of regional industry and commerce by means of publications, competitions, and prizes (Gullino 1996, pp. 38–40). Santini's and, even more so, Cittadella Vigodarzere's nomination demonstrated to all that, when there was a need for scientific-technological expertise and involvement, the Austrian authorities casted a searching glance toward Padua.

The membership of the *Istituto*, too, increasingly fell under Paduan influence. In the 1840s, 44% of its fellows (soci effettivi) were from Padua, but this fraction rises to nearly 80%, if we include Cittadella Vigodarzere and those who had spent at least part of their professionally formative years at Padua's university. The role of science and expert knowledge in the political and administrative redistribution of power between Padua and Venice and between the city of Padua and its province deserves much more attention. Several illuminating analyses of the relationship between the administrations of Veneto's cities and the provinces (e.g., Agostini 2011; Calabi 2001) have hitherto mostly ignored this aspect, yet it is beyond the aim of this book, too. However, for Padua, establishing itself as Veneto's intellectual center was politically immensely expedient as well as a form of revenge toward the rival lagoon city, especially at a time when many thinkers considered science and technology as key tools for future progress and prosperity. As limited as the impact of the *Istituto* may have been, it permitted Padua to increase its political authority and independence, and to extend its influence over the Veneto provinces and into the countryside. Concomitantly, the Paduan associations *Società d'incoraggiamento per l'agricoltura*, *Società promotrice di giardinaggio* as well as Andrea Meneghini's journal *Il Tornaconto* explicitly addressed all of Veneto, and the reach of De Visiani's successful floral exhibitions even extended into Lombardy (Sect. 8.2 in Chap. 8).

As the city of Venice fell behind, so did the arts and the humanistic disciplines. They had their own institutions, for instance the *Istituto di Belle Arti di Venezia*, founded in 1750, but these, despite their cultural and artistic relevance, lacked close institutional bonds with the political sphere. During the period 1840–1848, the members of the *Istituto Veneto* indeed mainly belonged to scientific-technological disciplines: seven mathematicians (including one astronomer), five engineers (including one architect), four chemists, three physicists, three botanists (Meneghini, De Visiani and the Veronese Giulio Sandri (1789–1875)), three physicians, three agronomists, two naturalists, one geologist, four men of letters, four historians, one philosopher, and one jurist. The scientific fellows also attended the (obligatory) meetings most assiduously: 89% of the natural scientists participated in all assemblies, but only 11% of the humanists (Gullino 1996, pp. 48–49).

A similar dominance of Paduan scholars and of the scientific-technological disciplines is evident at the two *Riunioni degli scienziati italiani* that took place in Veneto. The Meetings of Italian scientists were yet another kind of congregation initiated in the pre-1848 period. They were held in Pisa (1839), Turin (1840), Florence (1841), Padua (1842), Lucca (1843), Milan (1844), Naples (1845), Genoa (1846), and Venice (1847), and were expressions of the general spirit of associationism that pervaded these years, albeit on an extended geographical level. In fact, in addition to the local and regional level, these meetings afforded a unique occasion for networking on the national (Italian) and international level, and people like Giuseppe Meneghini made the most of this opportunity. These itinerant meetings, sometimes also called congresses, were inspired by the ideal of scientific universalism and by analogous occasions started in other European countries some years earlier. Lorenz Oken was indeed present at the first congress in Pisa to reinforce the ideal of an international confraternity of knowledge (Pancaldi 1983).

However, as has been stressed by Maria Pia Casalena (2007a), the Italian congresses differed considerably from their European counterparts. Indeed, every single congress had significant peculiarities, in particular with respect to organization and the make-up of the participants. The German meetings, the *Versammlungen Deutscher Naturforscher und Aerzte*, for instance, admitted only naturalists and physicians, and only those who demonstrated a certain level of professionality, that is, university professors or authors of important scientific monographs. The main criterion for admission to the Italian congresses was membership in an Italian or foreign academy, society, or educational institution. The fact that Italy possessed a particularly rich spectrum of academies of all types and levels created a large pool of possible attendees.

Attendees met in six distinct sections: Physics, chemistry, and mathematics; Geology, mineralogy, and geography; Botany and plant physiology; Zoology and comparative anatomy; Medicine; and Agronomy and technology. Subsequently, a section of surgery was added, and chemistry became a separate section. History, political economy, law, literature, philosophy, the arts, and similar disciplines suspected to be potential vehicles of political campaigning were officially excluded. Nevertheless, as members of multidisciplinary academies, many humanistic scholars received accreditation. Moreover, from the first congress, and in contrast to those taking place in other European countries, a section on agronomy was included, and became notorious for political discussions. Rather than concentrating on technical questions, it attracted numerous landowners, many with scientific interests, but few with publications. At Venice's congress, the last one before 1848, agronomy was the biggest section with 468 enrollees, while medicine had 412 and botany only 50. As general president of Padua's congress, Andrea Cittadella Vigodarzere had put particular emphasis on inviting a broad audience, and explicitly included women (Casalena 2007b, pp. 170–172; Fumian 2004, pp. 214–215). On other occasions, some participants even advanced proposals to open the congresses to anyone willing to participate (Pancaldi 1983). Instead of bringing together a carefully selected group of experts to exchange cutting-edge scientific research, the congresses were hybrid spaces of scientists, amateurs, notables, and administrators, and this was suspiciously and jealously, sometimes sarcastically noted by the excluded humanists and literates (Meriggi 2011). The particular make-up of the attendees also holds true for those from Padua. Cross-referencing with the members of Padua's associations indeed reveals significant overlap (Table 2.1) and shows that many attendees were not scientists in the strict sense, but members of the social elite. And yet another aspect dovetails with the social elitism of congresses and associations. Even though they embraced civil society, the congresses were not exercises in plenary democracy, but directed by a council, an inner circle of notable members of the scientific and civil community (Casalena 2018, p. 279).

Only about 40% of the 4800 Italian attendees of all congresses came from the same region where the congress took place. Despite initial restrictions, in particular from the governments of the Bourbon Kingdom of the Two Sicilies and of the Papal States, participation by scholars from all Italian states as well as from foreign lands grew steadily. More than 300, most of them subjects of Lombardy-Veneto, were particularly eager and took part in four or more congresses (Casalena 2007b, p. 163). For Italian scholars of the time, scattered as they were over many bigger and smaller centers, the congresses provided an extraordinary concentration of encounters and facilitated temporary but intense osmosis between local, regional, national, and international scientific realities and knowledges. They were a melting pot of professional or semi-professional scientists and civil society, of ecclesiastics and laymen, and of science and power.

The broad-ranging spectrum of participants, wide coverage by liberal newspapers, and the patronage by the governors and regents, keen to show their scientific treasures to their European neighbors, transformed the Italian scientific congresses into great social events, with dances, raffles, theatre plays, horse races at Prato del Valle, and excursions like the visit to the Festival of Flowers in Padua (Sect. 8.2 in Chap. 8). Venice even postponed the date of its traditional regatta to entertain the attendees (Anon. 1847p, p. 338). Every venue bestowed a different emphasis on its congress. Naples, for example, was keen to celebrate its archeologic sites and hence admitted archaeologists and even historians. Padua lived up to its botanical reputation and welcomed a high number of agronomists, gardeners, and apothecaries among the attendees (Fumian 1996, pp. 226–227).

Padua hosted the forth congress, and Venice was the ninth and last venue of the pre-1848 period. A look at the attendees of all nine congresses reveals that about 270 came from Padua. Of these, 149 attended the congress in Padua and 214 the one in Venice. Giuseppe Meneghini attended seven congresses. He skipped only those in Turin and in Lucca, though he participated through letters that were read before the audience. Nominations as secretary (in Padua and Genoa) or vice-president of the botany section (in Florence, Naples, and Venice) opened him doors to the Council of presidency and hence to the inner circle of decision makers. The only other equally active scholars from Padua were Roberto De Visiani and the physician Giuseppe Giannelli (1799–1872), who had moved to Milan in 1838, however. The physician Giacomo Andrea Giacomini attended six congresses, Achille De Zigno and six other Paduan scholars five, Giuseppe Bernardi, Giuseppe Clementi, Giovanni Zanardini, and nine other scholars four congresses. The congress of Padua, and even more so the congress in Venice, provided a unique opportunity to link Veneto's centers and peripheries of intellectual life, people with different levels of education—university, high schools, colleges, seminaries, even elementary schools—and the scientific and the economic elite. The congress in Padua hosted 514 attendees who belonged to 36 academic deputations (thirty-one from Italy and five from other countries). The congress in Venice welcomed 1478 attendees from 65 deputations (fifty-two from Italy and thirteen from other countries) and representatives from almost all Venetian urban and provincial learned institutions—universities, academies, societies, institutes, libraries, typographies, high schools, gymnasia, and Catholic, Armenian, and Jewish colleges (Soppelsa 2001, pp. 242–245).

Andrea Cittadella Vigodarzere was general president of the congress in Padua and president of the politically most delicate agrarian section of the congress in Venice. As many other attendees, he was not a scientist, nor did he want to engage in scientific disputes. For people like him, science was but one of many forms of social expression and a tool for educating people. He considered even agriculture primarily in its social rather than its technical dimension (Sect. 9.2 in Chap. 9). Both nominations were unique opportunities to demonstrate political leadership skills (Mogavero and Casalena 2020–2021). He therefore saw the congresses principally as social events and as venues for enhancing his socio-political position, and he was adamant to open them to the society at large. For networkers like Giuseppe Meneghini, the congresses in general and the two that took place in Veneto in particular, represented an unmissable scientific and professional platform. Indeed, on the occasion of the congress in Padua, he compiled a lengthy report to be published in Italian and, translated by Adalbert Bracht (1804–1848), in foreign journals like *Flora* (G. Meneghini 1844). The botanical section was the smallest one, but a glance at the attendees and their speeches reveals that it was the most international section and distinguished by a high scientific standard. In a letter to Meneghini, Hugo von Mohl, one of the most influential German-speaking botanists of the 1840s, wrote that he preferred them to the German meetings.[14] Andrea Meneghini attended only the congress in Milan and Venice, but played an important role at the agronomy sections (Sects. 9.2 and 9.3 in Chap. 9).

For a long time, historians have interpreted the Italian congresses principally as political pre-risorgimental events. Indisputably, the congresses had an explicit political dimension, yet a posteriori interpretations have distorted their significance by underplaying the heterogeneity of the participants, their aims, and contemporary effects. The congresses gave rise to broad coverage in German-speaking journals, which mostly emphasized how happy the Venetians were with the Austrian regents and their institutional innovations (Klemun 2021). The aforementioned Bohemian officer Adalbert Bracht illustrates the complex interplay between Austrian and Italian politics and botany, and the contiguity of being friends and being enemies. From 1825, Bracht was officer of the Austrian garrisons in various cities of Lombardy-Veneto (Bona 2012; Kiehn 2014). He married a Venetian noblewoman and participated in several of the Italian congresses. During the congresses, he passionately promoted nationalistic Italian projects, like the foundation of the Central Herbarium in Florence and the *Italian Journal of Botany*, and proposed (unsuccessfully) the creation of a central institution for the exchange of plants. He died in 1848 as commander of an Austrian brigade in the bloody battle of Custoza. Learning of his death, Paduan lichenologist and politician Vittore Trevisan (1848, p. 57) dedicated a new species of green algae to Bracht—even though it was the heyday of anti-Austrianism.

The congresses had a galvanizing effect on Paduan science and associations. Joint participation in numerous projects, committees, and institutions, initiated on occasion of the congresses, created lasting bonds between Paduan scholars and the intellectual elite of the peninsula. The Central Herbarium in Florence, a national Medical Dictionary, national, or at least supra-regional societies and associations like the Society for the Encouragement of Agriculture (Sect. 9.2 in Chap. 9), and journals like Andrea Meneghini's *Il Tornaconto* (Sect. 9.3 in Chap. 9) were important from a scientific as well as a political point of view. However, the case of botany shows that the marriage between science and civil society was incomplete. Despite commonalities, botany and agronomy developed in different directions. The botanical section was far more elitist, more international, had high scientific standards, and aimed to establish botany as a profession. Attendees of the agronomy section, on the other hand, sought a more inclusive and diverse audience, pleaded for less rigid scientific standards, and finally wanted to firm up their local sphere of influence (Casalena 2007b).[15] Moreover, as we will see in Chaps. 6 and 9, their interest in science became increasingly instrumental. Maria Laura Soppelsa, in fact, considers the event in Padua as the climax of the first phase of the congresses, which continued to echo the reform programs of the eighteenth century. This vision is exemplified by the figures of Roberto De Visiani, who called for unity between science and letters, and Cittadella Vigodarzere, who invoked a fraternal relationship between the physical and the moral sciences (see Sect. 9.2 in Chap. 9). At Venice's congress, in contrast, the center of gravity had shifted toward technology and the applied sciences (Soppelsa 2001).

Notes

1. Paolo Preto (1982b, pp. 62–63) reports that Andrea Cittadella Vigodarzere hurried to Venice with Carlo Leoni to implore the Austrian authorities not to rail against Niccolò Tommaseo and Daniele Manin. It is highly probable that they implored also for Meneghini and Guglielmo Stefani (1819–1861), who were in the same prison.
2. In 1838, its presidents were Giuseppe Cristina, Giordano Emo Capodilista, and Pietro Sagredo (G. 1838, p. 72).
3. The European context of reading cabinets is set out in Dann (1981). Similar cabinets were founded in 1847 at Este, south of the Euganean Hills, and, in 1857, in Monselice (Selmin 1997; Veronese 2012–2013).
4. Andrea Meneghini provides the number of inhabitants in *Il Tornaconto* II/5 (3 February 1848) pp. 37–38.
5. Carlo Fumian (1996) furthermore maintains that the ornithological, meteorological, and numismatic 'vocations' of some landowners were 'secondary and bizarre'.
6. These women were Sophie Prinzessin von Arenberg (1811–1901), Anna Basso Mussato (1798–1854), Antonietta Benvenuti Dal Cerè, Andrianna Bollani da Ponte, Rosalia von Bonar von Wüllerstorff-Urbair (b. 1815), Sofia Cabianca Onesti (m. 1904), Elisa Napoleona Camerata Baciocchi (1806–1869), Susan Maria Antonietta Chiereghin, Fulvia Chiericati Bissari Salvioni (n. 1802), Paolina Cittadella Dolfin (1822–1880), Arpalice Cittadella Papafava, Lucrezia Dolfin (1797–1871), Anna Maria Dolfin de Conninck, Lucia Emo Capodilista (1805–1892), Camilla Facciolati Buslacchi, Teresa Ferrari de Rocco, Costanza Ferri Patella, Loredana Morosini Gatterburg (1806–1884), Anna Gaudio Meneghini (d. 1867), Sofia Giacomini, Maria Giovanelli Buri (c1801–1884), Maria Felicita Hellman, Drusilla Loschi Dal Verme (1811–1885), Fanny De Luca Guolo, Aurora Manolesso Ferro, Laura Moretti, Anna Muchiutti Tomadini, Loredana Nievo, Therese Pallavicini zu Spaur und Flavon (1819–1902), Teresa Papadopoli Mosconi, Caterina Pasini Vandinelli, Chiara Paulucci Manin (1814–c1874), Angelina Piccini, Giulia Piovene Porto, Catterina Polcastro Querini (c1796–1869), Antonietta Pivetta d'Althan (m. 1871), Charlotte Reichlin-Meldegg von Dordi (b. 1815), Teresa de Rocco Mistruzzi, Luigia Rusconi (1797–1862), Elisa Salom, Linda Salvagnini, Andriana Santini (b. c1829), Angelina Sartori, Anna Schloissnigg (1803–1846), Enrichetta Treves dei Bonfili Consolo (1796–1851), Enrichetta Treves dei Bonfili Treves (1790–1858), Lauretta Turazza,

Maria Anna von Wimpffen, Marianna Zadra, Adele da Zara, Eloisa Zara, and Carolina Zucchetta Steyer.

7. Clara M. Lovett (1982, pp. 72–73) misunderstood Brofferio's story, thinking he was talking about real trees in the countryside.
8. In June 1870, Andrea Meneghini replaced a new copy of the Lion of San Marco, symbol of the Republic of Venice, destroyed by the French in 1797.
9. Tommaseo found the word *giardinaggio* 'inelegant'. See the letter from Tommaseo to De Visiani of 20 December 1845 (BOBP, Ar.B.13. https://phaidra.cab.unipd.it/detail/o:307111#?sortdef=created%20asc&page=1&pagesize=20&fr=bib_roles_pers_allroles_Niccol%C3%B2%20Tommaseo. Accessed 22 November 2019).
10. Dante, *Paradiso*, XXVII, 148: "e vero frutto verrà dopo il fiore".
11. Tommaseo 1846, p. 29: "possano gli uomini accorgersi finalmente, che le grandi cose s'intraprendono e compionsi, non aspettando il favore de' casi o invocando la carità de' potenti; ma sì le private forze, associando con ordine perseverante, e con fervore unanime volgendole a un fine. Di qui soltanto possono le nazioni aspettarsi rinnovellamento e grandezza."
12. Interestingly, his discourse of 1854 emphasizes the eminent role of Veneto not Italian scientists (De Visiani 1854).
13. De Visiani 1847b, p. 419: "[...] annestata felicemente in tal festa la utilità col diletto, fatta per essa sensibile, piacente e popolare la scienza, affratellate e confuse nell'unico sentimento delle naturali bellezze tutte le sociali disuguaglianze."
14. Letter from Hugo Mohl to Giuseppe Meneghini of 10 Dec 1845, BSNAP.
15. Casalena argues that the post-unity congresses failed because the coexistence of these two visions could not be sustained.

References

Agostini, Filiberto, ed. 2011. *Le amministrazioni provinciali in Italia: Prospettive generali e vicende venete in età contemporanea*. Milano: Franco Angeli.

Anon. 1842. *Guida di Padova e della sua provincia*. Padova: Coi Tipi del Seminario.

———. 1846. *Statuto della società promotrice del giardinaggio in Padova*. Padova: Tipi del Seminario.

———. 1846c. *Prima esposizione di piante e fiori, tenuta dalla Società promotrice del giardinaggio nell'I. R. Orto botanico di Padova nei giorni 7 ed 8 giugno 1846*. Padova: Tipi del Seminario.

———. 1846d. Manifesto. *Il Tornaconto* s.d. [October 1846].

———. 1847a. *Seconda esposizione di piante e fiori tenuta dalla Società promotrice del giardinaggio nell'I. R. Orto botanico di Padova*. Padova: Crescini.

———. 1847e. Prologo ai nostri articoli di agricoltura. *Il Tornaconto* 2 (14 January 1847): 9–11.

———. 1847k. Sull'intervento alle adunanze comunali e sociali. *Il Tornaconto* 17 (29 April 1847): 136–137.
———. 1847l. Riccardo Cobden a Padova ed a Venezia. *Il Tornaconto* 25 (24 June 1847): 201–202.
———. 1847m. Società d'incoraggiamento nella Provincia di Padova. *Il Tornaconto* 29 (22 July 1847): 239.
———. 1847n. Agricoltura. Società d'Incoraggiamento nella provincia di Padova. *Il Tornaconto* 34 (26 August 1847): 282.
———. 1847p. Spettacoli del Settembre 1847 in Venezia. *Il Caffè Pedrocchi* 2(40) (3 October 1847): 338.
———. 1848a. *Lista delle offerte fatte dai Cittadini al Comitato Dipartimentale di Padova pei bisogni attuali della Patria*. Padova: Tip. Crescini.
———. 1848b. Istituto medico-chirurgico-farmaceutico di mutuo soccorso in Padova. *Il Tornaconto* 2 (13 January 1848): 13–14.
———. 1854. *Atti della seconda distribuzione de' premj eseguita dalla Società d'incoraggiamento per l'agricoltura e l'industria in Padova il giorno XII giugno MDCCCLIII nella Gran Sala della Ragione*. Padova: Co' Tipi di Angelo Sicca.
———. 2019. *Das Florilegium Kaiser Franz I. Österreichische Nationalbibliothek*, on: http://www.bildarchivaustria.at/Pages/Ausstellung/Ausstellungen.aspx?AusstellungID=19246060. Accessed 12 October 2020.
———. 1827. *Il Monumento a Canova eretto in Venezia*. Venezia: Tipografia di Alvisopoli.
Antonielli, Livio. 2001. Venezia nel Regno Italico: Un'annessione di 'basso rilievo'. In *Dopo la Serenissima: Società, amministrazione e cultura nell'Ottocento veneto*, ed. Donatella Calabi, 123–151. Venezia: Istituto Veneto di Scienze, Lettere ed Arti.
Banti, Alberto Mario, and Marco Meriggi. 1991. Premessa. In *Élites e associazioni nell'Italia dell'Ottocento*, ed. Alberto M. Banti and Marco Meriggi, 357–542. Quaderni storici v. 26/n. 77.
Béguinot, Augusto. 1909. *Flora padovana, ossia prospetto floristico e fitogeografico delle piante vascolari indigene inselvatichite o largamente coltivate crescenti nella Provincia di Padova, con notizie storico-bibliografiche sulle fonti della Flora*. Vol. 1. Padova: Prem. Soc. Coop. Tip.
Berselli, Giovanni. 1847. Discorso d'inaugurazione. *Giornale Euganeo di scienze, lettere ed arti* 4 (1): 157–162.
Betri, Maria Luisa, and Elena Brambilla, eds. 2004. *Salotti e ruolo femminile in Italia tra fine Seicento e primo Novecento*. Venezia: Marsilio.
Bona, Enzo. 2012. Alberto (o Adalberto) de Bracht? lungimirante botanico e capitano boemo. *Notiziario Flora Alpina Bergamasca* 41: 20–23.
Breidbach, Olaf, Hans-Joachim Fliedner, and Klaus Ries, eds. 2001. *Lorenz Oken (1779–1851): Ein politischer Naturphilosoph*. Weimar: Böhlau.

Browne, Janet. 1995–2002. *Charles Darwin: A Biography*. Vol. 2 volumes. London: Jonathan Cape.
Busnardo, Giuseppe. 1998. Il ruolo degli *amateurs* veneti nello sviluppo delle scienze botaniche della prima metà del XIX secolo. In *Le scienze biologiche nel Veneto dell'Ottocento*, ed. Bruno Battaglia, Gian Antonio Danieli, and Alessandro Minelli, 207–232. Venezia: Istituto Veneto di Scienze, Lettere ed Arti.
Bussadori, Paola. 1988. Gli orti botanici privati padovani. In *Di sana pianta: erbari e taccuini di sanità: Le radici storiche della nuova farmacologia*, 47–54. Modena: Edizioni Panini.
Calabi, Donatella, ed. 2001. *Dopo la Serenissima. Società, amministrazione e cultura nell'Ottocento veneto*. Venezia: Istituto Veneto di Scienze, Lettere ed Arti.
Callegari, Marco. 2012. *Stampatori e librai a Padova nella prima metà dell'Ottocento*. Saonara: Il Prato.
Camurri, Renato. 2004. Il mutualismo e la diffusione dello 'spirito d'associazione' dopo l'Unità. In *Spazi laici: Strutture e reti associative tra Ottocento e Novecento*, ed. Renato Camurri and Marco Fincardi, 23–54. Venetica: Rivista di storia contemporanea 17: 1–221.
Casalena, Maria Pia. 2007a. *Per lo Stato, per la Nazione: I congressi degli scienziati in Francia e in Italia (1830–1914)*. Roma: Carocci.
———. 2007b. The congresses of the Italian scientists between Europe and the Risorgimento (1839–75). *Journal of Modern Italian Studies* 12: 153–188.
———. 2018. In Europa e ritorno: I congressi degli scienziati italiani tra modelli europei e via nazionale. *Mélanges de l'École française de Rome – Italie et Méditerranée modernes et contemporaines* 130 (2): 273–283.
Ciancio, Luca. 1994. La Repubblica de' naturalisti. Fortuna e declino della storia naturale nel Settecento veneto. *Atti dell'Accademia roveretana degli Agiati* VII/243: 57–110.
———. 2014. Il 'merito' dei veneti nella geologia: una riconsiderazione storico-critica. In *Esploratori del tempo profondo: Scienza, storia e società nella cultura veneta dell'età moderna*, ed. Luca Ciancio, 259–267. Verona: Qui Edit.
Clementi, Giuseppe. 1845. Giardinaggio. In *Enciclopedia italiana e Dizionario della conversazione, vol. VII, parte II*, 3094–3096. Venezia: Stabilimento enciclopedico di Girolamo Tasso.
Clementi, Moreno. 2017. *A cross-disciplinary study of the works and collections by Roberto de Visiani (1800–1878)* (dissertation thesis). Padova: Università degli Studi di Padova. http://paduaresearch.cab.unipd.it/10378/. Accessed 20 September 2019.
Cook, Alexandra. 2012. *Jean-Jacques Rousseau and botany: The salutary science*. Oxford: Voltaire Foundation.
Cox, Virginia. 2011. *The prodigious muse: Women's writing in Counter-Reformation Italy*. Baltimore: The Johns Hopkins University Press.

Curti, Luigino. 1993. Erbari e sviluppo della ricerca nell'800 presso l'Orto Botanico di Padova. *Webbia* 48: 501–514.
Curti, Luigino, and Fernanda Menegalle. 1996. Roberto de Visiani. In *Professori di materie scientifiche all'Università di Padova nell'Ottocento*, ed. Sandra Casellato and Luisa Pigatto, 77–85. Padova: Edizioni Lint.
D'Ezio, Marianna. 2012. Sociability and cosmopolitanism in eighteenth-century Venice: European travellers and Venetian Women's Casinos. In *Sociability and cosmopolitanism: Social bonds on the fringes of the Enlightenment*, ed. David Burow and Scott Brueninger, 47–58. London: Routledge.
Dal Cin, Valentina. 2019. *L'élite veneta fra rivoluzione e restaurazione (1797–1815)*. Venezia: Edizioni Ca' Foscari.
Dal Prete, Ivano. 2008. *Scienza e società nel Settecento veneto: Il caso veronese, 1680–1796*. Milano: Franco Angeli.
Dann, Otto, ed. 1981. *Lesegesellschaften und bürgerliche Emanzipation, ein europäischer Vergleich*. München: Beck.
Dante. 2007. *Paradiso*. A verse translation by Robert and Jean Hollander. New York: Anchor Books.
De Visiani, Roberto. 1840. *Illustrazione delle piante nuove o rare dell'Orto Botanico di Padova. Memoria I*. Padova: Tipi di Angelo Sicca.
———. 1842–1852. *Flora dalmatica, sive Enumeratio stirpium vascularium quas hactenus in Dalmatia lectas et sibi observatas*. 3 vols. Leipzig: F. Hofmeister.
———. 1847a. *Parole indirizzate il giorno 15 di Dicembre 1846 all'adunanza generale della Società promotrice del Giardinaggio in Padova*. Padova: Tipi del Seminario.
———. 1847b. Parole indirizzate alla Società promotrice del giardinaggio nell'adunanza generale del dì 5 dicembre 1847. *Il Tornaconto* a. I. n. 50 (16 dicembre 1847): 419–422.
———. 1854. *Delle benemerenze de' Veneti nella botanica: Discorso letto nella sala de' Pregadi del Palazzo ducale in Venezia nel dì 30 maggio 1854*. Venezia: Tip. Cecchini.
De Vivo, Filippo. 2008. Pharmacies as centres of communication in early modern Venice. In *Spaces, objects and identities in early modern Italian medicine*, ed. Sandra Cavallo and David Gentilcore, 33–50. Malden, MA: Blackwell Publishing.
Del Negro, Piero. 1992. Una società "per la lettura di gazzette e giornali" nella Padova di fine Settecento. *Archivio veneto* 138: 31–59.
Derosas, Renzo. 2000. Dati di stato, di flusso, di relazione. Un esempio di integrazione nello studio della proprietà fondiaria. *In Mélanges de l'École française de Rome. Italie et Méditerranée* 112 (1): 192–208.
Di Santo, Simonetta. 1993. *Nel giardino di Dante: Il sistema della vegetazione nella Commedia*. Chieti: Solfanelli.

Egmont, Florike. 2010. *The world of Carolus Clusius: Natural history in the making, 1550–1610*. London: Pickering & Chatto.

Ferrari, Giorgio E. 1958. L'attitudine di Padova verso Venezia nella crisi veneta del Quarantotto (esordio ad un bilancio bibliografico). In *Miscellanea in onore di Roberto Cessi*, vol. III, 183–234. Roma: Edizioni di Storia e Letteratura.

Filippini, Nadia Maria, ed. 2006. *Donne sulla scena pubblica: Società e politica in Veneto tra Sette e Ottocento*. Milano: Franco Angeli.

Filippini, Nadia Maria, and Liviana Gazzetta, eds. 2011. *L'altra metà del Risorgimento: volti e voci di patriote venete*. Sommacampagna: Cierre.

Fincardi, Marco. 2004. I luoghi possibili della sociabilità laica nel Veneto. In *Spazi laici: Strutture e reti associative tra Ottocento e Novecento*, ed. Renato Camurri and Marco Fincardi, 7–21. Venetica: Rivista di storia contemporanea 17: 1–221.

Franceschi, Franco. 1999. La bottega come spazio di sociabilità. In *Arti fiorentine: La grande storia dell'artigianato. Vol. II: Il Quattrocento*, ed. Franco Cardini, Franco Francesch, and Gloria Fossi, 64–83. Firenze: Giunti.

Freeden, Michael, and Javier Fernandez-Sebastián. 2019. European liberal discourses: Conceptual affinities and disparities. In *In search of European liberalisms: Concepts, languages, ideologies*, ed. Michael Freeden, Javier Fernandez-Sebastián, and Jörn Leonhard, 1–35. New York: Berghahn.

Freeman, Linton. 2004. *The development of social network analysis: A study in the sociology of science*. Vancouver: Empirical Press.

Fumian, Carlo. 1996. *Possidenti: Le élites agrarie tra Otto e Novecento*. Catanzaro: Meridiana Libri.

———. 2004. Il 'senno della nazione': I congressi degli scienziati e la cultura agronomica. In *Agricoltura come manifattura: Istruzione agraria, professionalizzazione e sviluppo agricolo nell'Ottocento*, ed. Giuliana Biagioli and Rossano Pazzagli, 203–251. Firenze: Olschki.

G., G. 1838. *Almanacco civile della R. Città e Provincia di Padova per l'anno 1838*. Padova: Tip. E Fond. Cartallier.

George, Samantha. 2005. The cultivation of the female mind: Enlightened growth, luxuriant decay and botanical analogy in eighteenth-century texts. *History of European Ideas* 31 (2): 209–223.

Ginsborg, Paul. 1979. *Daniele Manin and the Venetian revolution of 1848–49*. Cambridge: Cambridge University Press.

Giormani, Virgilio. 1988. Vincenzo Dandolo, uno speziale illuminato nella Venezia dell'ultimo '700. *Ateneo veneto* 175: 59–130.

Goody, Jack. 1993. *The culture of flowers*. Cambridge: Cambridge University Press.

Gullino, Giuseppe. 1996. *L'Istituto Veneto di Scienze, Lettere ed Arti: Dalla rifondazione alla seconda guerra mondiale (1838–1946)*. Venezia: Istituto Veneto di Scienze, Lettere ed Arti.

Hlavac, Christian, and Astrid Göttche. 2016. *Die Gartenmanie der Habsburger: Die kaiserliche Familie und ihre Gärten, 1792–1848*. Wien: Amalthea.

Hof- und Staatshandbuch des österreichischen Kaiserthums (HSHÖK). 1844. Wien: K.k. Hof- und Staats-Aerarial-Druckery.

———. 1845. Wien: K.k. Hof- und Staats-Aerarial-Druckery.

———. 1846. Wien: K.k. Hof- und Staats-Aerarial-Druckery.

———. 1847. Wien: K.k. Hof- und Staats-Aerarial-Druckery.

———. 1848. Wien: K.k. Hof- und Staats-Aerarial-Druckery.

Hof- und Staatsschematismus des österreichischen Kaiserthums (HSSÖK). 1840. Wien: K.k. Hof- und Staats-Aerarial-Druckery.

———. 1841. Wien: K.k. Hof- und Staats-Aerarial-Druckery.

———. 1842. Wien: K.k. Hof- und Staats-Aerarial-Druckery.

———. 1843. Wien: K.k. Hof- und Staats-Aerarial-Druckery.

Hyde, Elizabeth. 2002. Flowers of distinction: Taste, class, and floriculture in seventeenth-century France. In *Bourgeois and aristocratic cultural encounters in garden art, 1550–1850*, ed. Michel Conan, 77–100. Washington, DC: Dumbarton Oaks Research Library and Collection.

Jacobs, Joela M., and Isabel Kranz, eds. 2017. *Das literarische Leben der Pflanzen: Poetiken des Botanischen. Literatur für Leser* 2 (special volume): 1–121.

Jones, Peter M. 2016. *Agricultural Enlightenment: Knowledge, technology, and nature, 1750–1840*. Oxford: Oxford University Press.

Jütte, Daniel. 2012. Trading in secrets: Jews and the early modern quest for clandestine knowledge. *Isis* 103: 668–686.

Kelley, Theresa M. 2012. *Clandestine marriage: Botany and romantic culture*. Baltimore: Johns Hopkins University Press.

Kiehn, Michael. 2014. Die botanischen Aktivitäten des 1848 in Italien gefallenen Adalbert (von) Bracht, Hauptmann im k. k. Linienregiment Franz Carl. In *Erkunden, Sammeln, Notieren und Vermitteln – Wissenschaft im Gepäck von Handelsleuten, Diplomaten und Missionaren*, ed. Ingrid Kästner, Jürgen Kiefer, Michael Kiehn, and Johannes Seidl, 167–176. Aachen: Shaker Verlag.

Kirchner Reill, Dominique. 2012. *Nationalists who feared the nation: Adriatic multi-nationalism in Habsburg Dalmatia, Trieste, and Venice*. Stanford: Stanford University Press.

Kittelmann, Jana, ed. 2017. Botanik und Ästhetik. *Annals of the History and Philosophy of Biology* 22 (special volume): 1–335.

Klemun, Marianne. 2016. Franz Unger (1800–1870): multiperspektivische wissenschaftshistorische Annäherungen. In *Einheit und Vielfalt: Franz Ungers (1800–1870) Konzepte der Naturforschung im internationalen Kontext*, ed. Marianne Klemun, 15–92. Wien: Vienna University Press.

———. 2021. 'Spaces in between': Political, national and epistemological barriers and bridges. Meetings of German Naturalists, Austria and the Congresses of Italian Scientists in Padua (1842) and Venice (1847). In *Scienziati italiani a congresso nel Veneto asburgico (1842, 1847)*, vol. II, eds. Valeria Mogavero and Maria Pia Casalena. Venetica: Rivista di storia contemporanea 60(1): 35–60.

Lack, H. Walter. 2006. *Florilegium Imperiale: Botanische Schätze für Kaiser Franz I. von Österreich*. München et al.: Prestel.

Latour, Bruno. 2005. *Reassembling the social: An introduction to actor-network-theory*. Oxford: Oxford University Press.

Le Quellec, Jean-Loïc. 2008. L'arbre dans les mythes. In *Aux origines des plantes: Des plantes et des hommes*, ed. Francis Hallé and Pierre Lieutaghi, 416–439. Paris: Fayard.

Lovett, Clara M. 1982. *The Democratic Movement in Italy, 1830–1876*. Cambridge, MA: Harvard University Press.

Malatesta, Maria (ed.). 1988. *Sociabilità nobiliare, sociabilità borghese: Francia, Italia, Germania, Svizzera, XVIII–XIX secolo*. Cheiron 9-10 (special issue).

Maniero, Federico, and Elena Macellari. 2005. *Giardinieri ed esposizioni in Italia (1800–1915)*. Perugia: Ali&No Editrice.

Massaro, Martina. 2014–2015. *Giacomo Treves dei Bonfili collezionista e mecenate (1788–1885): La raccolta di un filantropo patriota*. Dissertation thesis, Università IUAV di Venezia, Venezia.

———. 2019. *Palazzo Treves dei Bonfili e il suo giardino*. Padova: Il Poligrafo.

Meneghini, Andrea. 1847. Ad un nostro associato di Venezia, autore del Piano di una Società di mutuo soccorso tra i possidenti per le Provincie Venete, pubblicato nel N° XII di questo Giornale. Il Tornaconto 19 (13 May 1847): 151–152.

Meneghini, Giuseppe. 1844. Bericht über die Arbeiten der botanischen Section bei der vierten Versammlung der italienischen Naturforscher und Aerzte zu Padua im September 1842. Flora oder allgemeine botanische Zeitung 27(1): 497–510, 556–566, 569–578.

Meriggi, Marco. 1992a. *Milano borghese: Circoli ed élites nell'Ottocento*. Venezia: Marsilio.

———. 1992b. Dalla restaurazione all'età liberale: Per una storia del concetto di associazione in Italia. In *I concetti fondamentali delle scienze sociali e dello Stato in Italia e in Germania tra Otto e Novecento*, ed. Raffaella Gherardi and Gustavo Gozzi, 87–106. Bologna: Il Mulino.

———. 2011. Prove di comunità: Sui congressi preunitari degli scienziati italiani. In *Storia d'Italia, Annali 26 (Scienze e cultura dell'Italia unita)*, ed. Francesco Cassata and Claudio Pogliano, 7–35. Torino: Einaudi.

Minelli, Alessandro, ed. 1995. *The Botanical Garden of Padua 1545–1995*. Venice: Marsilio.

———. 2008. Zoological vs. botanical nomenclature: A forgotten 'BioCode' experiment from the times of the Strickland Code. *Zootaxa* 1950: 21–38.

———. 2017. Die europäische Auswirkung der naturwissenschaftlichen Forschungen in Padua und Venetien: Antonio Vallisneri (1661–1730) und Giovanni Arduino (1714–1795). In *Padua als Europäisches Wissenschaftszentrum von der Renaissance bis zur Aufklärung*, ed. Dietrich von Engelhardt and Gian Franco Frigo, 109–124. Aachen: Shaker Verlag.

Minuzzi, Sabrina. 2016. *Sul filo dei segreti: Farmacopea, libri e pratiche terapeutiche a Venezia in età moderna*. Milano: Edizioni Unicopli.
Miotto, Onelio. 1942. *Andrea Meneghini: patriota inedito del nostro Risorgimento (1806–1870)*. Padova: Cedam.
Mogavero, Valeria. 2014. *Patrie patrizie. Una rete nobiliare alle prove dell'Otto–Novecento*. Doctoral thesis, Università di Verona, Verona.
Mogavero, Valeria, and Maria Pia Casalena (eds.). 2020–2021. Scienziati italiani a congresso nel Veneto asburgico (1842, 1847). *Venetica: Rivista di storia contemporanea* 58(1): 7–245 and 60(1): 1–352.
Moon, Francis C. 2016. *Social networks in the history of innovation and invention*. Dordrecht: Springer.
Nuovi Saggi della Imperiale Regia Accademia di Scienze, Lettere es Arti in Padova (NSIRAS). 1838. Vol. 4. Padova: Tipografia della Minerva.
———. 1840. Vol. 5. Padova: Coi Tipi di Angelo Sicca.
———. 1847. Vol. 6, Padova: F. A. Sicca e figlio.
Olmi, Giuseppe. 1991. 'Molti amici in varij luoghi': Studio della natura e rapporti epistolari nel secolo XVI. *Nuncius* 6: 3–31.
———. 2007. Per la storia dei rapporti scientifici fra Italia e Germania: Le lettere di Francesco Calzolari a Joachim Camerarius II'. In *Dai cantieri della storia: Liber amicorum per Paolo Prodi*, ed. Gianpaolo Brizzi and Giuseppe Olmi, 343–361. Bologna: CLUEB.
Paganelli, Arturo. 1988. L'*Horto Medicinale* dell'Università di Padova nella storia della medicina. In *Di sana pianta: erbari e taccuini di sanità: Le radici storiche della nuova farmacologia*, 41–46. Modena: Edizioni Panini.
Palmer, Richard. 1985. Pharmacy in the Republic of Venice in the sixteenth century. In *The medical renaissance of the sixteenth century*, ed. Andrew Wear, Roger K. French, and Iain M. Lonie, 108–118. Cambridge: Cambridge University Press.
Pancaldi, Giuliano, ed. 1983. *I congressi degli scienziati italiani nell'età del positivismo*. Bologna: Clueb.
Parker, John N., Niki Vermeulen, and Bart Penders. 2010. *Collaboration in the new life sciences*. Farnham: Ashgate.
Parrish, David. 2015. *Marketing Nature: Apothecaries, medicinal retailing, and scientific culture in early modern Venice, 1565-1730* (dissertation thesis). Durham: Duke University. https://dukespace.lib.duke.edu/dspace/bitstream/handle/10161/11326/Parrish_duke_0066D_13158.pdf?sequence=1&isAllowed=y. Accessed 1 March 2020.
Pederzani, Ivana. 2014. *I Dandolo: Dall'Italia dei lumi al Risorgimento*. Milano: Franco Angeli.
Plebani, Tiziana. 2004. Socialità, conversazioni e casini nella Venezia del secondo Settecento. In *Salotti e ruolo femminile in Italia tra fine Seicento e primo Novecento*, ed. Maria Luisa Betri and Elena Brambilla, 153–176. Venezia: Marsilio.

Pomian, Krzysztof. 1987. *Collectionneurs, amateurs et curieux: Paris, Venise XVIe–XVIIIe siècle*. Paris: Gallimard.

Preto, Paolo. 1982a. Un 'uomo nuovo' dell'età napoleonica: Vincenzo Dandolo, politico e imprenditore agricolo. *Rivista storica italiana* 94: 44–97.

———. 1982b. Andrea Cittadella Vigodarzere. In *Dizionario Biografico degli Italiani*, vol. 26, 62–63. Roma: Treccani.

Pugliano, Valentina. 2017. Pharmacy, testing, and the language of truth in Renaissance Italy. *Bulletin of the History of Medicine* 91 (2): 233–273.

———. 2018. Natural history in the apothecary's shop. In *Worlds of natural history*, ed. Helen Anne Curry, Nicholas Jardine, James Andrew Secord, and Emma C. Spary, 44–60. Cambridge: Cambridge University Press.

Rees, Katie. 2008. Female-authored drama in early modern Padua: Valeria Miani Negri. *Italian Studies* 63 (1): 41–61.

Riedl-Dorn, Christa. 1989. *Die grüne Welt der Habsburger: Botanik, Gartenbau, Expeditionen, Experimente (Ausstellungskatalog, Schloß Artstetten, Artstetten 1989)*. Wien: Naturhistorisches Museum.

Ronchi, Oliviero. 1906. L'albero della libertà a Padova. *Bollettino del Museo Civico di Padova* 9 (1): 26–31.

Rotenberg, Robert. 2002. *La pensée bourgeoisie* in the Biedermeier Garden. In *Bourgeois and aristocratic cultural encounters in garden art, 1550–1850*, ed. Michael Conan, 147–172. Washington, DC: Dumbarton Oaks Research Library and Collection.

Seaton, Beverly. 1995. *The language of flowers: A history*. Charlottesville/London: University of Virginia Press.

Selmin, Francesco. 1997. *Storia del Gabinetto di Lettura di Monselice 1847–1997*. Este: Società Gabinetto di lettura in Este.

Siegrist, René. 2013. On some social characteristics of the eighteenth-century botanists. In *Scholars in action: The practice of knowledge and the figure of the savant in the 18th century*, ed. André Holenstein, Hubert Steinke, and Martin Stuber, 205–234. Leiden/Boston: Brill.

Sittler, David. 2016. Medienpraktiken mit Blumen zwischen Protest und Revolution 1789–2011: Überlegungen zu einer politischen Botanik. In *Floriographie: Die Sprache der Blumen*, ed. Isabel Kranz, Alexander Schwan, and Eike Wittrock, 337–369. Paderborn: Wilhelm Fink.

Soldani, Simonetta. 2007. Il Risorgimento delle donne. In *Storia d'Italia, Annali 22: Il Risorgimento*, ed. Alberto B. Banti and Paul Ginsborg, 183–224. Torino: Einaudi.

Solitro, Giuseppe. 1927. Introduzione. In *Il Comitato Provvisorio Dipartimentale di Padova dal 25 marzo al 13 giugno 1848*, ed. Andrea Gloria, xii–xci. Padova: Tip. del Messaggiero.

———. 1930. *La 'Società di cultura e di incoraggiamento in Padova' nel suo primo centenario: Un secolo di vita padovana*. Padova: Società di cultura e di incoraggiamento.

Soper, Steven C. 2013. *Building a Civil Society: Associations, Public Life, and the Origins of Modern Italy*. Toronto/Buffalo/London: University of Toronto Press.

Soppelsa, Maria Laura. 2001. Immagini della cultura scientifica veneta nei Congressi degli scienziati italiani di Padova (1842) e Venezia (1847). In *Dopo la Serenissima: Società, amministrazione e cultura nell'Ottocento veneto*, ed. Donatella Calabi, 233–270. Venezia: Istituto Veneto di Scienze, Lettere ed Arti.

Soresina, Marco. 1985. Associazionismo e ruolo dei medici nel primo trentennio dello stato unitario. *Società e storia* 27: 85–118.

Stapelbroek, Koen, and Jani Marjanen, eds. 2012. *The rise of economic societies in the eighteenth century: Patriotic reform in Europe and North America*. New York: Palgrave Macmillan.

Stössl, Marianne. 1983. *Lo spettacolo della Triaca: Produzione e promozione della 'Droga divina' a Venezia dal Cinque al Settecento*. Venezia: Centro tedesco di studi veneziani.

Tomazzoli, Gaia. 2017. Le metafore della Commedia: tre modelli di lettura. In *Atti delle Rencontres de l'Archet. Morgex, 14–19 settembre 2015*, 180–187. Torino: Lexis Compagnia Editoriale.

Tommaseo, Niccolò. 1846. D'una nuova società istituita nel Veneto per promuovere la coltura dei fiori. *La Dalmazia* 2 (4): 28–29.

Tommaseo, Niccolò, and Bernardo Bellini. 1861–1879. *Dizionario della lingua italiana*. Torino: Pomba.

Trevisan, Vittore B.A. 1848. *Saggio di una monografia delle alghe coccotalle*. Padova: Coi Tipi del Seminario.

Veronese, Elisa. 2012–2013. *Storia del Gabinetto di Lettura di Monselice 1857–1939*. Tesi di Laurea specialistica. Venezia: Università Ca' Foscari di Venezia.

Vigodarzere, Antonio. 1823. *Per le nozze Gaudio-Meneghini: Terzine*. Padova: Dalla Minerva.

Weber, Max. 1978. *Economy and society: An outline of interpretative sociology*. Berkeley: University of California Press.

Wulf, Andrea. 2011. *The founding gardeners: How the revolutionary generation created an American Eden*. London: William Heinemann.

CHAPTER 3

Plants and the Social Ascent of the Meneghini Family

3.1 The Rise of the Meneghini Family

The family played a key role in the Italian Risorgimento. It performed crucial socializing functions, served as a source of political symbols, language, and role concepts of political behavior in the formation of almost all politically active young Italian men, and, finally yet importantly, provided insurrectionists with essential financial and moral support in difficult moments (Lovett 1982, p. 67–80). Contrary to Giuseppe Mazzini (1805–1872) and his manifesto *La Giovine Italia*, which electrified many young Italians in the early 1830s (Arisi Rota and Balzani 2012; Balzani 2000),[1] the Risorgimento of the 1840s and 1850s was not a generation conflict, at least not in Italy. Even less in Padua. Notwithstanding their frequent meetings and political discussions at the Caffé Pedrocchi, nothing points to the Meneghini brothers and their fellows as opposing their fathers as 'a new generation'. Their social self-consciousness continued to be predominantly defined through their family membership. The familial environment in which the Meneghini brothers grew up is therefore of great interest.

Unfortunately, we know almost nothing about the family's origin and the brothers' early years. The family name seems to originate from the prealpine plateau of Asiago (Mantese 1988). It is not clear whether there was any relation to the Meneghini families of Arzignano and Vicenza, a

A. Dröscher, *Plants and Politics in Padua During the Age of Revolution, 1820–1848*, Palgrave Studies in the History of Science and Technology, https://doi.org/10.1007/978-3-030-85343-3_3

Veneto city about thirty kilometers north-west of Padua, which became affluent in the eighteenth century through tannery. Andrea and Giuseppe's father, Agostino Meneghini, was born in Marostica, a small town about thirty kilometers southwest of Asiago and fifty kilometers northwest of Padua, where he moved at the age of fifteen.[2] The absence of any grandfather, uncle, or other parents in the records of Andrea and Giuseppe suggests that they did not have ramified ancestry in Padua. In the late 1840s, a notary Agostino Meneghini appears in several official documents. Most probably it is a homonym, because Agostino was of modest origins. The lyrical singer Adele Meneghini (d. 1838), who in 1826 married the Paduan writer and romantic poet Jacopo Crescini (1798–1848), director of the newspapers *L'Euganeo* and *Caffé Pedrocchi* (Fig. 2.1, F in Chap. 2), appears not to be a close relative of Agostino Meneghini, either. Some sources call Agostino an 'intelligent farmer' (Pasetti Medin 1996, p. 349; Galletto 1992, p. 52), Carlo Leoni (1812–1874) mentioned in his diary that he was a porter, who got rich selling military supplies during the Napoleonic Kingdom of Italy (Leoni 1976, p. 670–671). In 1816, during a serious agricultural crisis (see below), a governmental inquiry blamed him for having caused a food shortage by speculating and buying up great quantities of wheat (Monteleone 1969, p. 55). It is not clear whether this charge was true or whether it was an attempt to divert public anger from the government and toward a scapegoat, yet it seems certain that by that time he was already an important tradesman.

During the French domination that in Padua lasted from November 1805 to November 1813, many patrician families lost a good part of their property, in most cases due to accumulated debt. In addition, the Church alienated great parts of their property. Both developments cleared the way for a new group of landowners (Ventura 1989, p. 42–43). Some of these newcomers were rich Jewish families, like the Trieste and the Treves, who embraced the opportunity to invest after the opening of the Venetian Ghetto by Napoleon. Yet in the whole Veneto, Jews made up only about 9% of the purchasers of rural properties,[3] whereas about 49% were bourgeois families like the Meneghinis (Salvagnini 1841, p. 20). Similar cases of new wealth were those of the families of Fedele Lampertico (1833–1906) and Silvestro Camerini (1777–1866). Camerini died in Padua as the richest man in Veneto. The nature of the enterprises that lay behind the wealth of both families was obscure, and this gave rise to rumors and defamations (Fumian 1984, p. 118–141 and 142–149).[4]

In any case, a revolution of landed property took place in the first decades of the nineteenth century (Derosas 1987, p. 554). The revolution was cadastral, not political, because the emerging bourgeois bought the estates for the same reason as the wealthy nobility and the ecclesiastic institutions, that is, as a financial guarantee to maintain their standard of living. It was not a competition between nobility and bourgeoisie or a social revolution, but a simple replacement of old by new owners (Berengo 1963, p. 166–167). The social classes of Lombardy-Veneto were not neatly distinct. Marco Meriggi (1987, p. 115) has called the social elite a "fluctuating aristocracy of money", because affiliation was defined by financial resources, mostly property, in particular real estate. This made it much easier for people like Meneghini to become part of it. Yet, many of the new fortunes did not persist for long and soon crumbled. One such case was Agostino Meneghini, who in 1842 was forced to sell the villa and most of his properties.

In the 1810s, however, Agostino was one of those who had become wealthy and a *possidente* (landowner) by purchasing several plots, estates, and factories around Padua that he amplified and modernized (Maccalli 1974, p. 5–6; Zanetti 1989, p. 100). He soon became a Weberian notable and assumed several political-administrative positions, such as deputy of the Congregazione provinciale of Padua, a consultative Austrian organ introduced to learn "the desires and needs of the inhabitants of our Kingdom" (Tonetti 2000, p. 53). Agostino's social ascent was as rapid as his economic advancement. In 1814, he bought Pietro Selvatico Estense's famous Villa Selvatico (Fig. 3.1). Several reasons may have convinced him to make this considerable financial investment. The villa is situated at Battaglia, about seventeen kilometers southwest of Padua. The town was a center of traditional and emerging minor industries and as such since the twelfth century connected to three navigable canals. The Bisatto canal runs to Monselice in the south, the Battaglia canal to Padua in the north, and the Vigenzone canal to Chioggia and the Adriatic Sea in the east (Fig. 1.1 in Chap. 1). During times when the road network was in a more than desolate condition, these canals were of strategic importance. Moreover, the villa is located, together with many other villas and castles of the immediate neighborhood, in a most beautiful landscape, the Euganean Hills, which are of volcanic origin and have numerous thermal springs. The *Aquae patavinae* had attracted guests since Antiquity (Bassani et al. 2010). From the second half of the eighteenth century, in the context of a broader European revival of health resorts and thermal baths, the

Fig. 3.1 The villa of the Meneghini family on the top of colle Sant'Elena. In the background the Euganean hills. Lithograph by L.E. Petrovits. From Mautner (1883, p. 5). By courtesy of Ariane Dröscher

Paduan and Venetian governments frequently sent medical and academic commissions to control and improve the hot waters and associated infrastructure (Piva 1985). Villa Selvatico possessed its own thermal waters. They well up in caverns inside the Sant'Elena hill, which is therefore also called *monte della stufa* (stove hill). The over hundred publications on the Euganean baths, which appeared between the fifteenth and the nineteenth centuries, many of them scientific treatises (e.g., Fortis 1796; Ragazzini 1844), attest to their ongoing appeal and supra-regional economic importance. In fact, Meneghini was not alone in being attracted by the possible economic profit of the Euganean baths. Shortly after him, in 1823, also Moisé Trieste, member of one of the richest Jewish families in Padua, owners of several properties in the area, bought thermal springs in the nearby Abano Terme (Foscarini 1847, p. 16–17). Moisé Trieste purchased it from marquis Giovanni Antonio Dondi Dall'Orologio and continued to run the baths for the poor. He also bought the Orologio factories and restored them with the help of Giuseppe Jappelli. Yet, the project of creating an English garden around the factories was not realized (Piva 1985, p. 56). Instead, in 1835–1842 he created a Jappellian garden in his villa near Piazzola sul Brenta (see Fig. 1.1, no.13 in Chap. 1).

Apart from the economic potential, the purchase offered other important social opportunities. Villa Selvatico is a splendid and famous renaissance building, surrounded by a vast estate (Sect. 4.3 in Chap. 4). The villa was first mentioned in the late sixteenth century and had seen a series of architectural alterations up to very recent times. It had passed from the Carraresi family to the Dalesmanini and in the sixteenth century to the Selvatico family. Situated on the top of the small Sant'Elena hill, with a long stairway leading up to it, and its peculiar cubic structure with four crenellated towers, it catches one's eye from miles around (Fig. 3.1). The purchase guaranteed Meneghini visibility far beyond the immediate area and secured his admission to Padua's high society.

3.2 Marriage Politics

Agostino Meneghini was a clever social climber and successfully disposed of his daughters Angela (1802–1836) and Anna (†1867) in marriages to influential Paduan families. In the exclusive world of Veneto aristocracy, Padua's nobility was the most open to geographically and socially 'non-endemic' spouses (Derosas and Munno 2010, p. 249–252). Even if one may object that the apparent openness of the nobility only had the aim of conserving their hegemony, it provided opportunities for men like Meneghini. In 1819, Angela married the lawyer Giovanni Battista Cromer (1789–1846), only son of the known and in 1803 ennobled Venetian lawyer Giovanni Battista Cromer (1743–1809) (Schröder 1830, p. 282; Vedova 1832, p. 308). The Cromers lived in a villa at Monselice, a few kilometers south of Battaglia. Father Cromer had been a freemason, a very close friend of the poet Melchiorre Cesarotti (1730–1808) and was, like him—yet much less successfully—engaged in literary pursuits (Chiancone 2010, p. 29, 127, and 139). In 1786, Cromer and Cesarotti together undertook a *Grand Tour* in Northern and Central Italy (Leone et al. 2010, p. 285). In 1800, so long before Meneghini and Vigodarzere, Cromer had transformed the Buzzacarini Park into an English landscape with hills, lakes, a pergola, a Chinese pagoda, and a precious Canova statue (Ghidotti 2014). The decision to install a landscape garden was almost certainly influenced by Cesarotti, who owned what was probably the first English garden in Veneto (Finotti 2010) (Sect. 4.2 in Chap. 4). In 1805, Cromer commissioned the Venetian artist Teodoro Matteini (1754–1831) to draw a very sentimental English-style portrait of him sitting together with his family in the midst of his garden (Falconi 2010) (Fig. 3.2).

The Cromer-Meneghini wedding was a major social event. Giovanni de Lazara, an important Francophile offspring of the oldest Paduan aristocratic family (Caburlotto 2001; Preto 2005), gifted Agostino a small volume, *Raccolta per le faustissime nozze Cromer-Meneghini* (Collection for

Fig. 3.2 The Cromer family sitting in the landscape garden of their villa in Monselice, south of Padua. Canvas painting by Teodoro Matteini, 1805–1807. Angela Meneghini married in 1819 Giovanni Battista Cromer (on the left, playing the guitar). In the background: a pseudo-Roman aqueduct and the statue *Asclepius* di Antonio Canova. By courtesy of Matteo Ceriana and studio fotografico Claudio Giusti Firenze

the sumptuous Cromer-Meneghini wedding), that had been produced by several socialites for the occasion and contained forty-two poems, sonnets, madrigals, cants, and a list of plants (Anon. 1819b). He contributed Giuseppe Barbieri's (1774–1852) poem *I Bagni di S. Elena* annotated with his own historical notes. Antonio Vigodarzere dedicated Cesarotti's verses for the wedding (A. Vigodarzere 1819). The fact that the homages, in particular those of the then most prominent figures of Padua's political and cultural life, like de Lazara, Antonio Vigodarzere, Antonio Venturini (1781–1855), and Antonio Scapin (d. 1837), recalled their 'antique friendship' (Anon. 1819b, p. 5, 13, and 19) suggests that their bond was deep and dated back more than just a few years. They exalted Agostino's generosity and his 'beautiful heart'. The emphasis on his generosity is more revealing than it may seem at first glance. Jörn Leonhard argues in his essay on European liberalism that in the late eighteenth and early nineteenth centuries, in Britain the term 'liberal' was still pre-political and described an individual social quality, whereas in Germany and France it rather stood for an enlightened mind (Leonhard 2019). British Whig aristocrats distinguished themselves through their liberal attitude, that is, their social sensitivity and readiness to take responsibility. Anybody aiming to become part of their circle had to display a proper liberal education. Agostino's philanthropy and the eulogies of his guests indicate that similar canons of sociality and group dynamics were at work in Padua. This does not mean that Agostino's generosity was insincere, yet evincing a specific attitude, he represented himself as rich, socially sensible, philanthropic, and free-minded. The members of the Vigodarzere-de Lazara circle, on the other hand, proclaimed in front of all wedding guests and readers of the booklet that they had noticed and accepted that he possessed the necessary attributes to be one of them.

In Padua, the love for plants was another important part of the social performance. The celebratory wedding volume was much more than just a collection of personal homages. The publication reached an audience that went well beyond the wedding guests. Even Vienna's *Literarischer Anzeiger* mentioned it (Anon. 1819a, p. 344). It was Agostino Meneghini's *carte de visite* to Venetian and Austrian society. The repeated recall of plants and the frequent use of floral vocabulary in a print of such strategic importance for Meneghini's social advancement demonstrate that plants were an essential element of the set of local cultural principles of the group he wanted to belong to. Besides Barbieri and de Lazara's hymn on Meneghini's garden (Anon. 1819b, p. 7–10) and a register of the ornamental plants (Ibid., p. 70–80), the volume contained numerous other floral references. A long *verso sciolto* celebrated the virtue of Meneghini's garden as site of respite and peace ("In questo sì tranquillo e amabil loco") (Ibid., p. 47–52), a short sonnet extolled Cromer's garden (Ibid., p. 101), and numerous other poems and chants were replete with vegetal imagery, often mixed with Greek mythology. The groom Giovanni Battista Cromer relayed that in Monselice he had prepared a collection of plants for his bride Angela and praised her as loving "the beauty of simple nature" (Ibid., p. 69). As joyful as the wedding was, the happiness did not last. Angela died in 1836 at the young age of thirty-four. Her husband Giovanni Battista died in 1846, and their son Giovanni Battista (1824–1849) fell during the siege of Venice.[5]

In 1823, the second daughter Anna Meneghini married Francesco Gaudio, son of Luigi Gaudio (d. 1835), a well-known collector of historical artistic prints. In the 1810s, Luigi Gaudio, like Agostino Meneghini, had purchased numerous estates in Padua and in the province. The wedding was similarly sumptuous as the Meneghini-Cromer wedding. Luigi Gaudio commissioned a famous painter of the time, Giovanni De Min (1786–1859), to fresco the salons of his palazzo—evidence for the high social importance of the event (Ceccato 2013–2014, p. 51–55 and 85–97; Marsand 1823). Antonio Vigodarzere (1823) gifted a booklet. Giovanni Battista Cromer and a non-identifiable 'G.S.M.' commissioned two sonnets from the young and still unknown Niccolò Tommaseo (Michieli 2016; Tommaseo 1964, p. 588). However, the marriage seems to not have been happy and in 1835 became subject of public scandal (Anon 1852, p. 54–55). Giovanni Cittadella however, wrote an affectionate obituary (G. Cittadella 1867). Anna, too, cultivated botanical and floral interests, as testified by her membership in the Paduan Gardening Society (Sect. 2.3 in Chap. 2). During the events of the revolution, Francesco Gaudio was captain of the National Guard.

In 1832, Agostino's third daughter Marina (b. 1808) married Giovanni Nachich (also written as Nacchini, Nakić) from an important Croatian family. Little is known about this wedding. The marriages of the sons Andrea and Giuseppe were less sumptuous. Andrea married Catterina Fabris (1814–1871) of Udine in 1831, with whom he had two sons, Agostino (1833–1862) and Giuseppe (1835–1896). Giuseppe married Anna Lipparini (b. 1796) in 1861, the widow of the rich Paduan count Giacomo Negri.[6] The minor social visibility of these two wedding celebrations did not mean that Agostino neglected his sons. He invested into their education instead. A thorough school education was expensive, and a university degree even more so. As we will see in Chap. 5, he accorded much attention to their professional training and general education, supported their naturalistic passions, and sponsored Giuseppe's tour through North-Italian universities after his graduation.

All in all, it appears that the newly rich Meneghini family succeeded in achieving considerable social advancement and acceptance in a short lapse of time. In evident contrast to this, there is no obituary, nor a biographical record regarding Agostino. Andrea Cittadella Vigodarzere authored many obituaries, but evidently not for Agostino, 'the close friend' of his adoptive father Antonio. Maybe Agostino perceived his own failure as a shame, withdrew completely from public life, and rejected any commemoration. We can only speculate about the reasons of such a fall into oblivion. The relationship between Agostino's sons Andrea and Giuseppe Meneghini, and Antonio's adoptive son Andrea Cittadella Vigodarzere was still close, but no longer an intimate friendship.

Notes

1. Roberto Balzani (2000), however, stresses that even Mazzini's supporters did not define themselves in contradistinction to the 'old generation'. On the contrary, most fathers actively supported their sons.
2. Archivio di Padova, Anagrafe parrocchiale S. Francesco, register 68, Carta 735, numbers 1–20.
3. In 1841, 614 Jews lived in Padua, constituting about 0.08% of the city's population.
4. I thank Renato Camurri for this information.
5. Archivio di Padova, Anagrafe parrocchiale Cattedralle (Duomo), register 40, Carta 366, numbers 1–4. Giovanni Battista Cromer and Angela Meneghini also had two daughters: Paola (b. 1820), who married Basilio Saggini of Venice in 1842, and Elisabetta Rosa (b. 1822). The sister of Giovanni Battista, Rosa (d. 1844), in 1813 married the podestà of Padua, Andrea Saggini, who composed a short floral ode for Angela on occasion of the Gaudio-Meneghini wedding. On Cromer jr.'s death (Leoni, 1976, p. 302).
6. Archivio di Padova, Anagrafe parrocchiale S. Francesco, register 68, Carta 735, number 20. Almost nothing is known about her, except that she was born in Corfù and that her father was Giuseppe Lipparini. Her name does not appear in any of the obituaries of Giuseppe Meneghini. Maybe, she was a sister of the lyrical singer Caterina Lipparini (1792–1855), one of the great

prima donnas of her time, and daughter of the famous Buffo singer Giuseppe Lipparini. In the years 1817–1818, together with Caterina and Giuseppe, in the libretti appears a singer called Annetta Lipparini. In 1817, the Paduan police reported a scandal, when during the rehearsals for the world premiere of Giacomo Meyerbeer's (1791–1864) *Romilda and Costanza*, Caterina repeatedly appeared in public with the young count Giacomo Negri, probably Anna's future husband. After the Paduan affair, Caterina continued her successful career and later married a Sicilian count (Schuster 2003, p. 47).

References

Anon. 1819a. *Literarischer Anzeiger enthaltend die neuesten in- und ausländischen Bücher, Recensionen, Nachrichten von lebenden und verstorbenen Schriftstellern etc.* Vol. 1. Wien: Jakob Mayer und Comp.

———. 1819b. *Raccolta per le faustissime nozze Cromer-Meneghini*. Padova: Tip. Del Seminario.

———. 1852. *Carte segrete e Atti ufficiali della polizia austriaca in Italia dal 4 giugno 1814 al 22 marzo 1848*. Torino: Libreria Patria.

Arisi Rota, Arianna, and Roberto Balzani. 2012. Discovering politics: Action and recollection in the first Mazzinian Generation. In *The Risorgimento revisited: Nationalism and culture in nineteenth-century Italy*, ed. Silvana Patriarca and Lucy Riall, 77–96. London: Palgrave Macmillan.

Balzani, Roberto. 2000. I giovani del Quarantotto: profilo di una generazione. *Contemporanea* 3: 403–416.

Bassani, Maddalena, Marianna Bressan, and Francesca Ghedini, eds. 2010. *Aquae Patavinae: Il termalismo antico nel comprensorio euganeo e in Italia*. Padova: Padova University Press.

Berengo, Marino. 1963. *L'agricoltura veneta dalla caduta della Repubblica all'Unità*. Milano: Banca Commerciale Italiana.

Caburlotto, Luca. 2001. Private passioni e pubblico bene: Studio, collezionismo, tutela e promozione delle arti in Giovanni de Lazara (1744–1833). *Saggi e Memorie di Storia dell'Arte* 25: 123–217.

Ceccato, Stella. 2013–2014. *Collezionismo di stampe a Padova tra Settecento e Ottocento: Tre collezioni a confronto* (Tesi di laurea in Economia e Gestione delle Arti e delle Attività Culturali). Venezia: Università Ca' Foscari.

Chiancone, Claudio. 2010. *La scuola di Melchiorre Cesarotti nel quadro del primo romanticismo europeo* (doctoral thesis at the Université Stendhal—Grenoble III). Padova: Università degli studi di Padova.

Cittadella, Giovanni. 1867. *Anna Meneghini Gaudio*. Padova: Stab. Nazionale di P. Prosperini.

Derosas, Renzo. 1987. Aspetti del mercato fondiario nel Veneto del primo Ottocento. *Quaderni storici* 22 (65(2)): 549–578.

Derosas, Renzo, and Cristina Munno. 2010. La nobiltà veneta dopo la caduta della Repubblica: Verso la costruzione di un élite regionale? *Ateneo Veneto* 197: 233–274.

Falconi, Bernardo. 2010. Ritratto della famiglia dell'avvocato Giovanni Battista Cromer, 1805–1807. In *Da Canova a Modigliani. Il volto dell'Ottocento: Catalogo della mostra*, ed. Francesco Leone, Maria Vittoria Marini Clarelli, Fernando Mazzocca, and Carlo Sisi, 215–286, cat. 118. Venezia: Marsilio.

Finotti, Fabio, ed. 2010. *Melchiorre Cesarotti e le trasformazioni del paesaggio europeo*. Trieste: Edizioni Università di Trieste.

Fortis, Alberto. 1796. *Notizie storiche fisiche sui bagni di S. Elena ai colli Euganei con anedoti relativi alla loro celebrità, e virtù ed all'uso delle stesse acque. Al nobile signor marchese Pietro Estense Salvatico, patrizio padovano*. Venezia: Gio. Maria Bassaglia.

Foscarini, Giacomo. 1847. *Guida alle terme Euganee*. Padova: Tipi del Seminario.

Fumian, Carlo. 1984. Proprietari, imprenditori, agronomi. In *Storia d'Italia: Le regioni dall'Unità a oggi. Il Veneto*, ed. Silvio Lanaro, 99–162. Torino: Giulio Einaudi editore.

Galletto, Pietro. 1992. *Galantuomini padovani dell'Ottocento: cenni biografici ed istantanee*. Padova: Libreria Draghi Randi.

Ghidotti, Riccardo. 2014. *Parco Buzzaccarini, bosco dei frati*. Monselice: Tipografia Violato.

Leone, Francesco, Maria Vittoria Marini Clarelli, Fernando Mazzocca, and Carlo Sisi, eds. 2010. *Da Canova a Modigliani: Il volto dell'Ottocento*. Venezia: Marsilio.

Leonhard, Jörn. 2019. Formulating and reformulating 'liberalism': Germany in European comparison. In *In search of European liberalisms: Concepts, languages, ideologies*, ed. Michael Freeden, Javier Fernandez Sebasiàn, and Jörn Leonhard, 72–101. New York: Berghahn.

Leoni, Carlo. 1976. *Cronaca segreta de' miei tempi, 1845–1874; con prefazione e note di Giuseppe Toffanin jr*. Quarto d'Altino: Rebellato.

Lovett, Clara M. 1982. *The Democratic Movement in Italy, 1830–1876*. Cambridge, MA: Harvard University Press.

Maccalli, Lorenzo. 1974. *Storia di un insediamento industriale a Battaglia Terme: Dal XVI al XX secolo*. Padova: Erredicì.

Mantese, Giovanni. 1988. *Nota storica sull'arte della concia in Arzignano e sulle origini della famiglia Meneghini*. Arzignano: Grafiche A. Dal Molin.

Marsand, Antonio. 1823. *Il fiore dell'arte dell'intaglio nelle stampe con singolare studio raccolte dal signor Luigi Gaudio*. Padova: Tipografia della Minerva.

Mautner, Eduard. 1883. *Battaglia bei Padua. Mit 38 Illustrationen von L.E. Petrovits und J. Weber und einer Karte*. (Europäische Wanderbilder, 55 and 56). Zürich: Orell Füssli.

Meriggi, Marco. 1987. Il Regno Lombardo-Veneto. In *Storia d'Italia*, ed. Giuseppe Galasso, vol. 18. Torino: UTET.

Michieli, Fabio. 2016. Le poesie giovanili (1820–1833) di Tommaseo e la loro circolazione tra carteggi e stampe rare. In *Tommaseo poeta e la poesia di medio Ottocento*, ed. Mario Allegri and Francesco Bruni, vol. II, 423–451. Venezia: Istituto Veneto di Scienze, Lettere ed Arti.

Monteleone, Giulio. 1969. La carestia del 1816–1817 nelle province venete. *Archivio Veneto* 86: 23–86.

Pasetti Medin, Alessandro. 1996. Nuovi documenti per la villa sul colle di Sant'Elena ed i suoi giardini nell'Otto e Novecento. *Bollettino del Museo civico di Padova* 85: 349–371.

Piva, Raffaella. 1985. *Le "confortevolissime" terme: Interventi pubblici e privati a Battaglia e nelle terme padovane fra Sette e Ottocento*. Battaglia Terme: La Galiverna Editrice.

Preto, Paolo. 2005. Giovanni de Lazara. In *Dizionario Biografico degli Italiani*, vol. 64, 171–174. Roma: Treccani.

Ragazzini, Francesco. 1844. *Nuove ricerche fisico-chimiche ed analisi dell'acque termali euganee*. Padova: Tipografia del Seminario.

Salvagnini, Antonio Augusto. 1841. *Statistica della città e provincia di Padova*. Padova: Tipi di Angelo Sicca.

Schröder, Franz. 1830. *Repertorio delle famiglie confermate nobili e dei titolati nobili esistenti nelle provincie venete, contenente anche le notizie storiche sulla loro origine e sulla derivazione dei titoli, colla indicazione delle dignità, ordini cavallereschi e cariche di cui sono investiti gl'individui delle stesse*. Venezia: Alvisopoli.

Schuster, Armin. 2003. *Die italienischen Opern Giacomo Meyerbeers. vol. 2: Von 'Romilda e Costanza' bis 'L'esule di Granata'*. Marburg: Tectum Verlag.

Tommaseo, Niccolò. 1964. In *Memorie poetiche*, ed. Marco Pecoraro. Bari: Gius. Latera & Figli.

Tonetti, Eurigo. 2000. La rappresentanza politica nel Veneto e l'amministrazione delle città. In *Il Veneto austriaco 1814–1866*, ed. Paolo Preto, 53–67. Treviso: Fondazione Casamarca.

Vedova, Giuseppe. 1832. *Biografia degli scrittori padovani*. Vol. I. Padova: Minerva.

Ventura, Angelo. 1989. *Padova*. Bari: Edizioni Laterza.

Vigodarzere, Antonio. 1819. *Per le nozze Cromer Meneghini. Sonetti dell'abate Cesarotti*. Padova: Crescini.

———. 1823. *Per le nozze Gaudio–Meneghini: Terzine*. Padova: Dalla Minerva.

Zanetti, Pier Giovanni, ed. 1989. *Battaglia Terme: originalità e passato di un paese del Padovano*. Battaglia Terme: Comune di Battaglia Terme: La Galiverna.

CHAPTER 4

Garden Politics

Along with the marriage politics and the splendor of the weddings, another item of evidence bears out Agostino Meneghini's membership of Padua's cultural and political elite as well as how plants played a central role in the socio-cultural theatre of those years, namely the garden design at his villa Sant'Elena. The English garden style enjoyed particular success in Padua's liberal circles and became an effective means of identification and representation. The 'garden politics' displayed in these parks therefore provide precious insights into Paduan society, its political and social convictions, and the cultural and intellectual environment in which the Meneghini brothers grew up.

4.1 The Symbolic Role of Gardens

The eighteenth century was the 'century of the garden'. Even if the origins of gardens are antique and multiple, during this period in particular Europe was covered with a manicured green mantle (Assunto 1982, p. 4). Gardens and parks are places where humans transform nature according to their gusto and needs. Beyond that, the variety of garden types is great and their purposes manifold. Green areas may be places of leisure and consumption and, in times of turmoils and insecurity, safe havens of silence, private retreat, and respite. They are hybrids of nature and artifice, and

A. Dröscher, *Plants and Politics in Padua During the Age of Revolution, 1820–1848*, Palgrave Studies in the History of Science and Technology, https://doi.org/10.1007/978-3-030-85343-3_4

objects of varied and multi-layered natural, historical, social, aesthetic, scientific, practical, and symbolic meanings.

Sociologists argue that garden art was "a sign in a system of signs of status" of the aristocratic and bourgeois classes, and a statement of one's cultural, social, and political allegiances and ideals. Recent studies have indeed promoted the view of gardens as implements of a broad spectrum of political and religious intents, meanings, and influences (e.g., Kingsbury and Richardson 2005; Saguaro 2006; Spary 2000; Wickham 2012), even as the battlefields of international diplomacy (Callahan 2017; Hyde 2016). Andrea Wulf (2011, p. 2–3), for instance, contends that "it's impossible to understand the making of America without looking at the founding fathers as farmers and gardeners". An even more famous example are the Gardens of Versailles. Chandra Mukerji (2012) shows convincingly how in the late seventeenth century Louis XIV empowered the French state and his own position by means of a new vision of French grandeur and its Roman legacy. In his park, a lavish exposition of art and spectacular hydraulic engineering effectively displayed and conveyed a new image of France-as-Rome. Over a century later, Napoleon implemented a different type of 'plant policy'. His was not concerned with private gardens but with gardens for the people. The newly organized public green spaces were a demonstration of political and administrative skill and power. He transformed Turin from a military fortress into a peaceful and serene city with public parks and aesthetic boulevards (Ambrosoli 1998). These and similar botanical and arboricultural interventions of public utility also left a tangible sign of the new political order. Another, rather different attempt to use nature and plants to impose order was proposed by Robert von Mohl, jurist, professor of political science, and brother of botanist Hugo von Mohl. Born into a family that practiced extended natural and social walking tours as an important part of German bourgeois education, in 1844–1845 in his treatise *Die Polizei-Wissenschaft* [Police Science] he pushed for the installation of public parks and promenades as a way of disciplining the people (Lempa 2007, p. 179; Gamper 2005, p. 36).

The design of parks itself was a vehicle for many different kinds of socio-cultural messages. In the *Biedermeier* garden, baroque layouts and the dominance of old trees and weathered buildings, statues, and other objects provided a patina of old age, which for conservatives was an iconic representation of the 'good old times'. A predilection for fashionable trends, in contrast, communicated openness for change (Rotenberg 2002, p. 152–155). The most successful of the eighteenth and early nineteenth-century fashions was English landscape design. Developed in England around 1730, it became popular in Germany and Austria in the second half of the eighteenth century, and then spread across Italy. In Padua, it became a favorite form of expression of the new would-be intellectual and economic elite.

Landscape design started as an intellectual movement. As Marie Luise Gothein (1928, p. 279) famously said, "around its cradle stood poets, painters, philosophers, and critics" intensively discussing about its purposes, ideological meanings, and the best ways to implement them. One recurring aspect of all English gardens was a new love for nature and freedom, and the explicit opposition to barriers and fixed forms. The new feeling for nature was spiritual, but not necessarily theological. The early movement praised the wild lawless beauty, sometimes even deified untouched nature, and therefore completely did away with paths. In later years, paths were reintroduced, but never laid out in straight lines. This dislike of regular forms, and still more of symmetry, was rooted in a new philosophy of nature. This is exemplified by the exclamation "Nature abhors straight lines" by architect William Kent (c. 1685–1748), one of the originators of the landscape garden (Gothein 1928, p. 284–285). It signaled that a profound shift had taken place. Whereas French design espoused a concept of nature as an expression of a superior and precise mathematical plan, the prevailing concept of nature in landscape design was one of a composite and dynamic, historically and organically grown entity. Nature as a whole behaved in the same way as its single parts, the ever-developing and ever-transforming organisms.

The new style was not anti-scientific or anti-rational, but anti-mechanist. It gave expression to a different kind of science. The romantic sentiment yearned harmony with nature, by being an integral part of it. For Alexander Pope (1688–1744), French gardens forced nature into corsets, whereas English gardens sought to express the ideals of liberalism and free development. Along with this new garden philosophy came a new epistemology. Pope famously declared that "persons of genius, and those who are most capable of art, are always most fond of nature, as such are chiefly sensible, that all art consists in the imitation and study of nature" (Pope 1752, p. 354). Landscape gardens anticipated romantic *Naturphilosophie* and the emergent science of biology, which both valorized aspects such as the dynamicity, the singularity, and the metamorphic qualities of life. Romantic students of nature still looked to discover natural laws, yet without neglecting the almost endless diversity of forms and functions.

Giuseppe Meneghini was in fact such a naturalist. He always took account of the numerous exceptions to the rule (Chap. 7). This new form of naturalism was not limited to garden design and to philosophy of science. On the contrary, it permeated the (not exclusively romantic) political, economic, and juridical schools of thought. As we will see in Chap. 6, frequent reference to 'natural laws' was typical for many economic and political treatises written in Padua, as it was in scientific essays.

Another common feature of many landscape gardens was opposition to the French ideal of dominion over nature. English gardens—apparently—conceded primacy to nature, subordinating human design to the irregularity and spontaneity of life. Idyllic pastoral landscapes replaced the clear lines and absolute center of geometric parks with hills and clusters of trees. Meandering paths were used instead of the extended allées of royal processions. Geometric and enclosed pools gave way to irregularly shaped lakes while winding streams replaced fountains, formerly a sophisticated French symbol of controlling water. Yet, the love for wilderness had its limits. No landscape architect actually fancied the anarchy of nature unbound. Landscape gardeners certainly conceded more space to nature's peculiarities and banned, for example, topiary plants. Yet, every outgrowth was under their control. The art of the gardener consisted in manipulating nature such as to conform to the human sense of beauty, while preserving the illusion of wilderness. In the words of Andrea Cittadella Vigodarzere (1854, p. 167), "the limbs are natural, the composition is artificial. [...] the trickery is artfully mixed with the truth, and one employs trickery in foreseeing the future."

A third fundamental aspect of English gardening was sentimentalism. Visitors craved emotions. Therefore, certain landscape elements were deliberately laid out to impress. Some garden architects even staged elements of drama, installing hoary mountains instead of smooth falling hills, aged forests instead of harmonious clumps of trees, and roaring waterfalls instead of gentle streams. The second function of lakes, hills, groves, and winding paths was to create discontinuities. Instead of opening a complete conspectus that included even the smallest corner, as was common in French gardens, they produced ever-changing views and sceneries that instilled surprise and mystery. Visitors rambled from one picture and from one invitation to introspection to the next. In a later phase, landscapes were enriched by the placement of a long list of stimulating garden fabriques, like ruins, bridges, grottoes, pagodas, monuments, and temples. In Germany and in Italy in particular, romantic gardens often became places of time travel and exoticism, of fantasies of pagan worlds and Christian values, and of enlightened ideals and of nostalgia for a better world.

It may seem as if such fanciful gardens betrayed the original naturalistic idea, yet they were only extreme expressions of the same underlying idea. As such, they demonstrate that the early versions of landscape gardens, with their allegedly free play of nature, too, were illusions and green utopias. English landscaped pseudo-nature followed the vogues and tastes of its creators the same way as any other garden style (Munro 2002). A significant example is Horace Walpole (1717–1797), who defended the highly influential thesis that (English) garden design expressed the naturalization of a (Whig) political message. In a period when scholars were looking to nature as eternal truth, it was tempting to call on nature's authority and to see it as the basis for one's own social and political claims. Yet, the nature in Walpole's garden at Strawberry Hill was a pseudo-nature, with false rocks, artificial lakes, and imported exotic plants (Harney 2013).

Caution with symbolic interpretations of garden designs is therefore advisable, especially with those of the creators themselves. Moreover, it is impossible to generalize a uniform political profile or orientation of the landscape gardener. In fact, the owners of landscaped gardens in Padua were a heterogeneous group, comprising old-established Paduan families such as the Vigodarzere and Papafava, both by far the biggest landowners in the province, rich newcomers from outside Padua such as the Giacomini and De Zigno, rich Jews such as the Treves and Trieste, impoverished nobles such as the Cesarotti, ennobled bourgeois such as the Cromer, and newly rich low bourgeois such as the Meneghini. Some were progressivists, others were conservative. Still, in the context of the villas and parks of the Veneto, and notwithstanding the differences and contradictions within the liberal movement, by the 1810s and in the self-understanding of many adherents, English gardens had become a transcontinental symbol of republican and reformist convictions. For liberals, the landscape park was a symbol of these new sensibilities and the love for freedom, of common identity, and of the search for new forms of representation.

4.2 Romantic Landscapes in Padua

In Italy, the reception of the new garden fashion started rather late. Unsurprisingly, the first genuine theoretical and architectural take-up was in Padua, because Paduans had a long tradition and a broad and vivid interest in gardening. As much as plants have been an important part of Paduan and Veneto culture, gardens played a special role in its urban reorganization during the sixteenth and seventeenth centuries (Bussadori 1988). In that period, two opposite developments brought both along the creation of gardens. In the Veneto countryside, most of the villas and gardens arose when the spaces around the main buildings were cleared from defensive needs (Viviani 1975, p. 175). Instead, Padua's urban garden spaces were byproducts of the city's fortification. After the war of 1509, Padua became part of the *Renovatio securitatis*, the Venetian defensive pact. As a secondary effect of the erection of the city walls, the building of new houses was from then on limited to the enclosed area. This favored the construction of *insulae,* residential blocks composed of several buildings. The blocks surrounded open spaces, used as courtyards or small gardens. Later, rich patricians and aristocrats purchased several of these blocks and fused them together, creating palazzi and transforming the inner spaces into private contemplative and botanical gardens.

Between the sixteenth and eighteenth centuries, at least nine urban and eight suburban private botanical gardens were created in this way (Bussadori 1988, p. 51). Many of these housed noteworthy tree and flower species (De Toni 1887). Within a few decades, the city of Padua became botanically so rich and esthetically appealing that it attracted visitors from other European regions. In the late eighteenth century, Giovanni Marsili (1727–1795) compiled a historical list of at least sixteen patrician Veneto families who had installed important gardens in Padua, displaying plants and herbs mentioned by famous scholars like Leonard Fuchs (1501–1566), Luigi Anguillara (1512–1570), Conrad Gesner (1516–1565), and Gaspard Bauhin (1560–1624) (Marsili 1840; see also De Visiani 1840, p. 4). Among these celebrated urban gardens (Fig. 2.1 in Chap. 2), we find those of Daniele Barbaro (1514–1570), senator Lorenzo Priuli (not the Doge of the same name), superintendent Filippo Pasqualigo (1549–c1615), Monsignor Torquato Bembo, and of senator Gasparo Gabrieli in the sixteenth and seventeenth centuries, and Gian Francesco Morosini's famous botanical garden in the early eighteenth century (Lamon 2013; Zabbeo 2013; Zaggia 1993–1994). In 1840, De Visiani (1840, p. 4) added to this list the names of Jacopo Antonio Cortuso, Bernardino Trevisan, and the brothers Gianfrancesco, Alessandro, and Luigi Mussato.

Prior to the installation of the first English gardens in Padua, the new movement began, as it had about eighty years earlier in England, as a theoretical discourse. Between 1792 and 1798, a series of lectures on landscape parks were held at the *Accademia di Scienze, Lettere ed Arti* (Pietrogrande 1995). A few years later, one of the lecturers, Pier Luigi Mabil (1752–1836), authored a treatise on gardens that greatly favored the diffusion of the new design (Mabil 1801). A most striking aspect of all four talks was the "almost frenetic determination to establish Italian origins for the new landscaping style" (Hunt 2010, p. 14–15). The claim was grounded in a feeling of unjust marginalization in the contemporary European artistic debate; on the other hand, English landscape, ancient Roman, and Italian renaissance gardens actually shared many features (Azzi Visentini 2004; Ribouillault 2020). Setting aside the question of priority, the aesthetic affinity of late-English and Italian garden styles greatly favored the implementation oflandscape parks in Italy. Nevertheless, the 'English' gardens of Veneto were not simply copies of the northern models, but mixtures of both—not forgetting French and Chinese influences. The romantic, 'unbound wilderness' was embellished with clear-cut neo-classic and neo-Palladian buildings as well as with Italian fountains and grottoes and Venetian pergolas. Also the choice of most of the *leitmotivs* of the garden sceneries was Italian-inspired, by Virgil, Tasso, and other Latin and Italian poets and, not least, by local plant species.

The principal mediator between the English and the North Italian movements was Melchiorre Cesarotti, author of two of the academic lectures mentioned above. Around 1790, he created one of the first English gardens in Italy at Selvazzano, south-west of Padua (Finotti 2010; Pietrogrande 2010) (Fig. 1.1, no.9 in Chap. 2). Compared to Strawberry Hill or Chiswick Park, it was rather small (about two hectares). It was an intimate garden, meant as a retreat for the last years of his life, and conceived as an inseparable part of his poetry. However, Cesarotti's 'vegetal poem', as he called his park, provoked admiration but also criticism for its excessive fancifulness, especially in the Paduan circles, which favored concreteness and austerity rather than effusiveness. His close friend count Giovanni de Lazara, a man known for his sobriety, wrote to a common friend that he had hardly been able to hold back his laughter when Cesarotti emphatically presented his 'magnificent trees', which were actually rather young and not very tall (Caburlotto 2010, p. 135). His garden, replete with classical romantic architectural elements, resembled a 'take away landscape' (Venturi Ferriolo 1998, p. 72–78) rather than a rambling parkland. Equally harsh was the opinion of Cesarotti's disciple and successor Giuseppe Barbieri, tutor of the young Andrea Cittadella Vigodarzere and other offspring of the local nobility. Barbieri (1821, vol. 2, p. 146) wondered whether noble ladies living within Padua's city walls wished to see magnificent installations such as groves, temples, mountains, and lakes in their small urban gardens.

Notwithstanding de Lazara's and Barbieri's derision, English parks became a fashionable form of representation among the members of Padua's elite, mainly due to Antonio Vigodarzere and the people who came together around him. Cesarotti played an important part in the development of English gardens in Padua and, as we will see, in the establishment of the Meneghini garden. Yet, the cultural context of the 1810s and 1820s was not that of the 1790s. The protagonist of the 'Anglicization of the Paduan gardens' at the beginning of the Austrian period was Giuseppe Jappelli (Pietrogrande 2010). From 1816 onwards, Jappelli introduced new styles and fashions to Padua's architecture (Mazzi 1982, 2005, 2007). His most famous opus was the *Caffè Pedrocchi*, in the 1840s the meeting place of liberal Paduan aristocrats and students. That he was a 'foreigner' may have been one reason why the emerging liberal elite, in search of new directions and innovative styles of representation, preferred him to traditional Paduan architects. His contacts with influential Paduan figures had started around 1810, partially through masonic networks. Jappelli shared their liberal *visio mundi*, and, though working on commission and thus being obliged to adhere to the tastes and ideas of his commissioners, undertook his projects with sincere enthusiasm.

The principal reason for his success were his evocative scenographic arrangements (Baldan Zenoni-Politeo 1997). Although most of the gardens in Padua, like that of Cesarotti, were rather small, he skillfully integrated 'external' landmarks within the 'internal' prospect (Fig. 4.1). In this way, he created a spectacular view as well as the illusion of expanse and grandeur. He represented the material expression of the Paduan reformist spirit and enjoyed great success. His landscaped gardens and architectural work were innovative, yet did not radically break with previous styles, thus reflecting the desire for reform as well as the political moderation of the liberal circles. The gardens displayed an idyllic harmony of nature combined with elements of historical and literary fiction, and often masonic symbols and messages, for instance realizing Cittadella Vigodarzere's vision of a perfect world (Sect. 9.2 in Chap. 9).

Jappelli was the main figure responsible for the English transformation of Veneto gardens. His first park project was the garden of Antonio Vigodarzere at Saonara, a small town to the east of Padua (Fig. 1.1, no.

Fig. 4.1 The garden of the Treves family in Padua. Garden architect Giuseppe Jappelli integrated the roofs and tower of the Sant'Antonio Church into the scenography. Lithograph by G. B. Cecchini. From Anon. (1842, p. 274). By courtesy of Biblioteca Capitolare di Padova, Coll. 700.AA5.7

11 in Chap. 1). Vigodarzere commissioned the project in 1816, when Jappelli was still an outsider. One year earlier, Jappelli had come to Padua to install a scenography inside the great hall of the medieval *Palazzo della Ragione* (Palace of Reason) on occasion of the visit of Francis I of Austria. The scenography aroused general admiration. Vigodarzere was impressed and invited the architect to design a scenographic masonic initiation park. It contained typical landscape elements like artificial hills, lakes, and islands, bridges, and serpentine paths, as well as a series of fantastic and pseudo-historical elements like a temple and a statue of Hercules. The centerpiece of the fourteen-hectare park was the pseudo-Norman-Arabic Chapel of the Templars (Fig. 9.2 in Chap. 9) with its two scenographic grottos, the sepulcher of the knights, and the statue of Baphomet (e.g., Autiero 2006; Azzi Visentini 1990; Cittadella 1838). Vigodarzere's as well as Cesarotti's installations thus demonstrate that the influence of the English model was important but tempered. Most Paduan gardens had

more in common with Alexander Pope's (1688–1744) esoteric garden in Twickenham, than with Walpole's expansive landscapes at Strawberry Hill. They resembled Central European esoteric gardens (Fagiolo 2007) and recalled the Purification or Salvation Path of the seventeenth-century symbolic garden of Villa Barbarigo Pizzoni Ardemani in the Euganean Hills, a few kilometers west of Meneghini's villa (Fig. 1.1, no.6 in Chap. 1). The preference for pseudo-historical mainly pseudo-medieval elements in Paduan landscape gardens fostered the birth of the neoromantic movement in Italy (Bordone 1993; Chavarria and Zucconi 2016).

Over the following decades, many other Jappellian gardens were created in and around Padua (Agostinetti 2006; Levorato 1997). Almost all commissioners, or members of their families, like the Treves brothers, Achille de Zigno and Giacomo Andrea Giacomini in Padua (Fig. 2.1, no.9 in Chap. 2), Luigi Loschi (d. 1848) in Rosà near Vicenza, and Francesco Gera (1803–1867) in Conegliano, were later in one way or another involved in the reform movement of the 1830s and 1840s. Meneghini's garden project was the second, after that of Vigodarzere. Yet, before describing the circumstances of this garden, it is helpful to have a brief look at the history of the Treves garden (Fig. 2.1, no.7 in Chap. 2; Fig. 4.1), because, although the latter was an urban garden and created about fourteen years later, they show interesting parallels. The Treves brothers were much richer and they established themselves more successfully than the Meneghinis, yet both families were initially outsiders who in the beginning of the nineteenth century managed to become part of Padua's elite.

When Napoleon opened the Ghetto of Venice, the Treves family, like other Jewish families, quickly expanded their financial and real estate investments. In 1810, still facing the strong prohibitions as well as prejudice against Jews in Venice, the wealthy merchant Giuseppe Iseppo Treves (1759–1825) chose Padua as the center for his financial enterprise and bought an urban palazzo for his two sons Giacomo (1788–1885) and Isacco (1790–1855), from where they would run the Paduan activities. The moderate-liberal currents, dominant in Padua, were congenial with their convictions and activities. The relationship with Andrea Cittadella Vigodarzere was particularly strong. He praised their social and cultural patronage and sincere philanthropy toward the poor (Cittadella Vigodarzere 1862).[1] Like Agostino Meneghini, they provided important relief to the population during the famine of 1815–1816 (Monteleone 1969, p. 53).

The social commitment, visibility, and socio-cultural inclusion of the Treves brothers in Padua had also botanical roots (Massaro 2019). Like their aunt Enrichetta (Sect. 2.1 in Chap. 2), both brothers had a passion for plants. They established a famous urban garden, which was open to the public. Moreover, they were founding members of the Society for the Promotion of Gardening. In 1836, they commissioned Giuseppe Jappelli, who devised the hydraulic plan and came up with a romantic design with meandering paths, a small temple, a statue of a knight, an alchemist's grotto, a monument to friendship, a Chinese pagoda, and more (Dandolo 1836; Massaro 2015–2016, 2018, 2019). One peculiarity of their garden was its scientific-botanical aspiration. It contained a noteworthy collection of exotic palms, for which specific infrastructure, that is, glasshouses and a heating system, was built. As Martina Massaro (2014–2015, p. 101–102) rightly emphasizes, in realizing such an English park in line with the fashion of the emerging Paduan elite, the Treves family visibly signaled its membership in the Vigodarzere–de Lazara circle. The fact that the local administration, in particular the *Civica Deputazione all'Ornato*, approved their plans further attests to their admission to Padua's inner circle.

As we will see in the following chapters, Padua's high society was anything but homogeneous. On the one hand, the shared garden design was a visible expression of affiliation, on the other hand, it may have veiled the absence of a common political program. The gardens—and the discussions and events that evolved in and around them—reflect congruence in basic creeds, not a distinct political strategy. They were a public display of an aspiration to cultural leadership, a desire for innovation and a rejection of conventions paired with a marked sense of tradition and conservatism, an openness to foreign influences paired with patriotism, sentimental nostalgia paired with sober concreteness, beauty paired with usefulness.

4.3 The Meneghini Garden

It is difficult to say whether Agostino Meneghini actually shared the moderate liberal vision or whether he just wanted to become part of Padua's most influential economic and political circle. Agostino's social engagement and the later activities of his sons Andrea and Giuseppe suggest the former, yet there is no surviving recorded statement by Agostino himself to undergrid such a claim. However, his decision to engage, as early as 1816, the services of Jappelli and to install a garden that was similar to that of Antonio Vigodarzere surely demonstrates his sense of social parity. Both gardens indeed generated a sensation in Paduan society.

Meneghini's garden was situated in Battaglia, on the eastern side of the Euganean Hills. These hills are particularly densely covered in historical manors, villas, and parks (Fig. 1.1 in Chap. 1). Located within five kilometers of Villa Meneghini were the previously mentioned seventeenth-century garden of Villa Barbarigo, the castle of the Corinaldi family (today castello Sgaravatti), villa Emo, and the royal Catajo castle with its Renaissance gardens. In Monselice, about eight kilometers south of Battaglia, Giovanni Battista Cromer, the future father-in-law of Agostino's daughter, had established an English garden as early as around 1800. The apparent absence of pathways in this garden (Fig. 3.2 in Chap. 3) suggests that he followed the English style more closely than his friend Cesarotti and the later owners of landscape parks in Padua. The garden of Villa Benvenuti, at the southern end of the hills, benefitted from a complete re-design by Jappelli around 1848. Another vast English, but not Jappellian, park was that of Villa Papafava, located on the northern slopes. It received its design by count Alberto Papafava dei Carraresi (1832–1929) in 1860, son-in-law of Andrea Cittadella Vigodarzere.

Besides their predilection for English gardens, Agostino Meneghini and Antonio Vigodarzere shared a socio-economic vision. When Agostino purchased the villa at Sant'Elena di Battaglia and its surrounding estate, both properties were in a poor state. Like most of the urban landowners of his time, the previous owner count Selvatico had neglected his lands, renting them out and handing their management over to local administrators. This had devastating effects on the estates and their profitability (Sect. 9.2 in Chap. 9). The condition of the Selvatico estate further deteriorated due to inheritance disputes. In 1808, the family offered the property on the public market (Piva 1985, p. 50–51). Six years later, it was purchased by Meneghini, and in 1816, he began to transform the rugged land into fields for agriculture, revamped the thermal springs, improved the hospitality facilities, and laid out the English garden (Cittadella Vigodarzere 1842, p. 505). It is easy to understand why Vigodarzere was happy to see Agostino Meneghini, a newcomer, buying and improving the place. As we have seen in the previous chapter, the Veneto aristocracy was not as exclusive as their Austrian counterparts. Their social status depended on their landed property and, in a quasi-bourgeois manner, on their ability to generate wealth from real estates. People like Selvatico, who neglected their estates, actually endangered the social standing of the rural nobility in general, whilst newcomers, even if of modest origin, if wisely integrated into the system, could corroborate and continue the prevailing social order.

Agostino's actions surely had entrepreneurial aims, but he still emulated Vigodarzere's example. Vigodarzere had also undertaken his refurbishments to provide employment for local people, who suffered through a great famine in 1816, the 'Year without a Summer'. In 1817, the province of Padua alone counted 19,374 deaths (Monteleone 1969, p 56–75).[2] Thus, contrary to Horace Walpole's creation of a rural idyll at Houghton Hall that required the demolition of existing cottages and villages, and the forced eviction of most peasants (Burrell and Dale 2002, p. 118–119), Vigodarzere's as well as Meneghini's scheme made sure that the local population benefitted from the project. Following Gibson Burrell and Karen Dale's argument that Walpole's garden was part of an utopian order of the English aristocracy and wealthy bourgeoisie, whereby they treated the countryside as their garden and expelled anything unwanted from their property (Burrell and Dale 2002, p. 120), it is tempting to see Vigodarzere's and Meneghini's version as a utopian version of paternalism. It did not question private property, yet it had an eye on the wellbeing of the local population. The philanthropic employment did not challenge their own social dominance, but, on the contrary, underpinned their leadership as *patres familias*. As we will see, the bonds with the peasants established by this kind of activity were the basis for their claim for local social leadership and would be a crucial element during the revolutionary years (Chap. 10) as well as during the post-unification elections in Padua (Sect. 9.2 in Chap. 9).

Still other elements of the gardens provide precious information about prevailing cultural and socio-economic beliefs. Antonio Vigodarzere's Hercules statue and the 'Norman-Arab' chapel of the Templars—the first instance of gothic revival in Veneto (Levorato 1997, p. 98)—show his membership of Enlightenment freemasonry as clearly as Andrea Vigodarzere's marriage, in 1839, with Arpalice Papafava Antonini dei Carraresi (b. 1820), member of one of the most recognized enlightened-masonic families in Padua. During the French period, freemasonry and Jacobinism flourished in Padua, as elsewhere in Lombardy-Veneto. Even the bishop of Padua Francesco Scipione Dondi Dall'Orologio (1756–1819), notorious for his Francophile and democratic sentiments, was a member of a lodge. The situation changed radically with the arrival of the Austrians, who considered freemasonry as subversive (Mariutti 1930, p. 17–18). They dismantled all lodges and closely watched any person or activity that was suspected of reactivating them. Yet, some new and old lodges appear to have continued their activities. Vigodarzere and Jappelli were suspected to be adepts of the lodge *La Pace* (The Peace), though Jappelli, urged by the Austrian authorities, stepped down in 1826.

It is uncertain whether Meneghini was member of a lodge. Some elements of his garden project suggest an affiliation, others do not. Indeed, the interpretation of his garden is complicated by the fact that Agostino never actually implemented it. All that exists are Jappelli's sketches from 1816 and a poem composed on the occasion of the Cromer-Meneghini wedding in 1819 by Abbot Giuseppe Barbieri, Cesarotti's favorite disciple and later his successor to the university chair of philology (Barbieri 1819). The poem described the (imaginary) scenography of the garden, most likely according to Jappelli's plans. Similar to the itinerary through Cittadella Vigodarzere's park in Saonara, the *leitmotif* in Sant'Elena di Battaglia was an initiatory voyage culminating in purification and knowledge. The ascent from the darkness of grottos and caverns toward the light was to become a classic of Paduan liberal-masonic gardens. In the 1840s, for instance, Vincenzo Paolo Barzizza (1805–1873) created a similar serpentine ascending and descending route, probably devised by Jappelli, in the garden of his villa Belvedere at Mirano, halfway to Venice (Stefani Mantovanelli 1989) (Fig. 1.1, no. 15 in Chap. 1). Jappelli's ambitions for the Meneghini garden contained numerous architectural elements alluding to the myth of Aeneas and his voyage to the underworld as described in the sixth book of Virgil. Among the members of the cultural elite, knowledge and the love for classical mythology were as taken for granted as a passion for sprouting nature. Classical authors had been the sources of inspiration for leisure gardens since the late Renaissance. The idea of reconstructing the flora of classical literature, a sort of vegetal Greek revival, was therefore a trend in the gardens of the first decades of the nineteenth century (Becker 2017). Yet, achieving a fitting combination of classical philology and modern botany—of Ovid and Linnaeus, of *Arcadian* sensitivity and taxonomic rigor—was no mean feat. To avoid incongruences, most gardeners resorted to topographic partitions.

For the project in Battaglia, the misty and sulphurous thermal lakes (Fig. 4.2) and the small streams traversing the estate became Jappelli's *leitmotiv*. He planned two circular routes. The small lakes to the north of the Sant'Elena hill were to form the basis for his scenography of the underworld. Starting at the Avernus (the volcanic lake representing the mythological entrance to the underworld) and passing along Acheron (the

Fig. 4.2 The villa of the Meneghini family on the top of colle Sant'Elena. In the foreground a fountain with thermal water. Lithograph by L.E. Petrovits. From Mautner (1883, p. 5). By courtesy of Ariane Dröscher

river that divides the reign of life from that of the shadows), the itinerary was to lead to Lethe (the river of oblivion) along the western side of the hill. Thereafter, the path was to run through the oak woods, containing sceneries of the sanctuary of Miseno, toward the Elysium, south of the villa, finally terminating at the Gates of Horn and Ivory (the gate through which Aeneas and Sybil returned to the real world) (Mazzi 1997; Pietrogrande 2005). Even the plant species—horse chestnuts, ash trees, elms, and holm oaks—were taken from Virgil (Levorato 1997, p. 101).

The Virgilian project is also reminiscent of the garden of Melchiorre Cesarotti, which was inspired by the third book of the Aeneid. Yet, English gardens have certainly served as a model as well. Moreover, the project is indicative of Meneghini's affinity to masonic lodges. Nino Agostinetti (2006), in fact, included Meneghini's park among the masonic gardens. Likewise, Barbieri's hymn and its commissioning by Giovanni de Lazara as well as Meneghini's close relationship to Vigodarzere point toward membership: all three of them were being monitored as freemasons by the Austrians, Barbieri as a fanatic even (Anon. 1851–1852, II, p. 307–308).

Nevertheless, Meneghini's allusions might have been only a façade. His name did not appear in Austrian lists of persons suspected of freemasonry from 1814, 1824, and 1832, nor was he mentioned among those who (like Papafava, Gaudio, Selvatico, Treves, Trevisan, Trieste, and Vigodarzere) financially supported the anti-Austrian conspiracy of 1831 (Mariutti 1930, p. 132–169 and 184–193).[3]

Why Agostino Meneghini did not realize the garden project remains an open question. Apparently, it was not a question of money, because in 1828 he was able to discharge all remaining debt on his property (Maccalli 1974, p. 5–6). Maybe he was anxious not to be put under observation as a freemason; maybe he was, like de Lazara, too serious to install a Virgilian 'Disneyland'; or maybe he was not convinced that his guests would have been sufficiently keen to enter an (imaginary) underworld while enduring the (real) stench of sulfur. Most probably, Agostino's main objective in Sant'Elena was to realize the economic benefit of the thermal baths, which in fact reopened on 1 June 1818. While the garden project remained a paper sketch, Agostino invested a great deal of money and his considerable managerial skills into the enterprise of the thermal baths. The parts of Jappelli's design that were implemented were the hydraulic works, a general modernization of the surrounding fields and the sanitary infrastructure. In 1842, after the implementation of several alterations, the physician Gaspare Morgagni described two main complexes—one at the bottom of the hill, the other near the canal—with two hotels, stalls, a total of twenty-two Roman-like marmoreal baths, thirteen small rooms for mud therapy, one bath with mineral vapors, *frigidarium*, *tepidaria,* and *calidaria* (Morgagni 1842, p. 22–23). Moreover, from 1820, the baths offered a rich program of events and facilities—like horses, dances, a billiard room, restaurants of a high standard, a coffee shop, a garden for pleasure walks—that were superior to those of other baths nearby (Piva 1985, p. 58). All this is evidence that Agostino had an eye not only on the sanitary but also on the touristic opportunities associated with his investment. As early as in 1817, he revived the San Bartolomeo Fair in Battaglia, and in the following years, he regularly printed flyers to publicize his baths (Anon. 1823).[4]

Nevertheless, it is evident that with his enterprises he also claimed—and attained—social recognition and advancement. The spread of English gardens in Padua was certainly a question of individual tastes, yet even more so one of (real or intended) group identity construction. Michel Conan (2002, p. 3–7) suggests that this kind of behavior can be seen as conspicuous consumption, a concept introduced in 1899 by the sociologist

Thorstein Veblen to describe public displays of wealth that go far beyond the satisfaction of immediate needs but are rather aimed at attaining social status. Like the collection of objects of art or the patronage of artists, gardens can be means of emulation as well as distinguishing oneself from other social classes. In Padua, this was certainly a sign of not only a social but also a political sense of belonging.

This stroll through Agostino Meneghini's garden project has revealed some significant features about the position of the Meneghini family in the context of Padua. Firstly, it has shown that the group of owners of landscaped gardens in Padua was heterogeneous. Some were bourgeois, others aristocratic or even royal, some tended to fanciness, others were more sober. Likewise, the range of types of English gardens was broad: Cesarotti's garden reflected his poetry; Cromer's garden was modeled on the early picturesque English landscape garden style; Vigodarzere's garden was spectacular and spiritual-masonic; and Treves's garden distinguished itself by a strong botanical component. Likewise, their interests in the plant world varied, ranging from the historical (the reconstruction of a Virgilian flora) to the floristic-scientific (Treves's palm collection, Alessandro Papafava's conifer collection, Parolini's botanical garden in Bassano) and the ornamental (Treves, Vigodarzere, Giuseppe Cristina's camellia collection).

Secondly, putting aside the artistic and postponing for a moment the naturalistic-scientific aspects of Paduan gardens, the brief analysis has provided a series of insights into the general ambiance of the social circles with which the Meneghinis were associated. As any other garden or park style, the Paduan English gardens were status symbols. Planning, realizing, and maintaining a garden project required at least some hectares of land and a considerable sum of money, and was thus a sign of economic wealth. In addition, the choice of novel and foreign designs was as a visible and tangible expression of a break with traditional styles and their replacement with something unprecedented, and non-provincial and open to contemporary European movements. The gardens thus underpinned a claim to progress and cultural innovation.[5] Yet, they by no means represented a radical break with the past. Elements of continuity with Italian Renaissance and Baroque gardens were numerous and more important than the stylistic imports from England. Indeed, explicit references to Italianness were noticeable during the early theoretical debate at the Academy as well as in the actual garden installations. In the light of the emerging Risorgimento movement, these were early signs of increasing national pride. Many of the owners of landscape gardens were in fact financial supporters of the aforementioned anti-Austrian conspiracy of 1831. The display of Italian elements, however, did not necessarily reveal patriotic sentiments. This is demonstrated by the story of the subsequent owner of villa Meneghini, Maria Anna von Wimpffen. The Austrian countess was member of the Paduan Gardening Society, but also wife of the commander of the anti-Italian armies, Feldmarschall Franz von Wimpffen. In 1851, during the heyday of Austrian repression, she installed a geometrically designed garden *all'italiana* for her famous European guests, among them Heinrich Heine (1797–1856) and Alexander von Humboldt (1789–1859) (Pasetti Medin, 1996, p. 353–356).

Finally, and more than anything else, the Paduan English gardens, especially those of the 1810s and 1820s, were vehicles of identification. Much more than in England or Germany, they were a way to show one's affiliation to a specific group and to demonstrate a shared identity. The stories of 'newcomers' like Meneghini and Treves furthermore confirm that Padua's social elite was open to new members, including Jews and commoners, as long as they conformed to their precepts and were sufficiently wealthy to own and create a landscape garden. The canon of shared values was broad and varied, and encompassed general cultural beliefs (romanticism, the love of sentimental poetry, naturalism), political inclinations (moderatism, freemasonry, national and liberal ideals), and socioeconomic convictions (the pursuit of economic renewal and agronomic advancement, paternalistic social responsibility). This is especially true for the generation of Antonio Vigodarzere, Giacomo Treves, and Agostino Meneghini. Their sons grew up in this environment, yet later also created new forms of expression and congregation, namely, influential institutions like associations and journals discussed in earlier chapters. However, gardens and botanical metaphors continued to shape the social views of both generations (Chap. 6).

Notes

1. Their nephew Emilio Morpurgo (1836–1885), son of Fiorina (Flora) Treves dei Bonfili (1814–1873), daughter of Raffaele Vita Treves dei Bonfili (1792–1845), brother of Giacomo and Isacco, in 1859 wrote his doctoral thesis on *The proletariat and the mutual aid societies*. He was also a member of the Reading cabinet and the Società d'Incoraggiamento.

2. Starting in the last years of the Republic of Venice, Veneto suffered an economic and agricultural crisis, aggravated by six major military conflicts and entrepreneurial backwardness. Severe flooding in summer 1816 destroyed almost all of the wheat harvest, causing poverty, famine, epidemics of pellagra and typhus, and a worrying population decrease. Monteleone (1969, p. 55) also reports that Meneghini had been blamed for having caused a shortage by speculating and buying up great quantities of wheat.
3. Of interest is the note assessing Antonio Vigodarzere as a "Liberal who is said to have given money in favour of the revolt. He is very rich, of liberal political principles, but he was not censored because of his countenance, and is not fearsome" (Mariutti 1930, p. 192).
4. It showed the Sant'Elena hill, terraced by Agostino, and on its top villa Meneghini. The newly planted vegetation was still low. Today, there is no direct access to the principal entrance, the Battaglia canal.
5. Interestingly, for Genoa's patron, the doge Agostino Lomellini (1709–1791), the transformation of his estate into the city's first English garden represented not a claim for leadership, but, on the contrary, an admission of political failure and, like Cesarotti's almost coeval garden, melancholic withdrawal (Magnani 2002, p. 74–76).

References

Agostinetti, Nino. 2006. *Giardini massonici dell'Ottocento veneto*. Padova: La Garangola.

Ambrosoli, Mauro. 1998. Alberate imperiali per le strade d'Italia: La politica dei vegetali di Napoleone. *Quaderni storici* 33 (3): 707–738.

Anon. 1823. *Avviso di apertura dei bagni termali per il giorno 3 maggio 1823*

———. 1842. *Guida di Padova e della sua provincia*. Padova: Coi Tipi del Seminario.

———. 1851–1852. *Carte segrete e atti ufficiali della polizia austriaca in Italia dal 4 giugno 1814 al 22 marzo 1848*. 3 vol. Capolago: Tipografia Elvetica.

Assunto, Rosario. 1982. Teorie del giardinaggio nell'estetica romantica. In *Giuseppe Jappelli e il suo tempo*, ed. Giuliana Mazzi, vol. 1, 3–23. Padova: Liviana.

Autiero, Carlo. 2006. *Un percorso semiotico nel parco romantico jappelliano*. Roma: Meltemi editore.

Azzi Visentini, Margherita. 1990. Il giardino Cittadella Vigodarzere a Saonara. In *Il giardino italiano dell'Ottocento nelle immagini, nella letteratura, nelle memorie*, ed. Alessandro Tagliolini, 177–193. Milano: Edizioni Angelo Guerini e Associati.

———, ed. 2004. *Topiaria: Architetture e sculture vegetali nel giardino occidentale dall'antichità ad oggi*. Treviso: Fondazione Benetton Studi Ricerche.

Baldan Zenoni-Politeo, Giuliana, ed. 1997. *Il giardino dei sentimenti: Giuseppe Jappelli architetto del paesaggio*. Milano: Edizioni Angelo Guerini e Associati.
Barbieri, Giuseppe. 1819. *I Bagni di S. Elena. Versi per le faustissime nozze Cromer-Meneghini*. Padova: Tipografia del Seminario.
———. 1821. *Veglie tauriliane*. Padova: Valentino Crescini.
Becker, Marcus. 2017. *Leonicera caprifolium L.* im zauberischen Hain: Empfindsames Sprachspiel und botanische Nomenklatur im Frühen Landschaftsgarten. *Annals of the History and Philosophy of Biology* 22: 107–121.
Bordone, Renato. 1993. *Lo specchio di Shalott: La invenzione del Medioevo*. Napoli: Liguori.
Burrell, Gibson, and Karen Dale. 2002. Utopiary: Utopias, gardens and organization. *The Sociological Review* 50 (S1): 106–127.
Bussadori, Paola. 1988. Gli orti botanici privati padovani. In *Di sana pianta: erbari e taccuini di sanità: Le radici storiche della nuova farmacologia*, 47–54. Modena: Edizioni Panini.
Caburlotto, Luca. 2010. Un amico in visita al 'Selvaggiano': Giovanni de Lazara (e un seguito con Giuseppe Barbieri). In *Melchiorre Cesarotti e le trasformazioni del paesaggio europeo*, ed. Fabio Finotti, 129–144. Trieste: Edizioni Università di Trieste.
Callahan, William A. 2017. Cultivating power: gardens in the global politics of diplomacy, war, and peace. *International Political Sociology* 11 (4): 360–379.
Chavarria, Alexandra, and Guido Zucconi, eds. 2016. Medioevo fantastico: L'invenzione di uno stile tra fine '800 e inizio '900. *Archeologia dell'Architettura* 21: 13–85.
Cittadella, Giovanni. 1838. *Il giardino di Saonara descritto*. Venezia: Tipografia di Alvisopoli.
Cittadella Vigodarzere, Andrea. 1842. Colli Euganei. In *Guida di Padova e della sua provincia*, 463–506. Padova: Tipi del Seminario.
———. 1854. Elogio di Giuseppe Jappelli, letto nella tornata 14 maggio 1854. In *Rivista periodica dei lavori della I. R. Accademia di Scienze, Lettere ed Arti di Padova vol. II, Trimestre primo e secondo del 1853–54*, 163–181. Padova: Sicca e figlio.
———. 1862. Il cav. Isacco Treves de Bonfili. In *Memorie funebri antiche e recenti offerte per la stampa dall'ab. Gaetano Sorgato*, 30–32. Padova: Randi.
Conan, Michel. 2002. Introduction: gardens into cultural change, 1550–1850. In *Bourgeois and aristocratic cultural encounters in garden art, 1550–1850*, ed. Michel Conan, 1–24. Washington, DC: Dumbarton Oaks Research Library and Collection.
Dandolo, Tullio. 1836. Un giardino nell'Euganea. *Il Gondoliere* 4 (46): 183–184.
De Toni, Giovanni Battista. 1887. *Intorno ad alcuni alberi e frutici ragguardevoli esistenti nei giardini di Padova*. Padova: Tipografia Gio. Batt Randi.

De Visiani, Roberto. 1840. *Illustrazione delle piante nuove o rare dell'Orto Botanico di Padova. Memoria I.* Padova: Tipi di Angelo Sicca.

Fagiolo, Marcello. 2007. *Architettura e massoneria: L'esoterismo della costruzione.* Roma: Gangemi.

Finotti, Fabio, ed. 2010. *Melchiorre Cesarotti e le trasformazioni del paesaggio europeo.* Trieste: Edizioni Università di Trieste.

Gamper, Michael. 2005. Garten als Institution: Subjektkonstitution und Bevölkerungspolitik im Volksgarten. In *Der andere Garten: Erinnern und Erfinden in Gärten von Institutionen*, ed. Natascha N. Hoefer and Anna Ananieva, 35–54. Göttingen: Vandenhoeck & Ruprecht.

Gothein, Marie Luise. 1928. In *A history of garden art. Vol. 2: From the Renaissance to the present day in France*, ed. Walter P. Wright. London and Toronto: J.M. Dent.

Harney, Marion. 2013. *Place-making for the imagination: Horace Walpole and Strawberry Hill.* London and New York: Routledge.

Hunt, John Dixon. 2010. Transformations in landscape circa 1800: Landscapes of the bard, of the *promeneur solitaire*, & of the *pater familias.* In *Melchiorre Cesarotti e le trasformazioni del paesaggio europeo*, ed. Fabio Finotti, 7–23. Trieste: Edizioni Università di Trieste.

Hyde, Elizabeth. 2016. Of monarchical climates and republican soil: French plant and American gardens in the revolutionary era. In *Foreign trends in American gardens: a history of exchange, adaption, and reception*, ed. Raffaella Fabiani Giannetto, part 2. Charlottesville: University of Virginia Press.

Kingsbury, Noel, and Tim Richardson, eds. 2005. *Vista: the culture and politics of gardens.* London: Frances Lincoln.

Lamon, Roberta. 2013. Il palazzo del Bembo in via Altinate. *Padova e il suo territorio* 161: 35–40.

Lempa, Heikki. 2007. *Beyond the gymnasium: educating the middle-class bodies in classical Germany.* Lanham: Rowman & Littlefield Publishers.

Levorato, Margherita. 1997. Giuseppe Jappelli e l'arte del giardino: la variabilità del gusto. In *Il giardino dei sentimenti: Giuseppe Jappelli architetto del Paesaggio*, ed. Giuliana Baldan Zenoni-Politeo, 96–112. Milano: Edizioni Angelo Guerini e Ass.

Mabil, Pier Luigi. 1801. *Teoria dell'arte de' giardini.* Bassano: Remondini.

Maccalli, Lorenzo. 1974. *Storia di un insediamento industriale a Battaglia Terme: Dal XVI al XX secolo.* Padova: Erredicì.

Magnani, Lauro. 2002. *The rise and fall of gardens in the Republic of Genoa, 1528–1797.* In *Bourgeois and aristocratic cultural encounters in garden art, 1550–1850*, ed. Michel Conan, 43–76. Washington, DC: Dumbarton Oaks Research Library and Collection.

Mariutti, Angela. 1930. *Organismo ed azione delle società segrete del Veneto durante la seconda dominazione austriaca (1814–1847).* Venezia: R. Deputazione Editrice.

Marsili, Giovanni. 1840. *Notizie inedite dei patrizii veneti dotti nella cognizione delle piante e dei loro orti botanici più rinomati*. Padova: Tip. Cartallier e Sicca.

Massaro, Martina. 2014–2015. *Giacomo Treves dei Bonfili collezionista e mecenate (1788–1885): La raccolta di un filantropo patriota*. (Dissertation thesis). Venezia: Università IUAV di Venezia.

———. 2015–2016. Il palazzo, le adiacenze e il giardino per li nobili fratelli Giacomo Isacco Treves dei Bonfili nella Regia città di Padova. *Atti e Memorie dell'Accademia Galileiana di Scienze, Lettere ed Arti già dei Ricoverati e Patavina, parte II: Memorie della Classe di Scienze matematiche, fisiche e naturali* 128: 105–150.

———. 2018. Visualizing the Treves botanical garden in Padua: from documentary research to laser survey and 3D modeling. In *Visualizing Venice: mapping and modeling time and change in a city*, ed. Kristin L. Huffman, Andrea Giordano, and Caroline Bruzelius, chap. 7. Abingdon and New York: Routledge.

———. 2019. *Palazzo Treves dei Bonfili e il suo giardino*. Padova: Il Poligrafo.

Mautner, Eduard. 1883. *Battaglia bei Padua. Mit 38 Illustrationen von L.E. Petrovits undJ. Weber und einer Karte*. (Europäische Wanderbilder, 55 and 56). Zürich: Orell Füssli.

Mazzi, Giuliana, ed. 1982. *Giuseppe Jappelli e il suo tempo*, 2 vol. Padova: Liviana.

———. 1997. Un giardino per le terme: Il progetto di Giuseppe Jappelli per sant'Elena di Battaglia. In *Il giardino dei sentimenti: Giuseppe Jappelli architetto del paesaggio*, ed. Giuliana Baldan Zenoni-Politeo, 150–163. Milano: Edizioni Angelo Guerini e Associati.

———. 2005. Giuseppe Jappelli (1783–1852). In *Storia dell'architettura italiana: L'Ottocento*, ed. Amerigo Restucci, 590–605. Milano: Electa.

———. 2007. Giuseppe Jappelli. In *Padua felix: Storie padovane illustri*, ed. Oddone Longo, 281–292. Padova: Esedra editrice.

Monteleone, Giulio. 1969. La carestia del 1816–1817 nelle province venete. *Archivio Veneto* 86: 23–86.

Morgagni, Gaspare. 1842. *Del progressivo andamento sanitario e medico delle terme padovane*. Padova: Tipi della Minerva.

Mukerji, Chandra. 2012. Space and political pedagogy at the gardens of Versailles. *Public Culture* 24: 509–534.

Munro, Rolland. 2002. The consumption of time and space: utopias and the English romantic gardens. *The Sociological Review* 50 (S1) (Special issue: Utopia and organization, ed. Martin Parker): 128–154.

Pasetti Medin, Alessandro. 1996. Nuovi documenti per la villa sul colle di Sant'Elena ed i suoi giardini nell'Otto e Novecento. *Bollettino del Museo civico di Padova* 85: 349–371.

Pietrogrande, Antonella. 1995. Il dibattito padovano sui giardini all'inglese tra Sette e Ottocento. *Padova e il suo territorio* 54: 35–37.

———. 2005. Il progetto di Giuseppe Jappelli per il giardino di Villa Selvatico-Meneghini. *Padova e il suo territorio* 116: 23–27.

———. 2010. Un'interpretazione veneta del nuovo giardino europeo: Selvaggiano, il ritiro campestre di Cesarotti. In *Melchiorre Cesarotti e le trasformazioni del paesaggio europeo*, ed. Fabio Finotti, 57–90. Trieste: Edizioni Università di Trieste.

Piva, Raffaella. 1985. *Le "confortevolissime" terme: Interventi pubblici e privati a Battaglia e nelle terme padovane fra Sette e Ottocento.* Battaglia T: La Galiverna Editrice.

Pope, Alexander. 1752. *Guardians.* In *The works of Alexander Pope in nine volumes complete*, vol. VI, 308–356. London: printed for J. and P. Knapton, H. Lintot, J. and R. Tonson, S. Draper, and C. Bathurst.

Ribouillault, Denis. 2020. Garden architecture: from representation to transfiguration. *Le visiteur. Revue critique d'architecture* 25: 199–208.

Rotenberg, Robert. 2002. *La pensée bourgeoisie* in the Biedermeier Garden. In *Bourgeois and aristocratic cultural encounters in garden art, 1550–1850*, ed. Michael Conan, 147–172. Washington, DC: Dumbarton Oaks Research Library and Collection.

Saguaro, Shelley. 2006. *Garden plots: the politics and poetics of gardens.* Hampshire: Ashgate.

Spary, Emma C. 2000. *Utopia's garden: French natural history from Old Regime to revolution.* Chicago: The University of Chicago Press.

Stefani Mantovanelli, Marina. 1989. *Le ville e i parchi comunali di Mirano: Itinerari storico-artistici.* Mirano: Comune di Mirano.

Venturi Ferriolo, Massimo. 1998. *Giardino e paesaggio dei romantici.* Milano: Guerini.

Viviani, Giuseppe Franco, ed. 1975. *La villa nel veronese.* Verona: Banca Mutua Popolare di Verona.

Wickham, Louise. 2012. *Gardens in history: a political perspective.* Oxford: Windgather Press.

Wulf, Andrea. 2011. *The founding gardeners: how the revolutionary generation created an American Eden.* London: William Heinemann.

Zabbeo, Alessandra. 2013. Un giardino perduto a Piove di Sacco. *Padova e il suo territorio* 165: 33–35.

Zaggia, Stefano. 1993–1994. "L'Alcinoo d'Adria": Giovan Francesco Morosini e il suo orto botanico di Padova. Atti dell'Istituto Veneto di Scienze, Lettere ed Arti. *Classe di Scienze morali, lettere ed arti* 152: 371–390.

CHAPTER 5

Growing Up in a Progressive Environment

In conformity with Antonio Vigodarzere's agricultural reform project—later continued by Andrea Cittadella Vigodarzere's, project of 're-educating' the landowners (Sect. 9.2 in Chap. 9)—Agostino Meneghini did not give himself up to leisurely city life but stayed at Battaglia to watch over his estates. His sons, by contrast, were born and spent most of their youth in the city of Padua, about twenty kilometers north of Villa Meneghini. There is no evidence that Agostino contrived important marriages for them. Instead, he invested into their education. This was not unusual. Many families of the time accepted this financial sacrifice as the most accessible road to upward social mobility (Lovett 1982, p. 80–90). For both brothers, this implied considerable moral pressure to continue the family's socio-economic advancement. Like most of their peers, both brothers accepted this burden and soon displayed similar commitments and vigor as their father, yet not as entrepreneurs or farmers, but as scientists, professionals, and networkers.

For the most part, Andrea and Giuseppe's education followed the same path as that of other offspring of the Paduan social and intellectual elite and therefore also provides a general insight into the schooling of the city's rich. Families who could afford the expense engaged private teachers or tutors for the initial years. So did Agostino Meneghini. He paid two exceptional figures, the abbots Giuseppe Bernardi (1788–1851) and

A. Dröscher, *Plants and Politics in Padua During the Age of Revolution, 1820–1848*, Palgrave Studies in the History of Science and Technology, https://doi.org/10.1007/978-3-030-85343-3_5

Pietro Melo (1782–1829). Both deeply influenced the scientific and philosophical inclinations of his sons, the former in Padua, and the latter in Battaglia during the summer months.

5.1 Bleeding Polenta

Abbot Pietro Melo was the chaplain and botanically proficient guardian of the Cittadella Vigodarzere garden. His naturalistic skills are confirmed by Georg Matthias von Martens (1788–1872), who visited the garden in Saonara in the 1820s and received from Melo numerous Venetian plants for his publications (von Martens 1824, I, p. 178–181). Melo taught the young Meneghini brothers, as well as other offspring of Veneto notables (Lampertico 1882, I, p. 52 note 3), the techniques of collecting and identifying plants. It was in the park of Villa Meneghini and in the midst of the volcanic Euganean Hills that Giuseppe and Andrea learned to love the study of nature. Their father welcomed their passion and equipped them with microscopes and the most recent literature (Pirona 1889–1890, p. 683). With Melo, the brothers undertook field excursions into the surrounding area. Later, during his years as a university student, Giuseppe undertook, together with various friends, especially by his close friend the physician Giovanni Zanardini and the Venetian private scholar count Nicolò Bertucci Contarini (1780–1849),[1] other expeditions to study the algae of the Venetian lagoons and shorelines, and the flora of the Alps. Recent research has shed some light on the social and epistemic importance of expeditions (Klemun and Spring 2016; Thomas and Harris 2018), yet further work needs to be done on the multi-layered interactions of participants. In Meneghini's case, the influence of the private conversations during these days is poorly documented, but the excursions proved to be particularly intense and stimulating, scientifically as well as socially (G. Meneghini 1878–1879, p. 392).

Melo's scientific attitude had gained notoriety due to the curious and sensational episode of the 'bleeding polenta' (De Checchi 2008; Gaughran 1969; Meloni 1992). In early August 1819, at Legnaro, a little town near Padua (Fig. 1.1, no. 10 in Chap. 1), a peasant found his polenta, stored in a cupboard, sprinkled with a blood-like substance. He threw it away, but in the following days other pieces of polenta showed the same sprinkles. A priest was called several times to bless the house and the polenta, yet the phenomenon recurred, provoking such widespread turmoil that the Catholic and administrative authorities, and even the University of Padua, decided to send reliable local representatives to investigate the event. One of them was abbot Melo.

Melo's approach reveals much about his character. The Abbot went to Legnaro and ascertained the conditions inside the peasant's house. He then took a piece of polenta to reproduce the phenomenon in his own house. He excluded a miraculous nature, and rather saw the cause in spontaneous fermentation. With the results in hand, he hurried to be the first to publish on the topic (Melo 1819). His results were soon outdone by the extraordinary pre-bacteriological interpretation of Bartolomeo Bizio (1791–1862), who in 1823 described a bacterium he called *Serratia marcescens* (today *Bacillus prodigiosus*), that produces blood-red pigment in human food, preferably starchy ones like polenta. At any rate, this event offers a glimpse into the local situation of those days. It highlights the then still poor hygienic conditions in the rural centers of the Po Valley, and illustrates the prevailing coexistence of superstition and rationality. Last but not least, it underlines the role of clerics like Melo, who—in spite of their religious beliefs and widespread opinions, which also circulated in several newspapers, that such phenomena reflected a miracle or premonition—advocated personal verification, an experimental approach, and a chemical explanation. This explains why enlightened families like the Meneghinis appointed instructors like Melo who combined religious belief and scientific intellect. Andrea and Giuseppe, along with Christian values and botanical skills, soon also acquired the critical empirical spirit.

Very early, Giuseppe found particular delight in studying not the spectacularly ornamental plants and flowers, the object of desire of many gardeners and horticulturists, nor the majestic palms, the pride of the Treves garden, but the 'lower' plants, in particular the algae of thermal springs. In 1835, he listed 52 different species, 33 of which he introduced as new to science (Ragazzini 1844, p. 112–128). Likewise, the 18 halophytes (plants living in habitats of high salinity) of Giacomo Foscarini's (1810–1880) *Guida alle Terme Euganee* had been collected by Giuseppe Meneghini (Béguinot 1909, p. 73; Foscarini 1847, p. 39–41). Two years later in Prague, he presented a list of 179 species and 18 varieties of algae found in the Euganean hills, 54 of which he declared to be new (G. Meneghini 1837). The choice of algae is significant for at least two reasons. Firstly, it shows Giuseppe's openness toward research objects that others might have considered as ignoble. According to him (1844–1846, I, p. 314), we can admire "the mastery of organization" in the animated colossus as well as in the creeping worm. In 1843, he affirmed his passion for the tiny by starting his monograph on the cyanobacterium *Nostoc* with a quote from Saint Augustin *Deus autem ita est artifex magnus in magnis, ut minor non sit in parvis* (Now God is in such sort a great worker in great things, that He is not less in little things) (G. Meneghini 1843, p. 1).[2] Similar mottos, such as *Maxima in minimis*, were currently used in the physico-theological research of the seventeenth and eighteenth centuries (Ogilvie 2020; Stebbins 1980), and also in the early nineteenth century the naturalist and republican activist François Raspail similarly praised, at about the same year as Meneghini, the grasses as "these *pariahs* of vegetation" (Vienne 2017, p. 633–634). Secondly, as we will see in Chap. 7, precisely these tiny organisms introduced him into the debates that revolutionized botany as well as biology in general.

5.2 Bernardi's Lessons

A similar mixture of religiosity and rationality prevailed also in the education of his sons, which Agostino entrusted to Abbot Giuseppe Bernardi, first as a private instructor, then as teacher of the final triennium at the *I.R. Ginnasio di S. Stefano* in Padua, and finally as a life-long close friend of Andrea and Giuseppe. Agostino accommodated all three in an apartment in Borgo Vignali No. 3224 (today Via Galileo Galilei, because the astronomer lived here 1603–1610), close to Basilica Sant'Antonio (Miotto 1942, p. 8) (Fig. 2.1, C and D in Chap. 2). Although this is today often overlooked, Bernardi was without doubt a central figure in Padua's learned circles (Bellini 1951, p. 66–67; Fabris 1883, p. 30–31; Parazzi 1871). He was a man of manifold social and scientific interests and regular guest of many weddings and salons, such as the famous Bargnani-Dandolo salon, frequented also by Jappelli and Paduan professors like Gian Maria Zendrini and Giacomo Andrea Giacomini (Pederzani 2014, p. 206–207). Moreover, he was an intimate friend of Giuseppe Jappelli and many other eminent figures of his time. Between 1820 and 1826, he directed the *Tipografia del Seminario*, or typography of the Episcopal Seminar, one of the most important publishing companies of entire Veneto (Callegari 2016). In 1825–1826, he was president of the Paduan Academy of Science (Accademia Galileiana 2017), and he participated as its official representative in several of the Congresses of Italian scientists. In the early 1840s, he was also (unsuccessfully) put forward for membership of the *Istituto Veneto* (Gullino 1996, p. 40). Moreover, until his death he was prefect and teacher at the city's Lyceum-Gymnasium in the former convent Santo

Stefano, where many of the protagonists of Padua's public life and of the 1848 movements studied (Davi 2018) (Fig. 2.1, E in Chap. 2). Founded during the Napoleonic period, the Lyceum became the most prestigious school in Veneto due to its close ties with the university, with which it shared teachers and facilities. Many important families from all Veneto provinces sent their offspring to the Lyceum-Gymnasium to receive the best propaedeutic formation prior to enrolling at university. Despite its high reputation, the Lyceum-Gymnasium was not elitist but had a liberal outlook. In this regard, it is revealing that in 1850 the school's authorities resisted Austrian attempts to transform it into an exclusive aristocratic institution, and instead defended its ideals of being a vehicle for the legitimate social aspirations of the bourgeoisie (Davi 2018, p. 329).

The teaching at the Lyceum-Gymnasium was moderately liberal with a distinct rational-scientific orientation. Bernardi's pedagogical standpoints show a strong Rosminian influence. The liberal-moderate priest and philosopher Antonio Rosmini (1797–1855) was born in Rovereto, a Trentino city about hundred kilometers northwest of Padua. His main ties were, however, in Lombardy and Piedmont, but he also cultivated several contacts in Padua (Ambrosetti 1970a; Malusa 2011, p. 134–136). He was politically active in the Risorgimento and his works were placed upon the *Index* after the 1848 revolution (Traniello and Liermann 2020). He studied in Padua from 1816 to 1819, graduated in theology in 1822, and went on to make friends with several figures in Bernardi's Seminario (Fig. 2.1, B in Chap. 2), as testified by Niccolò Tommaseo and other common friends (Missori 1967–1969; Missori 1981, p. 226–228). Among other things, they shared a predilection for Virgil's poetry, which we have seen was a source of inspiration also in Cesarotti and Meneghini's garden projects. Mathematics and natural sciences, in particular anthropology, played an important part in Rosmini's teachings (see, e.g., Bonvegna 2013; Ottonello 1998).

Bernardi set out his own educational views in a small treatise, written on occasion of Giuseppe Meneghini's graduation and dedicated to Andrea Meneghini (Bernardi 1834). This was not an important doctrine, neither innovative nor well elaborated. Yet, it provides insights into the spirit of the education he gave to the Meneghinis and their fellow students. For Bernardi, similar to Rosmini's inductive conception of education (Bonafede 2019, p. 102–103; De Vivo 1970), the first step of any learning is the acquisition of data from sensory perception. In a second step, these data need to be rationally interconnected and then memorized. He thus emphasized empiricism and reflection as the primary sources of knowledge. He insisted on thorough scientific teaching, because mathematical reasoning was the best tool for discerning the links between the multitudes of facts, whereas natural history and geography provided the best methods for learning how to put these facts into a correct order.

Bernardi was a prolific scholar, yet only few of his talks were published. Moreover, in his few publications he failed to reveal his sources. It is therefore impossible to ascertain the origins of his inspiration. We have to make inferences based on spiritual kinship with influential philosophic currents of his time. He had three favorite topics: faith, physiology and psychology, and social inquiry. In his thinking, they were closely intertwined, a vision that prompted frequent charges of materialism. In 1845, he held a talk, *A doubt and nothing else*, at the Academy of Science, which was published a mere twenty-six years later by one of his numerous disciples, Girolamo Costantini (1815–1880), lawyer, philanthropist, politician, and owner of a big villa with garden in the hills of Vittorio Veneto (Bernardi 1871). Since it reveals at least part of Benardi's general philosophical views and some of these appeared in Giuseppe Meneghini's doctoral thesis, I will outline them in some detail.

The two most interesting aspects of Bernardi's reasoning were his historicist views and his recourse to analogies between the histories of individuals and of nations. Bernardi may have picked up his conception of history from Rosmini, who had an interest in the development of societies and assimilated basic ideas of Christian historicism (Ambrosetti 1970b, p. 98), yet it may also be the other way round. Another, probably stronger source of inspiration was the anti-Cartesian Neapolitan philosopher Giambattista Vico (1668–1744). Vico fully expounded his innovative vision of universal history and universal human nature in 1725 in his book *Scienza Nuova Prima*, or *The First New Science* and its rewritten versions of 1730 and 1744 (Vico 2002). It is uncertain if Bernardi's teaching already contained Vichian elements in the 1830s, when he taught the Meneghini brothers, but Vico's philosophy quickly became popular in particular in Venice and Padua (Cospito 2002, p. 74–75; Santinello 1982). As early as 1739, it was part of Jacopo Stellini's (1699–1770) university teaching and, one year later, of his famous essay *De ortu et progressu morum* (On the origin and progress of customs, 1740). Along with Stellini, several prominent Paduan savants like the poet Melchiorre Cesarotti and the naturalist Antonio Vallisneri (1661–1730) picked up and elaborated Vico's ideas. From the 1820s onwards, another, even more intense uptake of

Vico's ideas occurred in larger parts of Northern Italy, and Antonio Rosmini as well as liberals and socialists like Niccolò Tommaseo, Giuseppe Ferrari (1811–1876), and Vincenzo Gioberti (1801–1852) contributed to Vico's considerable posthumous success in Veneto (Ottonello 1989).

It is therefore plausible that Bernardi knew Vico's doctrine. Indeed, several fundamental aspects of Bernardi's short paper strongly recall Vico's conception of history that was—in the broadest terms—egalitarian, providential, lawful, emergent, genetic, cyclic, and recapitulatory. Vico's history focused not on individual 'great men' and their heroic (mainly belligerent) achievements, but on civil societies composed of many actors and their (mainly intellectual) activities. For him, humans not only made history, they were at the same time products of history. Humanity was historically conditioned, and its actions mirrored the historical context. At the same time, the historical sequence of events was governed by laws. Divine Providence guided the universal dimension of history and ensured that human actors, although endowed with free will and responsible for their individual deeds, always passed through the same stages of social, juridical, political, and religious order.

Vico's history of civilization thus assumed an underlying uniformity of human nature across historical and geographical settings. According to him, the common features of human nature gave rise to a fixed series of stages in the development of civil society, law, commerce, and government (Costelloe 1999). This began with the age of gods and heroes. This primitive state led to the subsequent stages: the age of aristocracy, and finally the age of democracy and rationality. Analogously, man's mental powers—the major product and expression of history—followed a sequence of stages, from the 'primitive mind' through the highest linguistic and rational capacities. Hence, for Vico, the overall historical process was not only ordered and lawful, but there existed a causal relationship and a genetic bond between the different ages. However, the process was by no means a continuous march toward perfection, but a cyclic succession of progress and decline. The final domain of rationality produced atheism and hence caused a regression to barbarity, which formed the starting point of a new cycle of history. Similar to the earlier philosophies of Niccolò Machiavelli (1498–1527), Francis Bacon (1561–1626), and Thomas Hobbes (1588–1679), Vico's idea of progress was thus limited in time, and general history was characterized by recurring patterns. A final crucial point of Vico's conception was the analogy between the ages of civilization and the corresponding stages—the formation of language, sociability, phantasy, and reason—in the history of humans.

Bernardi's essay was much less elaborate and innovative. However, his speculations underpin my argument that a particular historical-developmental view and organism-state analogy had gained a foothold in Padua's learned debates by the 1820s. The basic argument of Bernardi's disquisition was the relationship between feeling (*sentimento*) and reason, and the importance of a correct interaction—'a friendly association', as he called it—between them to produce positive effects (Bernardi 1871, p. 12). Like Rosmini, he thus identified these two domains as the fundamental and independent, partially cooperating and partially contrasting influences on human activity. Yet contrary to Rosmini (2001, p. 222), who understood the notion of *sentimento fondamentale* (fundamental feeling) as a form of bodily self-consciousness, for Bernardi *sentimento* meant affectionate emotion and passion. Both feeling and reason determine the histories of individuals and, in parallel, the histories of nations. Initially, a child's life is regulated by its immediate needs. After many sense perceptions, like pain, hunger, and pleasure, it starts to feel affections for those around him. As these feelings increase in intensity and variety, the young human individual starts to reason. In middle-aged individuals, both feeling and reason operate. Both are essentially positive as long as they mutually moderate each other, guiding the individual toward good behavior. In old age, however, feelings diminish and pure reason makes individuals cold and calculating (Bernardi, 1871, p. 10–14).

In analogy with the mental development of human individuals, nations pass through stages, from their initial appearance to their final decline. Blind virility reigns in the beginning. Those tribes that survive the initial adversity and the struggles between individuals and between peoples develop an equilibrium of affect and reason and reach the adolescence of a nation. A small number of men of high mind and great experience will now guide the nation. They will instill sane feelings and a collaborative spirit in the people, enforce family bonds, and institute order and inviolable laws. Yet, with each generation, the level of education (reason) will increase, while enthusiasm (sentiment) diminishes. An increase in comfort is accompanied by laziness, egoism, and dispassionate memory, ultimately leading to the nation's decline (Bernardi, 1871, p. 17–29).

Whatever one may think about Bernardi's rather simple account of the history and the destiny of humankind, his sketch contained several aspects that distinguish the Paduan conception of progress of these years as well as Giuseppe Meneghini's biological thinking, in particular his pre-Darwinian concept of evolution and his notion of the collaboration of parts. Bernardi's philosophy tightly linked the history of individuals to that of peoples and societies. Both pass through the same stages, and the forces that act upon the lower level (man) determine the destiny of the superior level (nation). His concept of history is characterized by pervasive progressivism. However, it was not a Hegelian linear-progressive conception of history, nor a Herderian universal *Selbstentwickelung*, which embedded human history in nature's history and followed a general plan from the birth of the universe to the creation of Earth, and from the first terrestrial life forms through to human civilizations (von Herder 1784–1791; Berlin 1976; Spahn 2015, p. 676–677). Bernardi's concept of history and development was cyclic like Vico's, and invoking individuals and single nations.

Another aspect in common between Bernardi, Rosmini, and Herder, and a basic premise in Giuseppe Meneghini's biological version (Chap. 7), was the interplay between history, development, and lawfulness. For Bernardi, the history of the individual as well as that of a single nation was one of steady advancement. He emphasized that "This is an inevitable law: that's how humanity is constituted" (Bernardi, 1871, p. 17). Yet, a romantic optimist he was not. For him, all development inevitably led to decline and demise. "That is no insult, but a history", he declared, and the law of history knew only few exceptions (Bernardi, 1871, p. 15). The causes of decline were not external, but intrinsic properties of individuals and nations. Intrinsic but not ontological. For him, there was no 'evil element'. Feeling and reason were both essentially positive. Rather, in the end all positive achievements were destroyed by the growing disequilibrium between them.

A further aspect worth noting is the (pre-Spencerian) notion of a 'struggle for life'. The individual parts of Bernardi's societies were heterogeneous. His concept of organization hence implied a certain degree of individuality. Societies were composed of different individuals in different stages of development. Equilibrium and cooperation were the key to advancing to a higher level of togetherness. But progress was by no means guaranteed. As long as the young prevail in a society, passion and 'arcane impulses' and hence selfishness, quarrel, and shortages dominate, in the end leading to extinction (Bernardi, 1871, p. 18–19). If internal quarrels are overcome and the struggle between individuals subsides, then quarrel and struggle shift to the level of peoples. Societies that have progressed have crushed many others (Bernardi, 1871, p. 20). If a society survives, its members grow older and these older elements, equipped with a higher degree of rationality and experience, exert a tranquilizing and pacifying effect. Now, finally, feelings and reason are in harmony. Progress is therefore not an automatic process. The basic impulse is there, but its concrete manifestation depends on many factors. Nor is it a democratic process, because a small number of leaders will conduct the multitude toward more affectionate family ties, more general laws and rights, and less individuality (Bernardi, 1871, p. 23). In the end, however, when the love of the fatherland becomes ever less passionate and ever more rational, and when leisure and convenience increase, society becomes egoistic and will soon decline.

Finally, a third aspect that appears significant in the Paduan intellectual milieu is religion. Notwithstanding all the differences between them, Bernardi as well as Rosmini and Vico advocated for a conciliation between faith and science. For all three, religion was the backbone of a prosperous society. Probably, Abbot Bernardi held a similar view as Vico, who saw the Divine Providence acting in the background and directing everything through general principles. Even more than for Vico, however, for Bernardi individual processes were caused by a multitude of factors and the analysis of these did not entail theological explanation. The word 'God' did not appear in Bernardi's treatise. Even more interesting, for him, religion itself was part of the developmental stages.

The relationship between Bernardi and the Meneghini brothers continued to be intimate, even in later years. As delegates of the Paduan Academy of Science, the Abbot and Giuseppe together traveled to the Congresses of Italian scientists, and Bernardi assisted Andrea as member of the *Consulta straordinaria* (Special Consult) in his post as head of the revolutionary government of Padua. The most important testament to the profound influence of Bernardi on Giuseppe Meneghini is their correspondence. In 1834, after his graduation in medicine, Giuseppe undertook a long trip to visit some of the main universities in Northern Italy. Between mid-May and mid-August, he wrote seventeen long letters to Bernardi, calling him "My friend" and continuously asking for his "much needed" advice.[3]

5.3 Seeking a Place in the Sun

In his youth, Andrea shared Giuseppe's passion for botany, but soon he followed a different path. He graduated in law in 1829. No professional trace of him can be found in the early 1830s. Possibly, he helped to run his father's businesses, but when these collapsed a few years later,[4] he had to stand on his own two feet and started his remarkable career in the public sector. In 1838, we find him, together with the then *podestà* (presumably Benedetto Trevisan) (president), Lodovico Menin (vice-president), Alessandro Papafava dei Carraresi, Marsilio Papafava, Pietro Selvatico Estense, Giuseppe Jappelli, and Nicolò Leoni, on the *Civica Deputazione all'Ornato* and the *Commission for Public Charity*, headed by the bishop (Anon. 1839, p. 44–45 and 78). The importance of both Commissions—the *Civica Deputazione all'Ornato* awarded building permits for palazzi and gardens—and the illustriousness of the other members is evidence that the Meneghinis participated in the elite's complex play of the distribution of public appointments. Still, Andrea took his tasks very seriously and distinguished himself in engagement, expertise, and independence. Giovanni Tomasoni (1821–1881) narrated that Andrea quarreled passionately with other delegates over proposals and opinions, a rarity in these years. He was "a man of opposition in an epoch in which the term and the fact were rare" (Tomasoni 1870, p. 4). Like for his father, a major concern of his were the rural poor (Gottardi 2009). He showed the same passion and organizational skills in his roles in Padua's associations, during the creation of an agronomic network in Veneto, in his journal *Il Tornaconto*, and in the events leading up and during the 1848 revolt. In 1841, he was elected as a delegate to the Provincial Congregation. In addition to his public duties, from 1842 to 1847, Andrea worked as notary in Padua and in Piazzola sul Brenta, a small town about twenty kilometres northwest of Padua.

Giuseppe graduated in medicine from Padua in 1834. He never practiced but was drawn to continue his botanical studies and became assistant at the Botanical Garden. The garden's prefect, Giuseppe Antonio Bonato (1753–1836), was already 81 years old, and Meneghini may have hoped to succeed him. Upon Bonato's death, however, the newly established chair of Botany was given to another former assistant, Roberto De Visiani. The Austrian authorities preferred him because he was Giuseppe's senior and a highly esteemed taxonomist with more expertise with exotic plants (Pirona 1889–1890, p. 685). Systematics enjoyed a high reputation at

Vienna at the time. Moreover, as we have seen in Sect. 2.3 in Chap. 2, taxonomic skills were indispensable in the upcoming restoration of the Botanical Garden (Klemun 2006). The Emperor's love for flowers redounded to De Visiani's advantage, too, whereas Meneghini's romantic conception of botany had many opponents in the imperial capital (Klemun 2016, p. 57–58). Despite any disappointment, the relationship between Meneghini and De Visiani remained amicable and collaborative. Meneghini was at the forefront of the committee that welcomed De Visiani upon his arrival in Padua (Galletto 1992, p. 41), and De Visiani played an important role in Meneghini's later appointment in Pisa (Chap. 10). Giuseppe continued as assistant at the Botanical Garden until 1838, but he had to think about alternatives for his future career. In Paris, the botanical garden had been successfully transformed into a center of general natural history, agricultural advancement, and imperial representation (Spary 2000). Maybe with this in mind, Giuseppe was keen to demonstrate his potential professional quality at various levels: scientific, cultural, and institutional. Networking was therefore fundamental, and we have seen the social dimension of his networks in earlier chapters.

In addition, Giuseppe became increasingly involved in scientific-botanical circles, and he did not restrict his activities to the local level. In Padua, he befriended other gifted botanists: Roberto De Visiani, the paleontologist Baron Achille De Zigno, the lichenologist Vittore Trevisan, Earl of San Leon, the agro-chemist Giuseppe Clementi, the paleo-botanist and lichenologist Abramo Massalongo, and the physician and phycologist Giovanni Zanardini, his closest friend. Many of these also became prominent in Padua's social and political life. De Zigno was *podestà* (mayor) of Padua from 1846 to 1856, Trevisan his counselor, De Visiani, president of the Gardening Society, and Clementi, co-founder of the journal *Il Tornaconto*. Looking beyond the local level, the role of these men as pioneers of the national botanical modernization is evinced by the fact that in 1843, four—Meneghini, Trevisan, Clementi, and De Visiani—of the ten founders of the *Giornale Botanico Italiano*, a monthly journal explicitly created to revitalize botanical studies in Italy, were Paduans (Moretti et al. 1844).

Indeed, Giuseppe had soon expanded his network beyond Veneto. Upon graduation, he visited the museums and botanical gardens of Bologna, Modena, and Florence, making lasting friendships especially with the brothers Paolo (1798–1871) and Pietro Savi (1811–1871), respectively professors of botany and zoology in Pisa. The *tour* was financed by his father. His principle aim was to introduce himself and become acquainted with other academic worlds. Yet Giuseppe was not merely attentive to the scientific-academic situation. In his letters to Bernardi, Giuseppe provided detailed accounts of the people he met and of his experiences during stays in Modena, Parma, Pavia, Milan, Turin, Genoa, Florence, and Livorno. He frequently included observations and reflections on the cities and their social situation. In Parma, for instance, he criticized that the rich indulged in luxury instead of helping the poor, concluding that "the people are left in ignorance, and the learned, perhaps out of jealousy, guard their knowledge".[5]

During the winter and spring of 1838–1839, De Visiani sent Giuseppe on a scientific mission to Vienna, where he came into close contact with many prominent exponents of the city's political and naturalistic elite. One of them was the previously mentioned rich military nobleman baronet Carl Alexander von Hügel. After participation in several wars on the Italian peninsula, Hügel had retired from the army and had become an avid traveler and naturalist. He shared many interests with the circle in Padua. The most important of these was horticulture. He possessed a magnificent garden outside Vienna where he established a thriving flower breeding enterprise (Sect. 8.2 in Chap. 8). In 1837, he became the first president of the *k.k. Gartenbau-Gesellschaft* (Imperial Horticultural Society). Giuseppe, in fact, met him to arrange an exchange of flowers and seeds with the Botanical Garden, yet had to admit that Padua had very little to offer in return.[6] Later, Hügel would again cross paths with Meneghini while attending the Ninth Congress of the Italian Scientists in Venice (Sect. 9.4 in Chap. 9), and, after the 1848 revolutions, as the Austrian ambassador to the Grand Duchy of Tuscany (Chap. 10). Other influential figures Meneghini met were Franz Anton von Kolowrat-Liebsteinsky (1778–1861), who in 1848 briefly succeeded von Metternich (1773–1859) as prime minister, Joseph Franz von Jacquin, director of Vienna's Botanical Garden, and his two successors Eduard Fenzl (1808–1879) and Stephan Endlicher.[7] The relationship to the latter must have been particularly intimate, because Endlicher dedicated a new genus of algae to him, *Meneghinia*. They also shared similar liberal political views (Sect. 9.4 in Chap. 9).

Giuseppe Meneghini's sphere of action also extended onto the international level. He sent his early plant-anatomical publications to many eminent European botanists, receiving complimentary letters from Hugo von Mohl, Ludolph Christian Treviranus (1779–1864), Alphonse de Candolle

(1806–1893), Adrien Henri de Jussieu (1797–1853), and others (Canavari 1889, p. 13–14),[8] many of whom he later met at the Italian scientific congresses (Sects. 2.4 and 7.4 in Chaps. 2 and 7 respectively). In 1837, he traveled to Prague, where Karel Boriwog Presl (1794–1852) read the results of his research on *Bryopsidum* and on the algae of the Euganean Hills to the *Versammlung Deutscher Naturforscher und Aerzte* (Sternberg and von Krombholz 1838, p. 160 and 173). The same year, he was appointed member of scientific academies and societies in Rostock, Regensburg, Breslau, and Moscow, and thereafter of seventeen Italian and European societies. Other naturalists met him in Italy. One of these was the famous phycologist Friedrich Traugott Kützing (1807–1893) in 1835, who visited, together with the pharmacist and phycologist Bartolomeo Biasoletto (1793–1858), founder of the first public botanical garden of Trieste, the thermal springs of Battaglia and Abano during his trip to Dalmatia, Northern Italy, and Switzerland. Meneghini hosted him and generously donated and explained to him the Euganean diatomeae (Kützing 1844, p. 10). In return, Kützing dedicated to him the two species *Rhipidophora meneghiniana* and *Cyclotella meneghiniana* (Kützing 1844, p. 50, plate 30, fig. 68; and p. 122, plate 11, fig. 2).

The Congresses of Italian scientists afforded the best platform for Meneghini's scientific ambitions and ideals. The botanical sessions attracted less numerous attendees than other sections; yet, they involved a rather tight group of scholars and attracted the largest number of non-Italian participants, illustrating the good level of interaction of Italian botanists with their European colleagues. Giuseppe was one of the most assiduous and active participants, and the fact that he corresponded with most of the others points to his possibile role in encouraging their attendance at the congresses. The frequent participation of eminent cryptogamists and pioneers of cell theory like Henri Dutrochet (1776–1847), Charles-François Brisseau de Mirbel (1776–1854), Charles-François Morren (1807–1858), Ludolph Christian Treviranus, Heinrich Friedrich Link (1767–1851), Robert Brown (1773–1858), Antoine Laurent Apollinaire Fée (1789–1874), Franz Unger, and Hugo von Mohl is especially significant. In fact, Meneghini repeatedly used this platform to stress the importance of the cellular approach for botany (Dröscher 1996, p. 36–37) (Sect. 7.4 in Chap. 7).

Hence, like his brother Andrea, Giuseppe Meneghini soon demonstrated a high level of expertise and keen organizational and networking skills. He used social relationships and his valuable botanical knowledge as a tool for establishing and enhancing his social and intellectual standing. In 1838, friends like Giacomo Andrea Giacomini, professor of theoretical medicine, anti-Austrian patriot, and owner of still another Jappellian masonic garden (Agostinetti 2006, p. 43–46; Anon. 2015–2017; Federspil and Martini 1991), endorsed Giuseppe's call onto the chair of propaedeutic sciences for students of surgery. This was indeed a professional advancement, since he became university professor with attendant privileges. However, he always remembered this period as one of great sacrifice, because it implied a heavy workload, low pay, little didactic satisfaction, and even less social recognition (Pirona 1889–1890, p. 685). His personal ambitions, fed by the international praise he had received for his scientific work, certainly meant that he was reluctant to content himself with this position.

By the mid-1830s, both brothers had thus set out on their professional careers. Through their father and the abbots Melo and Bernardi, they had received thorough moral, philosophical, and methodological instruction, and both showed an extraordinary level of social and professional commitment in two important fields, political economy and botany. Yet, both were still looking for an unfading place in the sun.

Notes

1. Contarini, descendant of an old and branchy family of Venetian doges, preferred naturalistic, in particular entomological and cryptogamic, studies to the study of law. He published numerous papers and maintained extensive correspondence with European scholars (Silvestri 1983). It was Contarini, who in 1834 introduced Meneghini to Pietro Savi and in 1835 to Zanardini; see the letters Contarini 1/20.22 (14 October 1834), Contarini 1/20.20 (11 July, 1835), and Contarini 1/20.19 (10 August 1835), preserved in the Library of the Natural History Museum of Venice.
2. The quote is from Saint Augustin, Liber XI, *De Civitate Dei*, cap 22; the English translation is from Augustin (1871, p. 462). It continues: "…for these little things are to be measured not by their own greatness (which does not exist), but by the wisdom of their Designer."
3. Letter from Giuseppe Meneghini to Giuseppe Bernardi of 27 May 1834. Biblioteca Civica di Padova (BCP) (C.A. 972.2, c. 1r).
4. In his letters, Giuseppe frequently asked Bernardi to assist his father, who was in need of support. This suggests that the financial squeeze started in the early 1830s. Letter from Giuseppe Meneghini to Giuseppe Bernardi of May 27, 1834. Biblioteca Civica di Padova (BCP) (C.A. 972.2, c. 1r).

5. Letter from Giuseppe Meneghini to Giuseppe Bernardi of 27 May 1834. BCP: C.A. 972.2, c. 2r.
6. Letter from Meneghini to Roberto De Visiani of 28 Nov 1838, BOBP. I thank Moreno Clementi for sending me the correspondence.
7. See the letters from Meneghini to Roberto De Visiani of 20 Nov 1838; 28 Nov 1838; 1 Feb 1839; and May 1839, BOBP.
8. Unfortunately, it was not possible to find the original letters.

References

Accademia Galileiana. 2017. Serie dei Principi, cioè Presidenti (17 March 2017). www.accademiagalileiana.it/elenco-dei-soci/fondatori-e-corso-storico/serie-dei-principi-cioe-presidenti. Accessed 20 April 2020.

Agostinetti, Nino. 2006. *Giardini massonici dell'Ottocento veneto*. Padova: La Garangola.

Ambrosetti, Giovanni, ed. 1970a. *Rosmini e il rosminianesimo nel Veneto*. Verona: Casa editrice Mazziana.

———. 1970b. Rosmini e il romanticismo politico e giuridico austriaci. In *Rosmini e il rosminianesimo nel Veneto*, ed. Giovanni Ambrosetti, 87–111. Verona: Casa Editrice Mazziana.

Anon. 1839. *Almanacco civile della R. Città e Provincia di Padova per l'anno 1838*. Padova: Tipografia e Fond. Cartallier.

———. 2015–2017. Giardino di Palazzo Giacomini Romiati. In *Italian Botanical Heritage: The Ultimate Guide of Italian Gardens & Landscapes*. https://luoghi.italianbotanicalheritage.com/giardino-di-palazzo-giacomini-romiati/. Accessed 14 July 2020.

Augustine, Aurelius. 1871. *The City of God, translated by Marcus Dods*. Edinburgh: T. & T. Clark.

Béguinot, Augusto. 1909. *Flora padovana, ossia prospetto floristico e fitogeografico delle piante vascolari indigene inselvatichite o largamente coltivate crescenti nella Provincia di Padova, con notizie storico-bibliografiche sulle fonti della Flora*. Vol. 1. Padova: Prem. Soc. Coop. Tip.

Bellini, G. 1951. *Sacerdoti educati nel Seminario di Padova distinti per virtù, scienza, posizione sociale*. Padova: Tipografia del Seminario.

Berlin, Isaiah. 1976. *Vico and Herder: two studies in the history of ideas*. New York: The Viking Press.

Bernardi, Giuseppe. 1834. *Alcune considerazioni intorno alla molteplicità delle materie d'insegnamento: Laureandosi in medicina Giuseppe Meneghini. Opuscolo dedicato al fratello Andrea*. Padova: Tipi della Minerva.

———. 1871. *Un dubbio e non altro: Pensieri*. Vittorio (Veneto): G. Longo.

Bonafede, Paolo. 2019. *L'altra pedagogia di Rosmini: Dilemmi, occultamenti, traduzioni*. Trento: Università degli Studi di Trento.

Bonvegna, Giuseppe. 2013. Rosmini naturalista? Note sul ruolo delle scienze naturali nell'antropologia filosofica rosminiana. *Rivista di Filosofia Neo-Scolastica* 1: 131–150.

Callegari, Marco. 2016. "La Stamperia si riscosse all'urto violento": la produzione libraria tra la caduta della Repubblica di Venezia e la nascita del Regno Lombardo Veneto. In *L'editoria italiana nel decennio francese: Conservazione e rinnovamento*, ed. Luigi Mascilli Migliorini and Gianfranco Tortorelli, 59–76. Milano: Franco Angeli.

Canavari, Mario. 1889. *Commemorazione di Giuseppe Meneghini fatta nell'aula magna dell'Università Pisana ai XXIV Marzo MDCCCLXXXIX*. Pisa: Tipografia T. Nistri.

Cospito, Giuseppe. 2002. *Il "Gran Vico": Presenza, immagini e suggestioni vichiane nei testi della cultura italiana pre-risorgimentale, 1799–1839*. Genova: CEIT.

Costelloe, Timothy M. 1999. The state of nature in Vico's New Science. *History of Philosophy Quarterly* 16 (2): 320–339.

Davi, Mariarosa. 2018. Le istituzioni scolastiche nel passaggio dall'Austria all'Italia: il caso del ginnasio-liceo di Padova. In *Il Veneto nel Risorgimento: Dall'Impero asburgico al Regno d'Italia*, ed. Filiberto Agostini, 320–333. Milano: Franco Angeli.

De Checchi, Franco. 2008. Il 'misterioso' caso della polenta porporina di Legnaro (1819), tra scienza e superstizione. *Saccisica: Studi e ricerche* 3: 203–217.

De Vivo, Francesco. 1970. Rosmini e l'Università patavina. In *Rosmini e il rosminianesimo nel Veneto*, 265–271. Verona: Casa Editrice Mazziana.

Dröscher, Ariane. 1996. *Die Zellbiologie in Italien im 19. Jahrhundert*. Halle: Deutsche Akademie der Naturforscher Leopoldina.

Fabris, Anton Maria. 1883. *Dei professori del Seminario di Padova morti dal 1821 al 1880*. Padova: M. Giammartini.

Federspil, Giovanni, and Chiara Martini. 1991. Giacomo Andrea Giacomini: a Padua clinician between metaphysics and science. *History and Philosophy of the Life Sciences* 13 (1): 73–95.

Foscarini, Giacomo. 1847. *Guida alle terme Euganee*. Padova: Tipi del Seminario.

Galletto, Pietro. 1992. *Galantuomini padovani dell'Ottocento: cenni biografici ed istantanee*. Padova: Libreria Draghi Randi.

Gaughran, Eugene R.L. 1969. From superstition to science: the history of a bacterium. *Transactions of the New York Academy of Sciences* 31 (1): 3–24.

Gottardi, Michele. 2009. Meneghini, Andrea. In *Dizionario Biografico degli Italiani*, vol. 73, 456–458. Roma: Treccani.

Gullino, Giuseppe. 1996. *L'Istituto Veneto di Scienze, Lettere ed Arti: Dalla rifondazione alla seconda guerra mondiale (1838–1946)*. Venezia: Istituto Veneto di Scienze, Lettere ed Arti.

Herder, Johann Gottfried von. 1784–1791. *Ideen zur Philosophie der Geschichte der Menschheit*. 4 vol. Riga—Leipzig: Johann Friedrich Hartknoch. Engl.

edition: Herder, J. G. 1803. *Outlines of a philosophy of the history of man* (transl. T. Churchill). London: J. Johnson (Luke Hansard).
Klemun, Marianne. 2006. Ausflüge in die Blumengefilde des Lebens. Leopold Trattinnicks "Flora des Österreichischen Kaiserthumes". In *Paradoxien der Romantik. Gesellschaft, Kultur und Wissenschaft in Wien im frühen 19. Jahrhundert*, ed. Christian Aspalter, Wolfgang Müller-Funk, Edith Saurer, Wendelin Schmidt-Dengler, and Anton Tantner, 433–449. Wien: Wiener Universitätsverlag.
———. 2016. Franz Unger (1800–1870): multiperspektivische wissenschaftshistorische Annäherungen. In *Einheit und Vielfalt: Franz Ungers (1800–1870) Konzepte der Naturforschung im internationalen Kontext*, ed. Marianne Klemun, 15–92. Wien: Vienna University Press.
Klemun, Marianne, and Ulrike Spring, eds. 2016. *Expeditions as experiments: practising observation and documentation*. London: Palgrave Macmillan.
Kützing, Friedrich Traugott. 1844. *Die kieselschaligen Bacillarien oder Diatomeen*. Nordhausen: W. Köhne.
Lampertico, Fedele. 1882. Delle scienze nel Veneto dal 1815 al 1866. In *Fedele Lampertico: Scritti storici e letterarii*, vol. I, 40–105. Firenze: Successori Le Monnier.
Lovett, Clara M. 1982. *The democratic movement in Italy, 1830–1876*. Cambridge, MA: Harvard University Press.
Malusa, Luciano. 2011. *Antonio Rosmini per l'unità d'Italia: Tra aspirazione nazionale e fede Cristiana*. Milano: Franco Angeli.
Martens, Georg Matthias von. 1824. *Reise nach Venedig*. 2 vols. Ulm: Stettin'sche Buchhandlung.
Melo, Pietro. 1819. Memoria sulla polenta rossa. *Giornale dell'italiana letteratura* II/18: 333–341.
Meloni, Giovanni A. 1992. I fatti di Legnaro e le osservazioni di Vincenzo Sette e Bartolomeo Bizio. In *Bartolomeo Bizio nel bicentenario della nascita (1791–1991). Atti del Convegno*. Rendiconti dell'Accademia Nazionale delle Scienze detta dei XL. Memorie di scienze fisiche e naturali 110: 87–94.
Meneghini, Giuseppe. 1837. *Conspectus algologiae euganeae, germanicis naturalium rerum scrutatoribus Pragae anno 1837 convenientibus*. Pataviae: Typis Minervae.
———. 1843. Monographia nostochinearum italicarum, addito specimine de Rivulariis. *Memorie della Reale Accademia delle Scienze di Torino* s. II/ v. 5: 1–144.
———. 1878–1879. Commemorazione di Giovanni Zanardini (1804–1878). *Atti del R. Istituto Veneto di Scienze, Lettere ed Arti* 37: 391–400.
Miotto, Onelio. 1942. *Andrea Meneghini: patriota inedito del nostro Risorgimento (1806–1870)*. Padova: Cedam.

Missori, Virgilio, ed. 1967–1969. *Niccolò Tommaseo, Antonio Rosmini: Carteggio edito e inedito*, 3 volumes. Milano: Marzorati.

———. 1981. Tommaseo a Rovereto. *Atti dell'Accademia degli Agiati* VI (20): 225–244.

Moretti, Giuseppe, Giorgio Jan, Giuseppe Meneghini, Giuseppe De Notaris, Filippo Parlatore, Adalbert Bracht, Giuseppe Clementi, Vittore Trevisan, Francesco Facchini, and Roberto De Visiani. 1844. Progetto. *Giornale Botanico Italiano* 1: 9–11.

Ogilvie, Brian W. 2020. Maxima in minimis animalibus: insects in natural theology and physico-theology. In *Physico-theology: religion and science in Europe, 1650–1750*, ed. Ann Blair and Kaspar von Greyerz, 171–182. Baltimore: Johns Hopkins University Press.

Ottonello, Pier Paolo. 1989. Rosmini e Vico: la 'filosofia italiana'. *Rivista Rosminiana* 83 (3): 265–274.

———, ed. 1998. *Rosmini e l'enciclopedia delle scienze. Atti del Congresso internazionale (Napoli, 22–25 ottobre 1997)*. Firenze: Olschki.

Parazzi, L. 1871. L'abate Giuseppe Bernardi da Noventa di Piave. *La Gioventù: Rivista dell'Istruzione pubblica in Italia, X/2, Quaderni* XXIII–XXIV: 812–815.

Pederzani, Ivana. 2014. *I Dandolo: Dall'Italia dei lumi al Risorgimento*. Milano: Franco Angeli.

Pirona, Giulio. 1889–1890. Della vita scientifica del prof. Giuseppe Meneghini. *Atti del R. Istituto Veneto di Scienze, Lettere ed Arti* 38: 683–700.

Ragazzini, Francesco. 1844. *Nuove ricerche fisico-chimiche ed analisi dell'acque termali euganee*. Padova: Tipografia del Seminario.

Rosmini, Antonio. 2001. *A new essay concerning the origin of ideas*. 2 vols. Transl. Denis Cleary and Terence Watson. Durham: Rosmini House.

Santinello, Giovanni. 1982. Vico e Padova nel secondo Settecento. In *Vico e Venezia*, ed. Cesare De Michelis and Gilberto Pizzamiglio, 77–89. Firenze: Olschki.

Silvestri, Daniela. 1983. Contarini, Nicolò Bertolucci. In *Dizionario biografico degli italiani*, vol. 28, 165–167. Roma: Treccani.

Spahn, Christian. 2015. Evolution. In *Oxford handbook of German philosophy of the 19th century*, ed. Michael Forster and Kristin Gjesdal, 674–694. Oxford: Oxford University Press.

Spary, Emma C. 2000. *Utopia's garden: French natural history from Old Regime to revolution*. Chicago: The University of Chicago Press.

Stebbins, Sara. 1980. *Maxima in minimis: Zum Empirie- und Autoritätsverständnis in der physicotheologischen Literatur der Frühaufklärung*. Frankfurt a.M.: Peter Lang.

Sternberg, Kaspar, and Julius Vincenz von Krombholz. 1838. *Bericht über die Versammlung deutscher Naturforscher und Aerzte in Prag im September 1837*. Prag: Gottlieb Haase Söhne.

Thomas, Martin, and Amanda Harris, eds. 2018. *Expeditionary anthropology: teamwork, travel and the 'science of man'*. New York and London: Berghahn.
Tomasoni, Giovanni. 1870. *Andrea Meneghini sindaco di Padova: Commemorazione*. Padova: Tipografia Luigi Penada.
Traniello, Francesco, and Christiane Liermann. 2020. Rosmini, Vico e la rivoluzione. In *Vico e la filosofia civile in Lombardia*, ed. Geri Cerchiai, 269–305. Milano: Franco Angeli.
Vico, Giambattista. 2002. *The First New Science*. ed. and trans. Leon Pompa. Cambridge: Cambridge University Press.
Vienne, Florence. 2017. Worlds conflicting: the cell theories of François-Vincent Raspail and Theodor Schwann. *Historical Studies in the Natural Sciences* 47 (5): 629–652.

CHAPTER 6

Organization, Cooperation, and Progress in Padua's Political Economy

In the previous chapters, I described the personal, social, and institutional entanglement of botany and politics in Padua prior to 1848. In this and the following chapter, I go further and illustrate at the theoretical-philosophical level, exemplified by the works of Andrea and Giuseppe Meneghini, the rich conceptual cross-fertilization that existed between these seemingly distant fields. The 1830s were characterized by general intellectual openness and curiosity toward recent scholarship from beyond the Alps, as evidenced by numerous translations, excerpts from, and reviews of foreign works. In particular, I argue that an 'organicist switch' took place in Paduan debates on political economy during the 1820s and 1830s. The organicist switch also entailed a 'botanical switch', with plants taking a prominent place as a source of metaphorical and explanatory inspiration.

It is not easy to gain access to Andrea Meneghini's (Fig. 6.1) private convictions. Like his brother Giuseppe, he was not prone to making effusive declarations nor did he spread popular mottos. Reasons for their restraint include caution in the face of ever-present Austrian censorship, the fuzziness of many organic concepts, and both brothers' sober character. Yet, this does not mean that their beliefs were not deeply and strongly held. On the contrary, the fact that both brothers maintained their basic visions into their later years shows that their convictions were sincere and

A. Dröscher, *Plants and Politics in Padua During the Age of Revolution, 1820–1848*, Palgrave Studies in the History of Science and Technology, https://doi.org/10.1007/978-3-030-85343-3_6

Fig. 6.1 Andrea Meneghini (1806–1870) in the late 1860s. Graph art by Till Claudius Dröscher. By courtesy of Till Dröscher Graphic & Design

profoundly rooted. In order to unveil their thinking, it is necessary, first, to survey their intellectual environment, and then to be attentive to the nuances in their soft-spoken words. The following sections provide a general sketch of political and economic thought in Veneto in the first half of the nineteenth century with special attention to the role of science, plants, and vegetal imagery. I will then transition to the ideas of Gian Domenico Romagnosi and of Andrea Meneghini and analyze how they reflected and influenced the organicist and progressivist philosophies of the nineteenth century.

6.1 Botany and Political and Economic Philosophies

Several scholars have posited the existence of intimate links between naturalism, natural philosophy, and the emerging economic thought of the late eighteenth and early nineteenth centuries (e.g., Clark 1992; Daston and Pomata 2003). The commonalities went beyond the complex concept of natural law. They ranged from questions of mathematical, descriptive, or empirical methodology to the use of metaphors and the more or less direct borrowing of models. Sergio Cremaschi (2002), for instance, offers a compelling analysis of the multitude of scientific and natural metaphors in Adam Smith's (1723–1790) discourse on the growth of wealth. Margaret Schabas (2005) goes further and locates the evolution of economic ideas within the history of natural science. She shows the works of economic theorists like David Hume (1711–1776), Adam Smith, and Thomas Robert Malthus (1766–1834) to be thoroughly infused with ideas and concepts from natural philosophy. Their discourses on money, trade, and commerce reflected the physical, chemical, and biological theories of their time. Only during the nineteenth century, economy, as a field of knowledge, began to divorce from general reflections about natural processes. It increasingly included bodies of knowledge accumulated by experts in commerce and public finance, and emerged as an autonomous discipline relying on social inquiry into human activity (Cremaschi 2017). During the first decades of the nineteenth century, however, the general belief that the same universal natural laws guide all processes still dominated—be they living or non-living, human or non-human. The body and its function continued to be a major point of reference. Yet the very metaphor of the body underwent several decisive modifications.

Plants were not absent in Enlightenment discourses. On the contrary, during the late eighteenth century, knowledge of botany, agriculture, and political economy, and the promotion of scientific societies and circles often went hand in hand. Young Herder intended to sketch in his *Ideen zur Philosophie der Menschheit* (1784–1791) a historicist *Philosophia anthropologica* inspired by botanist Carl von Linné's *Philosophia botanica*. His early writings were replete with botanical vocabulary and plant imagery (Goeth 2017; Janssens 1963; Janssens 1964; Simonis 2015). Conversely, Linné designed his 'new science', called *oeconomia*, as a natural history that encompassed information on the uses of nature's objects in technology (Koerner 1999; Müller-Wille 2003). Marc MacDonald similarly describes the British-Franco-Swiss network of the Delessert family, which included Jean-Jacques Rousseau, Thomas Robert Malthus's father Daniel (1730–1800), the diplomat Charles Pictet (1755–1824), and many others (MacDonald 2017). The network was fundamental for the European dissemination of the economic works of Malthus, Say, and others, and counted on their passion for plants and botanizing excursions. Botany was a popular interest in the Malthus household, and Daniel instructed Thomas Robert and his other children on nature walks and on the collection of plants. These and his later botanical studies also found their expression in his famous *Essay on Population* (MacDonald 2017, p. 7–10). Other notable examples of botany and natural history mingled with economic thought include the correspondence of Jean-Baptiste Say (1767–1832) and Augustin-Pyramus de Candolle (1778–1841) on natural history (MacDonald 2017, p. 17), and, in the 1830s, naturalist Charles Darwin who frequently dined with political economist Jean Charles Léonard Simonde de Sismondi (1773–1842), husband of Emma Darwin's (1808–1896) aunt Jessie Allen (1777–1853).

So far, historians of Italian political and socio-economic thought have mainly concentrated on the post-Risorgimento period. Whereas Veneto boasts several prominent thinkers in the second half of the nineteenth century—some of whom will touch on in this chapter—there are no equivalents in the first half. In his seminal history of political economy in Italy, Riccardo Faucci (2000, p. 113–115 and 127–128) considers Venetian economic scholarship of the pre-1848 period to be rather scarce in comparison with the rich literary movement of the time. Indeed, there is no leading political or economic scholar in Padua on whom my analysis could concentrate. Still, political and socioeconomic doctrines were not absent, and even unremarkable debates may sow the seeds for notable future developments.

At least three factors have to be considered in approaching the Paduan debate in the first half of the nineteenth century. Firstly, the strong Italian orientation. European writers were anything but unknown. Social and political concessions to the liberal bourgeoisie, even by the most reactionary regimes in Italy's south, provided new opportunities for ideas and debates, and for the reception and study of European authors (Gioli 2006, p. xlx–xlxi). Nevertheless, most of Padua's intellectual milieu evolved around Italian authors and their noteworthy and innovative concepts. The Italian Enlightenment was not a homogeneous movement, yet the ideas and works of these authors circulated more easily among the Italian states than the philosophical and political fragmentation of the time may lead us to suppose (Cremaschi 2020).

Secondly, the absence of neatly circumscribed theories. As elsewhere, the Italian political and economic writers of the period saw themselves as philosophers or general social scientists. This makes it difficult to discern their specifically economic and political ideas and, in addition, made them particularly receptive to the general intellectual *Zeitgeist*. One example is the complex influence of romantic ideas. German philosophy had no direct influence on economic thought, not even in Germany, although Hegel actually conceived a theory of the state. However, the absence of explicit references does not imply that certain general ideas did not represent powerful patterns of thought, and one of my arguments is that, along with the still strong Enlightenment tradition of Veneto's intellectual circles, romantic ideals actually grew in importance during the 1820s and 1830s. In the 1840s, the somewhat vague ideas of Giuseppe Bernardi, Antonio Vigodarzere, and other members of their generation continued to prevail, yet the new generation developed more concrete and effective instruments to implement these ideas and projects. The 1830s and 1840s were years of change in economic doctrine and civil participation. As outlined in Chap. 2, the city of Padua of the 1840s gave rise to lively brood of journals, associations, clubs, and congresses that aimed to realize the ideals of progress and social commitment. The pre-1848 period was thus a period of transition from general enlightened and romantic ideas to the more concrete and specific concepts of the mid-century. Nor did the new generation view itself as technocratic, however. They perceived political economy as a general discourse of 'modernization' rather than an analysis of the situation at hand. Their views went beyond local or regional boundaries and therefore tended to formulate abstract theories rather than concrete proposals with respect to local problems.

Thirdly, the tendency to maintain a low profile. The moderate spirit of the Paduan élite and the sober pragmatism of most its members as well as Austrian surveillance and censorship surely did not favor the pronouncement of explicitly political or, worse, revolutionary theories. In the 1830s and 1840s, for instance, Gian Domenico Romagnosi's works, object of the following section, in particular his book *On the Nature and Factors of Civilizing*, enjoyed considerable success in Veneto. Ettore Albertoni (1979) provides a complete list. The second edition appeared (posthumously) as early as 1834, that is, only two years after its first publication, and re-published in 1844 as the thirteenth volume of Romagnosi's complete works. This success notwithstanding, in the 1830s at least it was advisable to cite him with caution, because the Austrian government considered his liberal views dangerous. In 1821, he had been imprisoned and lost his teaching license, and subsequently he was kept under strict observation (Catellani 1934–1935, p. 483–484).

In the following, I will not dissect the details of particular economic and political questions, but rather describe a number of recurrent—and in fact inseparable—ideas to characterize the intellectual milieu and help to contextualize Andrea Meneghini's vision. In particular, I focus on the diffusion of historicism, organicism, cooperation, transformism, and progressivism, and the use of vegetal imagery in these contexts. The same categories will serve to analyze Giuseppe Meneghini's botanical works in Chap. 7.

6.2 Perfecting

Despite his repeated references to nature, Antonio Rosmini rarely used organicist or botanical metaphors. These were abundant, however, in the work of another most prolific and successful Italian writer, Gian Domenico Romagnosi. Born near Parma, he did not teach nor work in Padua, but his bonds to the city were close. Historians of economics (e.g., Romani 1988) have only considered his influence on economic thought in Veneto starting with the Veronese lawyer and politician Angelo Messedaglia (1820–1901), who taught political economy and statistics in Padua from 1858 to 1888. However, by that time, Romagnosi's tenets were not completely new to Padua. Rather, the institutional, editorial, and doctrinal success of Messedaglia as well as of Fedele Lampertico, Luigi Luzzati (1841–1927), and other exponents of the 'Lombardo-Veneto school' of political economy (see, e.g., Tusset 2016) was based on their ability to reflect the pre-existing, though still rather poorly defined creed of the local economic elites. Indeed, Andrea Cittadella Vigodarzere's endorsement was decisive for Messedaglia's university appointment (Cafarelli 2009).

Moreover, several supporters of Romagnosi's philosophy were already teaching at the University of Padua in the 1830s and 1840s. Less important than Messedaglia they were, but their activity demonstrates the early and broad diffusion of Romagnosi's ideas in Veneto. One of them was Baldassarre Poli (1795–1883), whose teaching of theoretical philosophy enjoyed considerable popularity between 1837 and 1852. He was also a main exponent of Kant's philosophy (Santinello 1990), and the first to sketch out a history of Italian philosophy. In 1827, he published his first book, whose long title reveals much about the intimate connection between philosophy and the biomedical sciences in Padua: *Essay on the School of the Modern Natural Philosophers, with an Analysis of Organology, Craniology, Physiognomy, Comparative Psychology and a Theory of Ideas and Feelings* (Poli 1827). Another Paduan student of Romagnosi in Padua was Alessandro Racchetti (1789–1854), professor of Criminal Procedure from 1815 to 1848 and dean of the university during the unrest in 1848. Tellingly, he possessed a small but noteworthy private botanical garden (De Toni 1909, p. 14) and attended the Venice congress not in the agronomical, but in the botanical section. Still another representative was Alessandro De Giorgi (1814–1878), an early Italian proponent of the German Historical School. He taught Roman law from 1849 to 1867, but he was active as assistant and editor years earlier. His fame was due to his Paduan edition of Romagnosi's collected works, published between 1841 and 1848 (Romagnosi 1841–1848). Finally, Gian Paolo Tolomei (1814–1893), Padua's professor of Natural Law and Philosophy of Law, was an opponent, yet connoisseur of Romagnosi's ideas (Novello 2020).

It is therefore most likely that Andrea Meneghini and his peers knew Romagnosi's works. Indeed, we find several of Romagnosi's basic assumptions in Andrea's writings. Both saw political economy as part of a universal science, and both intended to establish a science of economics based on scientific methodology. Both were moderate, progressivist, anti-subversive, and were attracted to constitutional monarchy as a guarantor of civil rights and the institution of universal suffrage. Even more interesting, Meneghini's *Social Economy* as well as Romagnosi's later works started their exposition 'from below', that is, by beginning with the needs of the individual and argued that law and politics cannot lead to progress without respecting the 'nature of things'. Both regarded man as naturally social and laid much emphasis on the cooperation between parts. Finally, both strongly defended the right to property and the fundamental role of education. Since Andrea Meneghini left many of his basic theoretical assumptions unstated, it is worth having a closer look at Romagnosi's philosophy.

Recent scholarship (Ciech and Soliani 2017; Ghiringelli 2012; Lanchester 2012) describes Romagnosi, a freemason, as an advocate of economic competition and liberalism, proponent of French constitutionalism, and strong opponent of British feudal constitutionalism. Born in 1761, he assimilated many classical Enlightenment ideas, yet also views increasingly contaminated by romantic currents. Like most of his contemporaries, he advanced a global vision and unitary conception of knowledge and science. In 1802, he observed the magnetic effect of electricity (Stringari and Wilson 2000). This discovery fell into oblivion, however, which may be one reason why he did not resume his engagement with natural philosophy. Nevertheless, for him, morals, law, economics, history, and the natural sciences were interconnected expressions of civilization, and he conceived his 'civil philosophy' as a combination of all kinds of knowledge of man. As it is impossible to do justice to the 12,000 pages of his truly polymathic work, I will only point to some features of his 'civil philosophy' that are particularly important for the purpose of this book.

For Romagnosi, political economy was the science of 'the social order of wealth'. Consequently, in typical enlightened-empiricist manner, he set out to discover the best form of this order, analyzing human nature and societies before drawing general conclusions. A distinctive and increasingly important characteristic of his analysis was his historicist approach. The influences of Vico and, after he became an Austrian subject in 1814, of the German historical school are evident, though rarely mentioned in recent scholarship. Together with Poli, in 1832–1836 he edited the Italian translation of Wilhelm Gottlieb Tennemann's (1761–1819), *Grundriss der Geschichte der Philosophie* (*A Manual of the History of Philosophy*) (Tennemann 1832–1836), and like his Paduan student De Giorgi, he knew and appreciated Savigny's history (De Giorgi 1863). Despite these influences, Romagnosi claimed to present a wholly new vision: "Open the books of Grotius, Selden, Hobbes, Pufendorf, Thomasius, Montesquieu, Rousseau, and all others, and you will find no account of it" (Romagnosi 1832, p. 265).[1] In 1805, he declared, "Societies and nations rise, grow, perfect themselves, as is well known" (Romagnosi 1805, II, p. 389).[2] He called the 'natural progress' of nations and human societies toward perfection *incivilimento*, or civilizing (ibid., p. 287).[3] And perfezionamento, or perfecting, was a natural drive for Romagnosi. However, it was not an all-encompassing natural force acting continuously and uniformly through time and space. Rather, societies and states developed in analogy to individual lives, which meant that they passed through birth, growth, maximum force, and eventually decline (Romagnosi 1815, I, p. 185–186 and 205), a concept that we also identified as basic in Giuseppe Bernardi's writing (Sect. 5.2 in Chap. 5).

Notwithstanding his emphasis on the necessity of organic organization and his accordance with Kant's philosophy, he did not consider an innate force or *Bildungstrieb* to be responsible for the civilizing drive. In more Herderian manner, for Romagnosi, the influence exerted by object and subject, internal and external factors was multidirectional. The magic words were harmony and equilibrium, both reminiscent of Enlightenment theories. He therefore did not oppose monarchy, but absolute power: "The organic theory of power arises from the necessity to temperate any absolute empire, in one man as well as in the union of many people" (Romagnosi 1815, I, p. 74–75).[4] In line with the civil movement of the Age of Revolution, his ideal was the wellbeing of the people. Only shared governance guaranteed all citizens the greatest level of rights and freedom. "[The double constitutional guarantee] must result from the simultaneous overall action of the governmental organism, as much as the robustness of temperament results from the simultaneous overall action of the animal organism" (Romagnosi 1815, I, p. 34).[5]

While the system of power Romagnosi set forth in 1815 was still hierarchically structured and exerted monarchically from the top, in 1832, he favored a system of shared power in the style of recent conceptions by German organismic political theorists (Sect. 6.4). He now emphasized the interplay between the parts and the whole. The single citizen was autonomous but could express himself most advantageously only as part of the entire system:

> Considering the effect of social development, it appears to us that ultimately the individual has only a rather small degree of personal attitude. Maximum splendour, maximum good, maximum power resides in the whole, and the more splendour, good, and power everybody obtains, the smaller is the fraction of individual power that remains isolated to him. (Romagnosi 1832, p. 24)[6]

Even classes, professions, and other civil categories could not survive in isolation:

> But every single one of these classes does not exist, nor is it able to act, not for itself nor for the others, if not by means of the whole. Moreover, each one is nothing but a shoot or a branch of the great social tree, each is sustained, nurtured, enforced by all others, so that the power and the action of all together will be unique, indivisible and supportive towards the universality of citizens as well as towards the private [citizens]. (Romagnosi 1832, p. 42)[7]

The foregoing three quotations show that, by the early 1830s, Romagnosi had replaced many of his mechanic state-body analogies with organic metaphors of cultivation and farming. He made extensive use of biological—mostly plant—analogies and tropes. Besides its greater organic expressiveness, the vegetal imagery was easily understood—and probably shared—in a rural society like Veneto and by an intellectual elite that almost exclusively consisted of landowners. The farming-civilizing analogy was furthermore congenial to the paternalistic views of *possidenti* like Andrea Cittadella Vigodarzere (Sect. 9.2 in Chap. 9), which may explain Romagnosi's success in Veneto throughout the nineteenth century.

In the opening pages of his *On the Nature and Factors of Civilizing*, Romagnosi (1832, p. 15–16) opined that "the concept of civilizing is a natural product of the health of the social body, as full flowering depends on good condition and the good soil on which a tree grows and on the healthy atmosphere." The tree metaphor was not accidental. Romagnosi (1832, p. 166, 169 and 182) repeatedly compared Italy, which he ardently desired to see become a unified nation, to a germinating plant. In an introductory chapter, Romagnosi provided a lengthy *excursus* on his conception of life and the usefulness of this concept for political philosophy. He held that life escaped any analysis, because "what we call life is nothing but the complex of its visible effects reduced to its most simple expression" (Romagnosi 1832, p. 21–22). He showed himself as well aware that life was a property of individuals and that its metaphorical application to societies was therefore only figurative. Societies were the sum of the successions of many lives. However, after having clarified this point, he continued to extensively use the trope of the living organism in the form of 'national vitality', 'the life of a State', and similar expressions (e.g., Romagnosi 1832, p. 35 and 88).

The process of perfecting depended on respecting nature's laws. Romagnosi warned that forcing reforms or aggregating non-homogeneous parts would result in delay and regression, just as "you cannot transplant nor graft before the plant has burgeoned the seminal leaves and is capable to nurture itself with robust roots" (Romagnosi 1832, p. 165). Corresponding ideas and similar vegetal imagery were also propagated in Herder's *Ideen zur Philosophie der Menschheit* (1784–1791) (Goeth 2017, p. 61). It is not clear if Romagnosi conceived *Incivilimento*, or the process of becoming civilized, as a sort of evolutionary process. Despite his almost manic use of terms like development, perfecting, perfectibility, and the like, as mentioned above, he did not conceive of a Kantian-Blumenbachian innate drive. For him, the cause of progress resided in man's ability to create complex wholes. An innate developmental drive existed within each organism, but a solitary human being was nothing. He possessed only the seed of the power of rationality. Man's great success derived from his sociability. His insistence on this aspect distinguishes Romagnosi from most contemporary authors. Only as part of a society was man able to realize his potential and build upon the achievements and traditions bequeathed by his ancestors (Romagnosi 1832, p. 23–24). In fact, Romagnosi continued to emphasize the influence of a complex of internal and external factors. Similar to Vico's standpoint, every society was thus the product of its specific space and time, and every expression of a society was historically and geographically contingent—an approach that opened the door to empirical research and to comparison between nations. The better a society succeeded in the aggregation of a harmonious body, the more efficiently it worked. Initially, in 1815, he proposed a geometrical "formola del perfezionamento politico", or formula of political perfecting (Romagnosi 1815, p. 405), further elaborated in his mathematical treatise of 1822 as the *principio di compotenza*, or principle of shared power, that aimed to mathematically determine the level of affinity between numbers (Romagnosi 1832, II, p. 443–447 and 454–498; re-published in 1841–1848, I, part II, p. 1310–1311 and 1317–1338).[8] In 1832, he instead likened the functioning—today we would say self-organizing property—of a society to a beehive (Romagnosi 1832, 23), another very common organismic analogy of this period (Mazzolini 1985; Wolfe and Kleiman-Lafon 2021). According to Romagnosi, the beehive was organized like a republic. Even the queen did not command or direct. Rather, every single bee was induced by its own very nature to cooperate in harmony.

6.3 Patterns of Social Organization

As previously mentioned, Andrea Meneghini's thinking parallels many of Romagnosi's basic assumptions. Given that he did not publish a lengthy theoretical treatise in the pre-1848 period, we must resort to his popular pocketbook *Elementi di economia sociale ad uso del popolo,* or *Elements of Social Economy for the Use of the People* (A. Meneghini 1851a) to gain insight into his basic convictions. Astonishingly, the book has been almost completely ignored by scholars of Italian economic history (e.g., Augello and Guidi 2007a). Meneghini's *Elements* surely did not show the theoretical depth of other post-1848 treatises. Nor was it innovative. Nevertheless, the lack of interest is curious given that the publishing house, the cousins Pomba of Turin, owned a successful and farsighted company that specialized in popular non-fiction books, series, and journals (Palazzolo 1997, p. 11–54). One of the numerous authors on economic questions published by Pomba in those years[9] was Antonio Scialoja (1817–1877) (Augello and Guidi 2007b). His books, the second edition of *Principi di economia sociale* (1846; *Principles of Social Economy*) and *Trattato elementare di economia sociale* (1848; *Elementary Treatise on Social Economy*), must certainly have overshadowed Andrea's *Elements*, even if the former addressed an academic and the latter a popular audience. Antonio Magliulo (2006, p. 5–6) emphasizes that the editorial output of this period represented and defended the viewpoint of the middle classes, rather than that of the aristocratic rural elite, and that in this context, Scialoja soon assumed a prominent role in the Italian debate. His university lectures in Turin attracted about 600–700 students and in the eyes of many of his peers this underpinned his claim for intellectual (and then political) leadership. In fact, he was appointed minister of agriculture in the 1848 revolutionary government in Naples, and after the Italian unification was several times minister of finance and of education.

Scialoja was one of the most important economists of the Risorgimento. At the age of twenty-three years, he wrote the first systematic treatise of economic science in Italy, published in 1840 in Naples (Scialoja 1840). Six years later a second, significantly augmented edition appeared in Turin, though the introduction was identical (Scialoja 1846). Roberto Romani (1992, p. 265) regards him as "the best expression of the peculiarly Italian way of reflecting on industrial development in the Risorgimento decades". Meneghini probably knew him personally when both were exiles in Turin.[10] No doubt he witnessed Scialoja's prolific output and the immediate public appreciation forthcoming from many liberal scholars in Europe, especially those with close contacts to Italy, such as Robert von Mohl and Carl Joseph Anton Mittermaier (1787–1867). Probably, Meneghini also shared Scialoja's sense of impatience about Italy's delay in responding to the social changes and his desire to make economic doctrine accessible to a broader popular audience (Gioli 1989). The fact that Meneghini's and Scialoja's works appeared in the same publishing house indeed points to some commonality of purpose, even if they addressed different audiences. Yet, they also disagreed in some respects. Certainly, Meneghini did not share Scialoja's staunch rejection of welfare policies and public charity. But the main difference between their worldviews concerns, I would argue, the position of the individual in the economic and social system, and hence the constitution of the social and the relation between part and whole. Contrary to Scialoja, who, following the tradition of the Milanese economist and philosopher count Pietro Verri (1728–1797), explicitly started with the basic fact of wealth and then discerningly moved to the needs of the individual and the society (Isabella 2005), Meneghini's treatise started from 'below' (A. Meneghini 1851a, p. 13–16). The first chapter was entitled *Needs and Demands of Man*.

The role of 'the people' was a central item on the social economy agenda. The early nineteenth century was imbued with a combination of multifaceted fears and hopes concerning social transformations, ranging from nostalgia for a supposedly quiet past to fervor for modernization and new opportunities. On the one hand, there was a growing demand for civil participation. The growing economic power of the bourgeoisie reinforced demands for more influence on political and administrative decisions. In addition, in Lombardy-Veneto the perception of Austrian governmental inaction and remoteness or even unwillingness to respond to the new social and political situation opened spaces for new forms of civil engagement (Camurri 2006). Associationism was but the most visible symptom of this growing desire (Chap. 2). On the other hand, most contemporary concepts of participation did not extend to all citizens. The ideals of the French Revolution lingered, yet they were accompanied by a fear of turmoil, unrest, and the 'anarchy of the masses'. Roberto Romani (1992, p. 261–263) argues that the fundamentally conservative ruling classes feared a modernization process outside their control and therefore proposed reform projects that alleviated social misery yet enforced their own privileged status quo. He regards the Lombard economist and politician Melchiorre Gioia (1767–1829) as exemplary for the general apprehension about upheavals. Gioia thought people were naturally lazy and enthusiastically advocated industrial production, yet not for its economic benefit, but because he saw it as a way to discipline people and create a well-ordered society (Romani 1992, p. 256).

Marco Meriggi (1987, p. 118) points to the absence of specific privileges for the nobility of Lombardy-Veneto. This meant that, from a juridical point of view, members of the aristocracy and the even most modest subjects were of equal standing. This did not give rise to egalitarian sentiments, but it did mean that, in some way, the modes and definitions of social distinction of the lower classes needed consideration. Socialist writers were not absent in Italy, yet a great majority of scholars were worried about rising socialism, and therefore always had an eye on the civilizing and disciplining effects of economic reforms. The marked overlap between the economic and the intellectual elite and the vague sense of being responsible for general local or regional wellbeing were reasons why the emerging economic and political sciences continued to stress the moral and civilizing dimension of political economy. In fact, political economy was more often called 'social economy' in those days. Another common response to the threat of socialism was a general insistence that the law must safeguard the individual's right to property. We have already seen this moderate liberal attitude in Meneghini as well as Romagnosi, and it had a lasting effect on Italian thought. In the second half of the century, it was a distinctive of the 'Lombardo-Veneto school' of economics, which included many scholars, who, like Messedaglia, went on the record against extreme *laissez-faire* and were deeply concerned about the social evil of industrialization.

The social dimension was also at the heart of Andrea Meneghini's writings. He was not a republican, even less a socialist, yet—and this distinguishes him from many other authors of his time—sincerely committed to the common people. In the premises of his book, he declared:

> I have taken into account the sad revelations made by socialism about the misery of the most numerous class, insisting on the importance to improve its condition, but I could not accept the major part of its proposals and I have fought against them with the evidence of the principles that it has developed. Rather than negating the evil, as many do, I have searched how far [economic] science could praise itself of providing a remedy. (A. Meneghini 1851a, p. 10)[11]

Andrea concurred with Romagnosi's plea for the benefits of equilibrium, social harmony, and philanthropy. These concepts generally fell on fertile ground in Veneto, yet the notion of equilibrium leaves much room for different conceptions of the relation between parts and whole. At the core of Meneghini's thinking and action was the dignity of the individual. Dignity was a recurrent theme in Veneto in the 1830s and 1840s. Around 1820, Romagnosi started to study the development of civil societies in depth. For him, any conception of social philosophy had to take into account that people were not an anonymous mass but that humans were agents that operated in a complex social reality (Moravia 1974, p. 34 and

37). Romagnosi explicitly referred to the works of Vico's disciple Abbot Jacopo Stellini, who had taught moral philosophy in Padua from 1739 and who had articulated proto-sociological ideas (Moravia 1974, p. 37). Societies continuously progress, but, as we have seen above, contrary to other European progressivists, for Romagnosi, this was not due to a (hidden) internal force or to natural laws alone. Rather, he regarded all humans as individual agents in a social whole. This conception prompted him to consider the many factors—social structure, religious beliefs, the state of agriculture, the condition of state power, the weight of public opinion, and so on—which lead to progress.

Meneghini thus did not start to develop his discourse from an abstract entity like 'wealth', as Scialoja did. For him, like for Romagnosi, any analysis had to start from the level of the individual and then proceed upwards to formulate general theories. Scialoja instead rejected any conception of economics as a study of the intricate congeries of particulars. He defended empiricism yet, in analogy to the approach of geometry, held that induction must be part of "the analysis of the *constant properties* of the things" (Scialoja 1840, p. 5).[12] In the best tradition of the Enlightenment, for him, there was only one eternal, universal, and ubiquitous science. Sergio Moravia (1974, p. 22–27), in fact, underlines that, despite all his empiricist avowals, Scialoja rarely started his analyses from concrete facts. He aimed to present a comprehensive system with a small number of universal laws, and therefore insisted on the necessity of a scientific approach. Like Galileo, who maintained that all objects fall with the same velocity when in a vacuum, even if this goes against observations under everyday conditions, Scialoja held that science must go beyond apparent phenomena.

Two further aspects of Meneghini's writing furnish evidence for the distinctiveness of his ideas about social organization and his regard for the role of individuals and individual data. In line with most liberal Italian writers, the right to property was a cornerstone of his beliefs. In his *Elements*, he maintained that man has social needs "among which we principally recognize education, the defence of one's own liberty and the protection of what he has obtained through work, hence the security of property" (A. Meneghini 1851a, p. 13–14).[13] Besides being clear opposition to socialist projects, the defense of the right to property places the rights of the individual above those of society as a whole. The attempt to promote capitalism among common people was characteristic for Andrea, as we will see below. The second aspect concerns Meneghini's passion for statistics and his conviction that it was the best tool for studying and improving society (Benvenuti et al. 1852). In this respect, too, he echoed Romagnosi's ideas and other European currents of the time (Petrarco 1996). Meneghini later made wide use of statistics when he studied the (negative) effect of Austrian taxes on the productivity of Veneto's economy (A. Meneghini 1859, 1860, 1863, 1865).[14] Yet, Andrea's preference also unveils another, more intimate conviction. In similar fashion as his brother Giuseppe, whose natural philosophy respected the immense diversity of living forms, Andrea's social philosophy respected the diversity of humans, because using statistics ultimately amounts to negating uniformity. Statistics is the favorite generalizing tool of the empiricist, because it produces results that accept and include exceptions and varieties.

Notwithstanding his defense of individualism, Meneghini did not maintain ultra-liberalist views. Like Romagnosi and most of his Venetian contemporaries, he pointed out that man is "by the laws of nature a social and perfectible being" (A. Meneghini 1851a, p. 13–14)[15] and that a functional society needed an equilibrium between the parts. He aimed to convince his readers that wealth for all could be achieved only through cooperation, and explained, "You will find that the highest of your well-being has to go in concert with those of the others, that happiness on Earth lies in solidarity" (A. Meneghini 1851a, p. 10).[16] Andrea's prominent involvement in Padua's emergent associationism, in particular his attempt to involve all members, the way he ran his journal, and his plans for creating a regional association of agronomists (Sects. 2.4 and 9.2 in Chaps. 2 and 9, respectively) demonstrate that his entreaties for the benefits of cooperation were more than theoretical lip service; they expressed his most intimate belief and formed a frame for his action.

This is also true for his support for mutual aid institutions. Today, mutualism is a key concept in the life sciences. It was introduced into the biological debate as late as 1873 when, in a university lecture, the Belgian parasitologist Pierre-Joseph van Beneden (1809–1894) distinguished mutualism, a type of interaction among organisms of different species with mutual benefit, from parasitism (van Beneden 1875, p. 6 and 11, 1876, p. 2 and 83). The phenomenon itself had been observed in all epochs and adduced as evidence of the harmony of nature. Most of the best-known examples of mutualism were first described between the 1840s and 1880s and principally involved organisms like fungi and algae (Bronstein 2015, p. 11–14), the preferred research objects of Giuseppe Meneghini. Douglas Boucher (1985, p. 9–11) maintains that the nineteenth-century interest in mutualistic phenomena expressed a general need of the time and that it was due to anti-Malthusian (and later anti-Darwinian) currents and the desire to develop counterpoints to the competition and progress-through-struggle themes in the natural as well as the social sciences. In fact, like many terms and concepts that cross the boundaries between fields, mutualism was originally a political-economic term. In 1840, the French anarchist Pierre-Joseph Proudhon (1809–1865) in his essay *Qu'est-ce que la propriété?* (What is property?) conceived an ideal of just and equal exchange that would eliminate the capitalist unfair advantage of the inheritance of property and replace it with mutual relationships (Anon. n.d.). In Lyon, he had shared the experience of the home weaver's organization, the *mutuellistes*, and around 1843 he developed the theoretical concept of *mutuellisme*. In 1848, he proclaimed that the paths to an 'organic revolution' and to 'scientific socialism' lay neither in social revolution nor in communization of property, but in a system of mutual credit for workers and the creation of cooperatives. According to George Woodcock (1993), Proudhonian mutualism built on the experiences of mutual aid societies, but likewise on a somewhat romantic image of peasant life and equitable exchange of products among them.

In France, mutual aid societies, mutual saving banks, working-class organizations, and the like were founded to counter exploitation and to create groups representing legitimate interests by pooling the forces and resources of the common man. In Italy, the idea fell on a fertile ground and the system of mutual credit banks enjoyed some success. However, the mutualistic movement in Veneto was a far cry from anarchism or socialism. In the 1840s, mutual aid was ideologically still a rather open concept, even if many of its complex and multilayered socialist, anti-capitalist, and anti-Malthusian undertones were present (Isenburg 1977). Like most Italians, Meneghini did not accept the Malthusian principle, because he did not consider the problems of the growth of populations to be imminent in Italy (A. Meneghini 1851a, p. 192–194). Moreover, as seen above, he refuted socialism. Nevertheless, he accepted that socialism "has boosted the development of the principle of association, which represents the basis of all civil advancement" (ibid., p. 63)[17] and fervently advocated the (timid) mutualistic movement in Padua. People like him contributed to developing this institution into a politically less suspicious activity. In fact, in Veneto these banks were headed by wealthy *notabili*—and Andrea did not omit to praise their benevolence (A. Meneghini 1851b, p. 300). Around mid-century, according to Renato Camurri, mutualism was one of the most important ideological elements of Veneto moderatism (Camurri 2002, 2004).

The apparent contradiction between an idea of mutualism tainted by French roots and socialism and the Venetian and Austrian version of mutual banking may find an explanation in Catholic social doctrine. The Veneto of the 1840s had no experience in industrial proletarianism worth speaking of. Still, the problem of poverty was omnipresent. In 1814, 3269 or about 10% of the 31,612 inhabitants of the city of Padua were indigent; in the province, it was 7942 people out of a population of 80,180 (Monteleone 1969, p. 37). These people depended on governmental assistance and private philanthropy. Due to its roots in Catholic charity, the Italian debate and action had a strong humanitarian tone, in particular when voiced by moderates (Woolf 1991). The historical compromise concerning the ambivalent relation between Catholic ethics and capitalist market economy (a term coined in the mid-nineteenth century that entered scholarly discussion only at the beginning of the twentieth century) was the notion of the common good (instead of the absolute good of capitalism). From the late Middle Ages onwards, Italy, with its long tradition of commerce and capital flows as well as its strong Catholic influence, produced a long list of authors who were inclined to offer reflections on charity, profit, and the common good (Zamagni 2010).

Veneto was the first part of Italy to literally inaugurate mutual saving banks (*cassa di risparmio*). Modelled on Austrian saving banks, they opened on 12 February 1822, the birthday of Emperor Francis I, in Padua and other Veneto cities. They were founded to counter rampant usury and oppressing debt (Bof 2018). An article in the *Tornaconto*, most probably written by Meneghini, praised saving banks as "one of the most beautiful terms of modern civilization" (Anon. 1847g)[18] and as a way to establish one's own business and to enable social advancement for one's children (A. Meneghini 1851b, p. 298). As is typical for his pragmatic and statistical approach, he provided instructions and tables of how one's savings would increase with 4% interest (ibid., p. 301). He thus saw mutual saving banks as instruments of popular capitalism rather than of anti-capitalism or anarchy.

Saving banks did not become popular before the end of the nineteenth century, so Andrea's early allegiance demonstrates his tendency toward the philosophy of self-help. Compared to the paternalistic vision dominant in Veneto (Sect. 9.2 in Chap. 9), he thus advocated a more associative and collaborative idea of society. He preferred self-help to charity, and in contrast to conservative liberals like Cittadella Vigodarzere, he extended his ideals of individual rights, economic participation, and the common good to the lower classes. Meneghini's *Elements* never addressed the state or governors. His central foci were the society (in the sense of *societas*), its constituents (the people), and their various economic interactions. For him, the society was the biggest of all associations (A. Meneghini 1851a, p. 252).

In this context, Meneghini's popular pocketbook takes on a different significance. Its principal aim was not to civilize his audience nor to develop innovative doctrines. Andrea followed a much more pragmatic strategy, namely to provide a practical *vade mecum* for common people, which would enable them to understand and partake in current economic language, terms, and conceptions—and possibly convince them to participate in this game. Prosper Paillottet (1804–1878), disciple and posthumous editor of the liberal economist Frédéric Bastiat (1801–1850), noticed and lauded this aspect explicitly in his commendatory review in the *Journal des Économistes*. He remarked that Meneghini's little opus was not innovative, but concise, very clear, and well structured, and he confessed to envying Italy for having authors and publishing houses able to popularize political economy (Paillottet 1853).[19] One may disregard Andrea's *Elements* as low-grade philosophy, yet one may also appreciate it as a purveyor of a tacit philosophy. Without formulating an explicit doctrine, he promoted the philosophy of popular participation through his actions.

6.4 Organic Frameworks

Conceiving of society as a togetherness of individual parts that jointly form a functional unit went hand in hand with a decisive refusal of absolutistic and feudal regimes. The latter receive their *raison d'être* principally from a single individual at the top, and this individual determines the political and social substructure. Yet, if not the supreme director or leader, who or what else holds the parts together? Associative conceptions require a different ideological answer to the question why liberal societies keep on working and do not fly apart like a heap of leaves in the wind. A justification is even more necessary when a considerable degree of individuality and independent rights are ascribed to the parts, as Meneghini did. Andrea did not offer a lengthy discourse to justify or compare his conception of social organization, but his terminology and several of his arguments can help to reveal the framework of his view.

Recent scholarship (e.g., Freeden and Fernandez-Sebastián 2019; Leonhard 2004) no longer propagates narratives of the homogeneity of liberal thought, and instead differentiates several contemporaneous currents and treats liberalism as a mottled cluster of basic cultural postulates rather than a monolithic theory. Moreover, several liberal postulates were vague, and, not uncommonly, we can find different, even seemingly contradictory conceptions and tropes in one and the same treatise. I therefore sketch several currents characteristic of the debate in Veneto, yet mainly focus on organicism which has been somewhat disregarded by historians of Italian economic and political thought.

In the wake of rationalism and the Enlightenment and owing to the close adjacency to natural history, a distinctive feature of nineteenth-century philosophy was the tendency to understand politics as a science. This attitude led to a growing methodological and theoretical convergence with conceptions and techniques coming from the natural and the life sciences. Initially, the cross-fertilization occurred on the level of general inspiration and the desire to learn from nature's apparent perfection. The pioneering ideas thus remained somewhat ambiguous. Later, some scholars went further and developed concrete models, for instance, of future societies shaped by conceptions of the constitution of biological entities. These models did not radically break with past conjectures, but they consisted of eclectic combinations of traditional and innovative elements (Guillo and Jacobs 2002), and they left much room for idiosyncratic theories. Natural and liberal explanations, for instance, did not necessarily exclude divine explanations, as long as one saw God as the originator of the natural laws.

Concepts of organization and self-organization swept across many domains of thought and human experience in Europe during the period from the late seventeenth to the end of the nineteenth century. Eighteenth-century European cultural thought, revolving around the question of order, concerned itself with a multitude of apparently ordered systems—natural, social, political, economic, cognitive, and others (Sheehan and Wahrman 2015). In Britain, ideas of spontaneous self-organization became a major current in several areas of social and economic theory in the 1720s. Nature was regarded as possessing its own logic that humans ought to understand but leave alone if things were to work well. Humans were not impotent, yet these conceptions implied less confidence in the power of reason and human intervention, and more humility toward the 'nature of things'. If seen optimistically, *laissez-faire* philosophy implied a faith that events would naturally turn out well. This view appealed to many religious thinkers who equated nature's laws with Divine providence. David Hume, for example, argued that the rules of property, as well as the rules and conventions of justice, money, and language, evolved spontaneously, yet were guided by a 'divine hand'. Adam Smith, too, believed in spontaneous social order enabled by providence (Dale 2018). Later, this current became increasingly imbued with elements of developmental, evolutionary, and progressivist ideas.

Social Newtonianism, that is, social and moral theories claiming to be modeled on Newton's *Principia* (1687), was an intellectual program that partly overlapped with the aforementioned current (Jacob 1977). Newtonian mechanics provided a church-endorsed and intellectually attractive rational model, consisting of a universal natural law and a marvelously harmonious and simple explanation of order within the enormity of space. Social theorists like John Locke and Adam Smith availed themselves of the laws of attraction and repulsion to conceive of social aggregation as a loose interplay of individual parts, similar to the spontaneous binding of atoms. In British liberalism, the tendency to think of society as a great self-equilibrating machine became a scientific and divine law, and *laissez-faire* theories conceived of economic forces as equilibrating themselves through markets, just like the universe stayed in equilibrium because of the gravity of celestial bodies (Cremaschi 2009; Fong 1988). In Italy, along with Newton, many economists made explicit reference to Galileo as the ideal universal scientist. Galilean economists like Angelo Messedaglia favored an empirical approach and promoted the use of observation, 'strict scientific' tools, and the apt use of mathematics (Romani 1988; Tusset 2018).

A third powerful influence came from the biomedical sciences and concerned body-state analogies. The conception of bodies as composites of parts is antique, as is the inclusion of this idea in social and biomedical philosophies. Plato, Aristotle, and Thomas Aquinas are just three of the foremost thinkers of states as bodies with corporate order. Early Christian conceptions teemed with *corpus*-theories, too. During the period of the French Revolution and German Idealism, such traditional conceptions underwent a decisive transformation. Inspired by contemporary debates in natural philosophy, in particular concerning the contemporaneously developed biological concept of 'organism', the idea of 'organization' was incorporated into theories of the state and soon became a basic socio-political term. Until the mid-eighteenth century, the term organism had mechanistic connotations, describing all kinds of bodies, organic and inorganic (Dohrn-van Rossum and Böckenförde 1978, p. 559). Around 1800, organism metaphors were still rare, although

> the attempt is discernible to identify and theoretically digest by means of these terms the new realities of form and organisation of political rule, as perceived in part with and after the French Revolution and thronging towards realisation, and to attribute to them a heuristic political value. (Ibid. p. 561)[20]

Supporters of mechanist conceptions, dominant in the seventeenth and eighteenth centuries, used bodies as tropes for rational and efficient organization, but these bodies were, like machines, rather static aggregations of parts. Thomas Hobbes, for instance, compared societies with human body machines in 1651, and made this analogy the basis of his social contract theory. In contrast with the organicist theories, which blossomed from the late eighteenth century onwards, especially in the circles of romantic *Naturphilosophie*, the parts were removable, exchangeable, and subordinate to the whole. Like a machine, the mechanist body-state needed a supreme director or at least a controller.

In the wake of Immanuel Kant's elaborate theory of the organism (see, e.g., Nassar 2016; Zammito 2018), from about 1815 onwards, several German political theorists began to take inspiration from organic bodies. Not uncommonly, they did this, as romantic natural philosophers, in explicit opposition to mechanistic worldviews. Kant based the logic of organization on the interplay between means and ends. In contrast to a machine, a living body was cause and end in itself. In a living being, he argued, the parts existed only in relation to the whole and the whole existed only in relation to its parts. It was an organic unit. Both sides determined each other. On this basis, Kant refuted absolutism and elevated the relevance of the individual while avoiding individualism or anarchy.

Today, organicist theories of the state have fallen into ill repute for a variety of reasons. In the second half of the nineteenth century, opponents emphasized the theoretical and practical shortcomings of the state-organism and other bio-political analogies. In the twentieth century, they were instead criticized for promoting fascist *Lebensraum* ideologies (e.g.,

Goggans 2004).[21] It is important to bear in mind that political organicism was not a well-defined theory. It meant quite different things in different historical, geographical, and philosophical contexts, and included a variety of currents. There were conservative-reformist organicists, like Friedrich Karl von Savigny (1779–1861) (Sect. 6.5) and Karl vom Stein (1757–1831), and progressivist organicists. Most organicist theories strictly opposed any type of absolutism or dictatorship and instead favored forms of governance based on the principle of subsidiarity and the ideal of interdependence. Yet, later in the nineteenth century hierarchical conceptions experienced a renaissance and theorists like the Italian constitutionalist Luigi Palma (1837–1899) followed Herbert Spencer (1820–1903) in arguing that precisely the organismic nature of states necessitated the existence of a dirigiste entity that all elements had to obey to assure the vitality of the whole body (Palma 1877, I, p. 122–123). The picture of organicism is further complicated by the fact that many participants in the debate did not scrutinize their tropes. Not uncommonly, even directly cited sources were misinterpreted. Many nineteenth-century German political theorists, for instance, adopted cellular metaphors, yet few of them correctly understood cell theory.

All organic theorists of the early nineteenth century conceived of bodies as composite collaborative entities. Individual parts were meaningless and unviable without natural context. The focus of this period was hence directed to sociality and cooperation, both indispensable for creating a dynamic living being. In the emancipatory context of the period around 1800, when a growing number of citizens understood themselves not as obedient subjects of a superimposed masterminding authority but rather as social partners participating actively and collectively for the common good, these ideas fell on fertile ground (Goggans 2004). Organicists rejected the traditional settings of feudalistic policies, and welcomed new ideas about organization in-between the individual elements (citizens) and the state. Between 1830 and 1870, in parallel to the formulation of cell theory (Sect. 7.1 in Chap. 7), the new organism concept became the *leitmotiv* of the constitutional discussion in Germany (Dohrn-van Rossum and Böckenförde 1978, p. 587). Adherents to political as well as biological organicist conceptions approached complexity from the bottom up and, combined with the historicist philosophies of the same period (Sect. 6.2), considered bodies as historically grown and therefore indissolubly fused together. The organic philosophy thus considered political constitutions not as unilaterally conceived and then conceded to the people, but as the product of the interplay of forces (Dohrn-van Rossum and Böckenförde 1978, p. 595–596). The monarch, if included in such conceptions of state order, did not stand above the state, but was an integral part of it—often the heart or the brain.

Interestingly, a vital materialist like Immanuel Kant frequently adopted 'the tree' as a model for living organized bodies. This is curious because, at the time, philosophers still saw plants mostly as purely mechanical entities, at most as inferior and only partly living objects. Yet, in §64 of his *Kritik der Urteilskraft* (1790), Kant referred to the tree to illustrate his idea of the essence of life as being a self-organizing whole, because "in such a product of nature each part is conceived as if it exists only *through* all the others, thus as if existing *for the sake of the others* and *on account of* the whole" (Kant 2000, p. 245). The tree illustrated the "special character of things as natural ends" and the fundamental organizing activities of reproduction and regeneration, hence the capacity of the living being to produce other bodies "of the same species", to produce itself as an individual (organic growth), and to reproduce its own parts through nature's self-help in case of injury (Kant 2000, p. 243–244). Tobias Cheung (2009) shows that Kant had in mind a very peculiar idea of 'the tree'. In particular, inspired by botanical grafting experiences and following the natural philosopher Charles Bonnet (1720–1793), he conceived of trees as assemblies of various levels of interrelated parts, each of which possessed a high degree of individuality.

Thereafter, other kinds of botanical metaphors and philosophies became increasingly popular among organicists. The young Herder, for instance, carefully chose his metaphors and his botanical imagery—in the first place that of organic growth—as they had a concrete explanatory function (Schick 1971). Although they stimulated the senses and invoked poetic imagery, these tropes did not work against but together with abstract linguistic thought, to yield deeper understanding. This holds true in particular for the romantic thinkers and the rapturous style of *Sturm und Drang*. For Lorenz Oken, originator of influential botanical-political analogies, plants provided intellectually plausible and emotionally powerful imageries for the growth and blossoming of vital organisms and societies. In Chap. 4, I argued that English gardens, too, expressed and conveyed ideas of harmonious growth and togetherness and of humans as a part of nature and subject to the same laws. French gardens were places of controlling nature; English gardens were places where one immersed oneself into nature.

The success of organic theories in Italy did not equal that in the German lands, not even in Lombardy-Veneto, the region most influenced by scholarship from north of the Alps. Likewise, it seems that the pragmatic and down-to-earth Paduans did not warm to the abundance of metaphors typical for the romantic currents. Nevertheless, the transition from French to Austrian domination also marked a gradual transformation of basic philosophical assumptions in Lombardy-Veneto. Romagnosi's aforementioned theoretical 'conversion', for instance, also involved a switch to agro-botanical metaphors (Sect. 6.4). Others, like Antonio Rosmini, favored historicist positions as early as in the late 1810s (Sect. 5.2 in Chap. 5). Giovanni Ambrosetti (1970, p. 103) has noticed that his socio-political conceptions shifted from a mechanic to an organicist framework around 1823–1824. Rhetorically, however, Rosmini persisted in using mechanic tropes. This becomes evident in his most important work on the nature of civil societies, *La società ed il suo fine* (Society and its Purpose), drafted around 1826–1827 and printed in 1837 as part of his *Philosophy of Politics* (Rosmini-Serbati 1837, 1839, 1994). Here, he repeatedly expressed his desire to see the social sciences attain the strictness of mathematical formulas (Rosmini-Serbati 1837, p. 85), a debt to his Augustinian beliefs. He sketched a procedure for the "art of directing civil society to its ends". He called the procedure 'social mechanics' and conceived of it, inspired by Saint Augustine's doctrine of God as the supreme mathematician, as a movement ruled by universal and immutable principles (Rosmini-Serbati 1837, p. 78–79, 1994, p. 2–3). Importantly, his analogy between developing societies and fluctuating equilibria recalled Klemens von Metternich's political theory of the *Gleichgewicht der Kräfte*, or balance of powers (Rosmini-Serbati 1837, p. 80–81, 1994, p. 3; for Metternich, see Sofka 1998). For Rosmini, societies never achieved perfection or, at the opposite extreme, destruction, but continually fluctuated between the end points.

However, around 1823–1824, Rosmini clearly broke with Metternich's opposition to popular sovereignty, and instead insisted on the dignity of every human individual. Rosmini's position became increasingly liberal, romantic, and organicist. In his 1837 treatise, the classical mechanist body-state metaphor underwent a decisive Kantian interpretation: "Society exists when all the individuals are united with a single common end, in the way that all our bodily limbs have the well-being of our whole body as their end, and the whole body has as its end the well-being of the limbs" (Rosmini-Serbati 1837, p. 102–103, 1994, p. 22). In the simplest terms, Rosmini conceived of societies as historically contingent bottom-up forms of organizations. Like for the advocates of organic theories, for Rosmini, the individual parts constituted the whole and citizens existed for the state, not *vice versa*. These beliefs dovetailed with the Catholic social ethic of the common good (Sect. 6.3), and found expression in declarations that every individual human being's purpose should be to act to support the common good, and not for personal benefit. The wellbeing of the whole and the wellbeing of the parts were mutually dependent. Organic theories of organization strengthened the focus on the parts. In return, the growing interest in the nature of human individuals opened the door to conceiving of social theory as a 'social physiology', as Romagnosi called his doctrine (Moravia 1974, p. 41). Romagnosi's Paduan supporters, in particular Baldassarre Poli and Giuseppe Bernardi, indeed often used bio-medical analogies (Sect. 5.2 in Chap. 5 and Sect. 6.2). This was probably one of the reasons why Bernardi recommended Giuseppe Meneghini investigate the human brain (Sect. 7.1 in Chap. 7).

Scialoja's works, too, initially contained organicist elements. Recent Italian scholarship emphasizes that his thinking was inspired principally by French and British economists, and that it "links the Italian tradition of eighteenth-century economists with the introduction of recent British and French writings, in particular those of Smith, Ricardo and mainly Say" (Faucci et al. 2007, p. 121–122). However, a look at his *I Principi della Economia sociale* (1840) shows that Scialoja had assimilated an organicist attitude. His introduction started with the sentence: "Society is an organized and living body: constant and emergent laws of its own proper nature must therefore regulate its life" (Scialoja 1840, p. 3).[22] He compared the task of a social politician to that of a physician—both aimed at preserving or re-establishing public health. Referring to physiologist Friedrich Tiedemann (1781–1861), who also had a deep impact on Giuseppe Meneghini's conception of anatomy (Sect. 7.1 in Chap. 7), he deemed political economy to be the science of the "history of those *organic manifestations* of society, which constitute the life of conservation" (Scialoja 1840, p. 4, his emphasis)[23] However, state-organism analogies or biological-economic metaphors were rare in the following chapters. Scialoja's second *Elementary Treatise of Social Economy* (1848) did not contain any organicist terminology, an indicator that his audience in Piedmont preferred a more positivistic attitude (Scialoja 1848).

Hence, organicist elements were common in Lombardy-Veneto in the 1830s and 1840s, though more often latent than explicit. They were part of tacit, basic assumptions, but did not amount to a distinct current.

Angelo Messedaglia's position shows that organicist views continued to be present in the second half of the century. Messedaglia was convinced that every development—whether social or natural—progressed from the simple to the complex. Moreover, he viewed populations as organisms or organic systems, "where every single part has its right to exist, in correspondence to all others" (Messedaglia 1877, p. 24).[24] Finally, he transformed statistics—his main field of study—from a mathematical to a bio-medical tool. Originally, William Petty (1623–1687) had conceived of statistics in the Cartesian terms of number, weight, and measure (Petty 1899, p. 244). For Messedaglia (1877, p.6), in contrast, it was "up to limits a kind of anatomy and physiology in mass, a double character: *organic* and *social*."[25]

The organicist attitude also fitted well with associationism. In fact, the role of non-governmental political entities moved nearer the top of the agenda of political theorists and policy-makers of the first half of the nineteenth century. Opinions on the topic diverged considerably. The French revolutionaries mistrusted any form of intermediary political power between the state and the people. Similarly, many British eighteenth-century theorists distrusted political organizations as subversive interest groups (Gregorio 2012, p. 2–4). Many liberals, who leaned toward organic theories of the state, instead considered viable bodies not as loose aggregates of individual parts, but—in analogy to cells, tissues, and organs of growing complexity—as functionally structured and hierarchically ordered. Romantic conceptions thus envisaged society as internally articulated with local intermediary divisions and institutions and therefore anticipated later consociational theories. This concept of an articulated state supported the creation, authority, and political empowerment of associations (and later of political parties), as long as these acted not selfishly or corporatively, but for the wellbeing of the whole (the state) and the parts (the individuals). As we have seen in Chap. 2, in Veneto these ideas enjoyed considerable success, especially among the moderate elite, which continued to win political power and electoral success well into the twentieth century.

In this context, we can better understand Andrea Meneghini's definition of associations as "that concourse of labour or capitals that renders them suited to obtain a greater utility, or to produce what would not have been accomplished if employed in isolation" (A. Meneghini 1851a, p. 52).[26] His basic epistemology can indeed be described as organic-holistic, because it focused on the individual elements and the interplay among them, and because it understood precisely these interrelations as the producers of emergent economic phenomena that were profitable for all. On the other hand, his emphasis on the common good and his declaration that God has created man as eminently sociable (A. Meneghini 1851a, p. 15) derive from his roots in the Catholic social ethic. A third aspect that emerges from Meneghini's treatise, and that complicates the assignment of his ideology to a specific current, is the centrality of human agency. The order of the content in his treatise as well as the fact that all items were systematically described from the perspective of the common man was surely chosen with the aim of providing easy and immediate access to the principles of social economy. At the same time, it discloses his intimate conviction. His readers did not find themselves awestruck by an exalted scholarly spectacle of abstract theories and geniuses, but as parts of a play that directly involved them.

His insistence on agency places Andrea Meneghini closer to French than to German thought. When describing cooperation between citizens, he often used the term 'solidarity'. Regarding the purposes of a society, he maintained, for instance, "Another purpose of the society is the assistance in accidents, because private charity provides insufficient relief. [...] public assistance can be said to be mandatory for the solidarity that is established among the members of the social body" (A. Meneghini 1851a, p. 253).[27] At that time, the term 'solidarity' was not in use in Padua, but fundamental in French social philosophy and biological discourse. It is a term that connotes a profoundly different concept of togetherness. 'Solidarity' assumes the possession of a social self-consciousness and a deliberate choice by each citizen, rather than an internal natural drive or intrinsic force as favored in German debates. French social thought was greatly influenced by Auguste Comte (1798–1857). In 1839, Comte (1839, p. 325) put forward the basic premise that each social element "is always exclusively to be conceived in relation to all others, with whom it forms a *fundamental solidarity*"[28] as the indispensable basis of any study of social order.

Later, this current became distinctive of the French debate. Michael Osborne (2017) regards the account of zoologist Edmond Perrier (1844–1921) as representative of the dominant French biological and political philosophy of the late-nineteenth century. In his major work, *Les Colonies animales et la formation des organismes* (*Animal Colonies and the Formation of Organisms*, 1881), Perrier flagged the lessons to be learned from biological association, division of labor, and parasitism, and how these lessons might be applied to human society via Solidarism and an organicist conception of society (Osborne 2017, p. 215–216). Nobel laureate, statesman, and leading member of the Radical Party Léon Bourgeois (1851–1925) published his book *Solidarité* (1896) in order to promote his political program of social harmony and to offer a contrast to the idea of a 'struggle for life' (Demko 2002; Thomas 2017). Solidarism also enjoyed some success beyond France. In Britain, 'solidarity' was a central theme of Herbert Spencer's *Principles of Sociology* (1876). In Italy, similar ideas were discussed by, among others, Roberto Ardigò (1828–1920), professor of History of philosophy in Padua from 1881 to 1909, who defended an "organicist community-based evolutionism" of societies (Lentini 1980–1981; Padovan 1998).

Hence, Andrea was one of the pioneers of solidarism in Veneto, a view that differed somewhat from that of his Paduan contemporaries as well as from that of his brother Giuseppe, who placed greater emphasis on intrinsic forces, an inevitable 'march of nature', and the guidance bestowed by a vital force (Sect. 7.3 in Chap. 7). A similar combination of organicism and solidarism only attained wider prominence several years later, when Giuseppe Toniolo (1845–1918), student of Messedaglia in Padua, proposed a banking system counter to socialism as well as capitalism and extreme liberalism that was grounded in 'organic democracy' and Christian solidarism (Fanfani 2005; Poggi 2005).

6.5 The 'Law of Progress'

The previous section sought to refute the scholarly view that the increasing dynamism inherent in political and economic doctrines of the early nineteenth century was exclusively due to the incorporation of elements from mechanics and sociology (Tusset 2009, p. 267). Organicist theories and recourse to the distinctive phenomena of life contributed more to this dynamism than the rather static mechanist models. Dynamic conceptions gained further momentum through the inclusion of a developmental dimension, in particular historicist perspectives. This current, too, explored socio-political as well as biological areas of inquiry and theory simultaneously. This vast intellectual development started around mid-eighteenth century, involved a broad spectrum of fields, and took shape in several varieties (Beiser 2011). Indeed, the term historicism carries a number of partly overlapping and partly contradictory meanings. Andrew Reynolds (1999) has distinguished five varieties: (1) mundane historicism, which stresses the necessity of considering historical contexts; (2) methodological historicism, characterized by the definition of history as a science together with objection to universalist (physicalist) and absolute claims of any kind; (3) Popperian historicism, that is, the proclivity to find general laws or patterns in history; (4) epistemic historicism, the rejection of fixed and eternal standards of rationality; and finally (5) total historicism, which roughly equates with the claims of social constructivism.

Of this list, only the first three are relevant to our understanding of the first half of the nineteenth century. In contrast to natural law philosophy, from a historicist perspective man and his culture were seen as historically contingent entities—contingent yet subject to general laws. Many historicists of the period indeed maintained the idea of a common overall pattern in all history. They rejected the ahistorical ideals of the physical and experimental sciences, and found inspiration in living nature, in particular in the developmental patterns therein. The idea of human social life as a lawful historical sequence that paralleled by human individual ontogenesis was very explicit in Giuseppe Bernardi's doctrine (Sect. 5.2 in Chap. 5). Such views culminated in 1844 in Comte's 'law of the three stages' (Comte 1844, p. 2). Basically, historicist scholars, whether of the social or the biological kind, believed that historical roots determine the present and, according to some proponents, also the future. They therefore sought to understand a phenomenon, process, or form through its antecedents (Thienemann 1909, p. 273–274). They emphasized the value of having introduced a new type of causality for explaining human practices and of considering institutions and events as being in a state of perpetual change, whereas opponents criticized the determinism and inherent conservativism of historicism. Disagreements circled, among other things, around the degree of lawfulness of historical development, around the degree and manner in which history influenced current events, and around the inevitability and predictability of the process.

Contrary to the views of most present-day scholars, who locate the Italian historicist debate in the second half of the nineteenth century, my analysis of Andrea Meneghini's philosophy and that of his peers demonstrates that historicist perspectives were already influential in the first half of the century. In Italy, in fact, historicism received an early and important boost through the works of Giambattista Vico (Sect. 5.2 in Chap. 5). The publications of Rosmini, Romagnosi, and Bernardi are but some examples of the diffusion and of the varieties of historicist reasoning in the 1830s and 1840s. Andrea Meneghini, too, incorporated the concepts of perfecting and civilizing into his *Elements*. He did this without further explanation, which suggests that he took for granted some general familiarity and the accord of his readers with these assumptions. However, Vico's cyclic conception of *incivilimento*, still present in Bernardi's account, had disappeared from Meneghini's as well as from the treatises of other authors of the 1840s and 1850s. The idea of progress had become more linear and constant. Expounding upon the purposes of societies, Meneghini (1851a, p. 253) maintained that "The second purpose of social association is the moral and intellectual perfecting of our species which must obey the law of perfectibility and progress that distinguishes us from the other beings."[29] Although this sounds supremely deterministic, the sentence immediately took a different turn, as Andrea continued "and which the governments must promote especially through education and public instruction". Like the Meneghinian arguments about individualism and solidarity, mentioned earlier, the passage highlights that Andrea maintained a subtly different view. Like Romagnosi, he considered internal *and* external factors.

In fact, a particular form of historicism prevailed in Italy. On the one hand, historicism had seen an early and favorable reception. Georg Wilhelm Friedrich Hegel's (1770–1831) *Geschichte der Philosophie* (1837), for instance, was translated several times during the 1840s (Hegel 1840, 1848). The German Historical School of Jurisprudence, too, in particular Friedrich Carl von Savigny's historical writings as well as those of Georg Friedrich Puchta (1798–1846), Eduard Gans (1797–1839), and others enjoyed great success in the late 1840s and early 1850s. One reason behind the Italian historicist outlook was the above-mentioned popularity of Vico's philosophy, which paved the way. Corrado Bertani (2015, p. 218–219) explains the positive reception with the growing role of history in Italian nation-building. History was fundamental to Italy's identity and mingled with patriotic reprisals of Italy's priority and supremacy in culture, science, and economic theory (Cazzetta 2013, p. 163–200, 341–384, and 385–423). This sentiment prompted many kinds of historical research. From the late 1820s, Italy was among the first European nations to engage closely with the path-breaking work of Savigny, who greatly contributed to the prominence of Roman law and Roman history (Furfaro 2016, p. 13–17; Moscati 1990). By far the greatest number of translations of his works, exactly 49 out of 103, were Italian (Rückert and Duve, 2015, p. 456–477).[30] It resonated with the longstanding Italian tradition in philological and antiquarian scholarship, which meant that these fields were particularly inclined toward history (especially Roman history) and historical reasoning. Moreover, Savigny and other like-minded scholars frequently visited Italy and contacted Italian colleagues in order to recover the historical sources of Roman and Germanic law. For instance, Friedrich Blume (1797–1874) visited Padua on 14 September 1823 (Blume 1824, I, p. 167–189). In addition, Savigny's book on the right to property, translated in 1839, made him popular in Italy's moderate circles (Savigny 1839).[31]

Geographically, the reception occurred mainly in Piedmont, Naples, and Tuscany. Still, his doctrines attracted an important audience in Padua, too. As early as 1818, the abbot Angelo Ridolfi (1752–1825) emphasized the importance of the historical dimension of Savigny's *History of Roman Law in the Middle Ages* in his university lectures on German literature (Moscati 1997, III, p. 250). Another key proponent was Luigi Bellavite (1821–1885), who had studied political sciences in Padua in the early 1840s and became professor of Roman and later Civil law in 1855 (Pertile 2011). Between 1850 and 1852, he translated, summarized, and commented on excerpts of Savigny's *International Private Law*, and in 1855 Rudolf von Jhering's (1818–1892) *Geist des römischen Rechts*. In addition, the poet and novelist Leone Fortis (1824–1896), frequent patron of the Caffé Pedrocchi and involved in Padua's 1848 student unrest, translated the preface for the—never completed—first translation of the eight volumes of the *System des heutigen römischen Rechts* that Paride Zajotti had begun in 1828 for a Venetian publisher.[32] However, he published the preface separately (di Savigny 1853). Hence, even if it was unsystematic, limited, and sporadic, the reception of Savigny's work in Padua in the first half of the century nevertheless points to the existence of a generally favorable milieu for historicist reasoning.

Yet the German Historical School was not assimilated uncritically. For all its success, the diffusion of Savigny's thought was incomplete and rather granular. In fact, more interesting than its diffusion is the significant modification of key terms. In Naples, his historicist mindset found many sympathizers, but his life metaphors and his state-body analogies did not. In 1847, the Neapolitan translators L. Lo Gatto and V. Janni systematically replaced Savigny's abstract philosophical terms with concrete ones, and, in the context of this book even more importantly, replaced organic and biological metaphors with more neutral terms. Corrado Bertani lists some of them. The "organische Zusammenhang des Rechts" (organic cohesion of law) of the original version became "questa natural dipendenza del diritto" (this natural dependence of law). Likewise, "in lebendiger Verbindung und Wechselwirkung" (in lively connection and interaction) was translated as "in intimo legame ed in azion reciproca" (in intimate link and mutual action), and "diese innere Fortbildung" (this inner growth) was translated as "questo interior progresso" (this inner progress), which gave it a marked progressivist touch (Bertani 2015, p. 246–247). Finally, Blumenbach's term 'Bildungstrieb' (formative drive), fundamental also for Kant's concept of the organism, was translated as 'organizzazione' (organization), a rather different term. These semantic shifts are evidences of Italian pragmatism and of the separation between philosophical and legal thought in Neapolitan jurisprudence. Since no Venetian translation exists, we can only surmise whether this would have been more organicist. On the strength of my argument in Sect. 6.3, I suspect the terms would have been more faithful to the original, yet nevertheless more sober and philosophically attenuated.[33]

A look at the second half of the century provides evidence of the persistence of a peculiar idea of historical causality in Veneto. In fact, debates about Darwinian and social Darwinist ideas show that North Italian social philosophy remained in thrall to the theses of Vico, Romagnosi, and romanticism. This is even more astonishing given that Charles Darwin rang in a profound rupture in concepts of progress and the assumed mechanisms of historical change. However, declared anti-Darwinians like Angelo Messedaglia in the first place rejected the Malthusian catastrophism and the capitalist-like 'struggle for life' ideology (Isenburg 1977; Romani 1988). Yet, he as well as supporters of Darwin such as the Vicentine novelist and politician Antonio Fogazzaro (1842–1911), who sought to reconcile Darwin's ideas with Catholic theology, adopted arguments that were still imbued with concepts of naturally lawful progress and harmonious equilibria (Marangon 2003–2004; Rossi 1977).

Transformist-progressivist versions of historicism also stand out at the national level. Two examples drawn from the fields of legal science and government policy will serve to illustrate this. The influential juridical school of the Sicilian lawyer Giuseppe Vadalà-Papale (1854–1921), who coined the term 'social Darwinism' in 1883, promoted concepts like 'juridical evolutionism', 'evolutionary historicism', and 'sociological school of law', which had little to do with Darwinian evolution, but directly referred to Vico's and Romagnosi's historism (Mazzarella 2012). Agostino Depretis's (1813–1887) political project of 'transformism' denotes a specific political period that lasted approximately from 1876 to

1892. During these sixteen years, the liberal Lombard politician first conceived and then actually presided over governments based on cross-party agreements (Sabbatucci 1998). His fiercely debated and later mostly repudiated approach to politics came close to abolishing political parties in favor of a dynamic and cooperative form of decision-making for the declared principle aim of the common good. At first glance, it seems to have nothing to do with the mid-century pre-Darwinian transformism. On closer examination, however, it turns out that Depretis's argumentation was based on his conviction that his approach was the consequence of an innate unstoppable historical progress of the modes of policy-making. He used chemical-paleontological metaphors to argue graphically that parties ought not crystalize or fossilize, but themselves transform and advance (De Mattei 1940, p. 13).[34]

Even if they did not amount to a distinct doctrine, historicist beliefs played an important role in the general intellectual environment of pre-1848 Padua. During the reform period, historicist optimism provided a new kind of justification for the ascent of the bourgeoisie and for claims to civil rights and national ideals, taken to be logical consequences of historical development. However, understanding themselves as parts of a continuous and natural, for some also inevitable, historical flow may have also seduced the proponents of progressivist currents into underestimating the incongruencies and shortcomings of their reform program and the possibility of failure.

Andrea Meneghini's social ideology differed somewhat from that of his fellow Paduans. Following Reynolds' distinction mentioned above, Andrea incorporated mundane and methodological historicist elements, and only a small amount of Popperian historism. His idea of progress rested on four pillars: Catholic ethics, historicist organicism, belief in human sociability and solidarity, and faith in the positive impact of education. All four were called into question when he considered the role of 'pleasures' (*piacieri*), or life's sources of happiness. He explained that pleasures—better houses, food, or clothes, but also culture and free time—were important and not impossible to those of small means. Yet for the latter class of persons, pleasure was confined to family life and to the exchange of affection and ideas. However, precisely these affections constituted "the truest and most satisfying pleasure of a being created by God as eminently sociable. We should thank Providence, and we should also search in science for a way to increase the sum and intensity of so holy pleasures, soliciting with education the development of those faculties and sentiments" (A. Meneghini 1851a, p. 15).[35]

Notes

1. Romagnosi (1832, p. 265): "Aprite pure i libri dei Grozj, dei Selden, degli Hobbes, dei Puffendorf, dei Tommasj, dei Montesquieu, dei Rousseau e degli altri tutti, e voi non la ritroverete in conto alcuno".
2. Romagnosi (1805, II, p. 389): "Le società e le nazioni nascono, crescono, si perfezionano coi secoli, come è noto".
3. Romagnosi (1805, II, p. 287): "il perfezionamento della società, che appellasi incivilimento".
4. Romagnosi (1815, I, p. 74–75): "La teoria organica dei poteri nasce dalla necessità di temperare ogni impero assoluto, sia in un uomo solo, sia nell'unione di più uomini."
5. Romagnosi (1815, I, p. 34): "Questa doppia garantia ... deve risultare da tutta l'azione complessiva e simultanea dell'organismo governativo, come la robustezza del temperamento risulta dall'azione complessiva e simultanea dell'organismo animale."
6. Romagnosi (1832, p. 24): "Considerando diffatti l'effetto del sociale sviluppamento ci par di vedere che in ultima analisi non ci lascia all'individuo che una più o meno piccola attitudine personale. Il massimo lume, il massimo bene, la massima potenza risiede nel tutto, e da questo tutto ognuno ritrae tanto più di lume, di bene, di potenza, quanto minore è la frazione di potere individuale che a lui rimane in senso isolato."
7. Romagnosi (1832, p. 42): "Ma ognuno di queste classi non esiste né può agire, né per sé, né per gli altri se non per mezzo del tutto. Oltrecchè ognuna non è che un getto, o sia un ramo del grand'albero sociale, ognuna è sostenuta, nutrita, afforzata dalle altre tutte, di modo che unica, indivisibile e solidale si è la potenza e l'azione di tutte, sia verso la universalità dei cittadini, sia verso i privati."
8. Rosmini-Serbati (1836, p. 333–340) criticized this view harshly as conceiving an 'occult force'.
9. Another work that provides evidence of the central role of Turin in economic thought in Italy in the late 1840s are the three volumes of Gerolamo Boccardo's *Theoretical-practical treatise of political economy* (Boccardo 1853). Other books that appeared during these years were Salvatore Majorana Calatabiano's *Wealth and misery: New treatise of political economy* (Calatabiano 1847); and Placido De Luca's *Elementary principles of economic science* (Napoli 1852).
10. In 1846, Scialoja moved from Naples to Turin to the chair of Political Economy, but he returned to Naples during the 1848 insurrection. After three years in Naples' prison, he moved back to Turin, this time as an exile, and was mainly politically active and became an influential politician after 1860. For Scialoja and the history of his books and their reception, see Gioli and Magliulo (2007).

11. A. Meneghini (1851a, p. 10): "Ho tenuto conto delle tristi rivelazioni fatteci dal socialismo sulla miseria della classe più numerosa, insistendo sulla importanza di migliorarne la condizione, ma non potei accettare la maggior parte delle sue proposte e le ho combattute coll'evidenza dei principii che andava svolgendo. Anziché negare il male, come è usanza di molti, ho invece cercato fino a qual segno la scienza poteva lusingarsi di porvi riparo."
12. Scialoja (1840, p. 5): "... l'analisi delle *proprietà costanti* delle cose." (his emphasis).
13. A. Meneghini (1851a, p. 13–14): L'uomo ha ulteriori bisogni sociali "principali tra questi noteremo l'educazione, la difesa della propria libertà e la tutela di quanto seppe procurarsi col lavoro, cioè la sicurezza della proprietà."
14. Pietro Giovanni Trincanato (2017) maintains that these publications on the (devastating) effects of Austrian taxes and confiscations were the result of propagandistic efforts by Piedmont's liberals to denounce Austrian despotism and disrespect of the right to property (see also Girardi 2017–2018).
15. A. Meneghini (1851a, p. 13–14): L'uomo è "per legge della stessa natura, un essere sociale e perfettibile."
16. A. Meneghini (1851a, p. 10): "Troverete di più come il maggior bene vostro debba andare di concerto con quello degli altri, come solidale sia la felicità su questa terra."
17. A. Meneghini (1851a, p. 63): Al socialismo va riconosciuto il merito di "dare sviluppo maggiore al principio d'associazione, nel quale sta la base d'ogni miglioramento civile."
18. Anon. (1847): "uno dei più belli titoli del moderno incivilimento."
19. I thank Gianfranco Tusset for this information.
20. Dohrn-van Rossum and Böckenförde (1978, p. 561): "erkennbar ist der Versuch, anhand dieser Begriffe die neuartige Wirklichkeit der Form bzw. Organisation politischer Herrschaft und Ordnung, die teils mit der, teils nach der Frz. Rev. erkennbar wird und zur Realisierung drängt, zu erfassen, theoretisch zu verarbeiten und ihr zugleich politische Richtpunkte zu setzen."
21. An early twentieth-century supporter of a fascist and nationalist organic theory of the state in Padua was Alfredo Rocco (1875–1935), politician and professor of Commercial law. His theory of the state as an ethic organism formed the ideological basis for an 'authoritarian state of the masses'. Interestingly, the personal and social composition of Rocco's Nationalist Group showed some parallels and continuities with the nineteenth-century manifestations of this current. Many of its members were young landowners of the Euganean province, and many noblemen endorsed but did not publicly participate: they were the 'grey eminences' (Simone 2012, p. 35–37).

22. Scialoja (1840, p. 3): "La società è un corpo organizzato e vivente: leggi costanti ed emergenti dalla natura sua medesima deggiono quindi regolarne la vita."
23. Scialoja (1840, p. 4): "... la storia di quelle *manifestazioni organiche* della società che ne costituiscono la vita di conservazione" (his emphasis).
24. Messedaglia (1877, p. 24): "dove ogni singola parte ha la sua ragione di essere, in corrispondenza con tutte le altre."
25. Messedaglia (1877, p. 6): "È, fra certi limiti, una specie di anatomia e fisiologia in massa, a doppio carattere: organico e sociale" (his emphasis).
26. A. Meneghini (1851a, p. 52): "Associazione è quel concorso o di lavoro o di capitali che li rende atti ad ottenere una maggiore utilità, od a produrre quella, che non avrebbero raggiunta impiegati isolatamente."
27. A. Meneghini (1851a, p. 253): "Altro scopo della società si è quello dell'assistenza agli infortunii, in quanto la carità privata riesca insufficiente al loro sollievo. [...] l'assistenza pubblica può dirsi obbligatoria per la solidarietà che si stabilisce tra i membri del corpo sociale."
28. Comte (1839, p. 325): "... soit toujours exclusivement conçu comme relatif à tous les autres, avec lesquels une *solidarité fondamental.*"
29. A. Meneghini (1851a, p. 253): "Secondo scopo dell'associazione sociale è il perfezionamento morale ed intellettuale della nostra specie, che deve ubbidire a quella legge di perfettibilità e di progresso che ci distingue dagli altri esseri, e che i governi promuovono specialmente coll'educazione e coll'istruzione pubblica."
30. The complete translation of the *Sistema del diritto romano attuale* was produced in the 1880s by Antonio Scialoja's son Vittorio (1856–1933).
31. However, after 1849 the Austrian government created two chairs for history of law in Padua and Pavia, hoping that these would function as a bulwark of tradition against the supporters of natural law, which was considered to be excessively liberal.
32. Fortis' mother, Elena Wollemborg, hosted a famous literary salon in her Paduan house, which attracted famous poets and intellectuals, among them many young insurrectionists from her native Trieste (see Luzzatto 1966).
33. The great success of the German historical school in the late nineteenth-century juridical debate in Italy may in fact also be attributable to the circumstance that by that time, terms like 'organ' and 'organization', which are so distinctive for the early period, had definitely entered the political and juridical vocabulary, yet not as stimulating organic metaphors and analogies, but as specific juridical terms independent from any biological meaning (see Dohrn-van Rossum and Böckenförde 1978, p. 614–621).
34. Interestingly, among those who contributed to the parliamentary discussion, many, even allies of Depretis' move, criticized the term transformism,

preferring ‘cooperation’, ‘combination’ (*connubio*), or, like the Venetian politician and naturalist Paolo Lioy (1834–1911), ‘palingenesis’. Giovanni Mussi (1835–1887) even declared to prefer ‘evolution’, because it implied a smoother and less active process (see Anon. 1883). Depretis’ main supporter from the right-wing parties was Marco Minghetti.

35. A. Meneghini (1851a, p. 15): “… il più vero e soddisfacente piacere di un essere creato da Dio eminentemente sociabile. Ringraziamone la Provvidenza, e cerchiamo anche nella scienza il modo di aumentare la somma e l’intensità di piaceri così santi, eccitando coll’educazione lo sviluppo delle facoltà e dei sentimenti”.

References

Albertoni, Ettore A. 1979. *La vita degli stati e l’incivilimento dei popoli nel pensiero di Gian Domenico Romagnosi, con la cronologia degli scritti e delle edizioni italiane ed il testo integrale del libro I Della Vita degli Stati*. Milano: A. Giuffrè.

Ambrosetti, Giovanni. 1970. Rosmini e il romanticismo politico e giuridico austriaci. In *Rosmini e il rosminianesimo nel Veneto*, ed. Giovanni Ambrosetti, 87–111. Verona: Casa Editrice Mazziana.

Anon. 1847. Casse di risparmio. *Il Tornaconto* 3 (21 January 1847): 24.

———. 1883. Tornata di sabato 19 maggio 1883. Atti parlamentari, Camera dei Deputati, Legislatura XV, 1ª Sessione, discussioni, 3226–3231. https://storia.camera.it/regno/lavori/leg15/sed118.pdf. Accessed 10 April 2019.

———. n.d. *Pierre-Joseph Proudhon*. In Encyclopédie Larousse, https://www.larousse.fr/encyclopedie/personnage/Pierre_Joseph_Proudhon/139696. Accessed 14 July 2020.

Augello, Massimo M., and Marco E.L. Guidi, eds. 2007a. *L’economia divulgata: Stili e percorsi italiani (1840–1922)*. Milano: Franco Angeli.

———. 2007b. I manuali di economia politica nell’Italia liberale: Primo bilancio di una ricerca. In *L’economia divulgata: Stili e percorsi italiani (1840–1922)*, ed. Massimo M. Augello and Marco E.L. Guidi, vol. I, vii–xxix. Milano: Franco Angeli.

Beiser, Frederick C. 2011. *The German historicist tradition*. Oxford: Oxford University Press.

van Beneden, Pierre-Joseph. 1875. *Les Commensaux et les Parasites*. Paris: Bibliotheque Scientifique Internationale.

———. 1876. *Animal parasites and messmates*. New York: D. Appleton & Company.

Benvenuti, Bartolomeo, Andrea Meneghini, and Domenico Berti. 1852. *Manuale del Cittadino degli Stati Sardi*. Edizione a beneficio dell’emigrazione. Torino: Tiprografia Economica.

Bertani, Corrado. 2015. Italienische Übersetzungen von Savignys Schriften im 19. Jahrhundert: Bestand, Charakteristika, Schlüsselworte. In *Savigny International?* ed. Joachim Rückert and Thomas Duve, 217–250. Frankfurt a. M.: Klostermann Verlag.

Blume, D. Friedrich. 1824. *Iter Italicum*. Vol. 1. Berlin and Stettin: Nicolaische Buchhandlung.

Boccardo, Gerolamo. 1853. *Trattato teorico-pratico di economia politica*. 3 vols. Turin: Società editrice della Biblioteca dei comuni italiani.

Bof, Frediano. 2018. Credito e risparmio nel Veneto dalla dominazione austriaca al Regno d'Italia. In *Il Veneto nel Risorgimento: Dall'Impero asburgico al Regno d'Italia*, ed. Filiberto Agostini, 217–234. Milano: Franco Angeli.

Boucher, Douglas H. 1985. The idea of mutualism, past and future. In *The biology of mutualism: ecology and evolution*, ed. Douglas H. Boucher, 1–28. Oxford: Oxford University Press.

Bronstein, Judith L. 2015. The study of mutualism. In *Mutualism*, ed. Judith L. Bronstein, 3–19. Oxford: Oxford University Press.

Cafarelli, Andrea. 2009. Angelo Messedaglia. In *Dizionario Biografico degli Italiani*, vol. 73, 779–784. Roma: Treccani.

Camurri, Renato. 2002. *Censimento storico delle società di mutuo soccorso del Veneto*. 2 vols. Verona: Regione del Veneto.

———. 2004. Il mutualismo e la diffusione dello 'spirito d'associazione' dopo l'Unità. In *Spazi laici: Strutture e reti associative tra Ottocento e Novecento*, eds. Renato Camurri and Marco Fincardi, 23–54. Venetica: Rivista di storia contemporanea 17: 1–221.

———. 2006. La 'seconda società': L'associazionismo borghese nel Lombardo-Veneto (1848–1866). In *Memoria, rappresentazioni e protagonisti del 1848 italiano*, ed. Renato Camurri, 249–276. Verona: Cierre edizioni.

Catellani, Enrico. 1934–1935. Gian Domenico Romagnosi. *Atti del Reale Istituto Veneto di Scienze, Lettere ed Arti* 94 (2): 467–513.

Cazzetta, Giovanni, ed. 2013. *Retoriche di giuristi e costruzione dell'identità nazionale*. Bologna: Il Mulino.

Cheung, Tobias. 2009. Der Baum im Baum: Modellkörper, reproduktive Systeme und die Differenz zwischen Lebendigem und Unlebendigem bei Kant und Bonnet. In *Kants Philosophie der Natur: Ihre Entwicklung im 'Opus postumum' und ihre Wirkung*, ed. Ernst-Otto Onnasch, 25–51. Berlin: De Gruyter.

Ciech, Edoardo, and Riccardo Soliani. 2017. Economics and 'civilization' in Gian Domenico Romagnosi. In *Economic thought and institutional change in France and Italy, 1789–1914: a comparative study*, ed. Riccardo Soliani, 3–36. Cham: Springer Nature.

Clark, Charles M.A. 1992. *Economic theory and natural philosophy: the search for the natural laws of the economy*. Aldershot: Edward Elgar.

Comte, Auguste. 1839. *La philosophie sociale et les conclusions générales*. Paris: Bachelier.

———. 1844. Discours sur l'esprit positif. In *Traité philosophique d'astronomie populaire*, 1–108. Paris: Carillan-Goeury et Vor Dalmont.

Cremaschi, Sergio. 2002. Metaphors in *The wealth of nations*. In *Is there progress in economics?* ed. Stephan Boehm, Christian Gierke, Heinz Kurz, and Richard Sturm, 89–114. Cheltenham: Edward Elgar.

———. 2009. Newtonian physics, experimental moral philosophy and the shaping of political economy. In *Open economics: economics in relation to other disciplines*, ed. Richard Arena, Sheila Dow, and Matthias Klaes, 73–94. London and New York: Routledge.

———. 2017. Early modern political philosophies and the shaping of political economy. In *Routledge Historical Resources. History of Economic Thought*, ed. Heinz Kurz, https://www.routledgehistoricalresources.com/economic-thought/essays/early-modern-political-philosophies-and-the-shaping-of-political-economy. Accessed 13 January 2020.

———. 2020. The Italian Enlightenment and the rehabilitation of moral and political philosophy. *The European Legacy* 25: 743–759. https://doi.org/10.1080/10848770.2020.1758411.

Dale, Gareth. 2018. 'Our world was made by nature': constructions of spontaneous order. *Globalizations*. https://doi.org/10.1080/14747731.2018.1498172.

Daston, Lorrain, and Gianna Pomata, eds. 2003. *The faces of Nature in Enlightenment Europe*. Berlin: Berliner Wissenschaftsverlag.

De Giorgi, Alessandro. 1863. *La filosofia del diritto e la scuola storica*. Padova: Tipi del Seminario.

De Luca, Placido. 1852. *Principii elementari della scienza economica*. Napoli: Pellizone.

De Mattei, Rodolfo. 1940. *Dal trasformismo al socialismo*. Firenze: Sansoni.

De Toni, Giovanni Battista. 1909. Una lettera inedita del botanico padovano Giuseppe Meneghini. *Bollettino del Museo civico di Padova* 12: 13–16.

Demko, Denis. 2002. *Léon Bourgeois: Philosophe de la solidarité*. Paris: Éditions Maçonniques de France.

Dohrn-van Rossum, Gerhard, and Ernst-Wolfgang Böckenförde. 1978. Organ, Organismus, Organisation, politischer Körper. In *Geschichtliche Grundbegriffe: Lexikon zur politisch-sozialen Sprache in Deutschland*, ed. Otto Brunner, Werner Conze, and Reinhart Koselleck, vol. 4, 519–622. Stuttgart: Klett-Cotta.

Fanfani, Tommaso. 2005. Solidarietà e cooperazione nel pensiero di Giuseppe Toniolo: credito e capitale. In *Giuseppe Toniolo: Il pensiero e l'opera*, ed. Romano Molesti, 95–104. Milano: Franco Angeli.

Faucci, Riccardo. 2000. *L'economia politica in Italia: Dal Cinquecento ai nostri giorni*. Torino: UTET.

Faucci, Riccardo, Giovanni Pavanelli, and Marco E.L. Guidi. 2007. Sulle opere di Antonio Scialoja: un dibattito. *Il Pensiero Economico Italiano* 15 (2): 117–129.

Fong, Peter. 1988. Ethics, politics and sociology as Newtonian sciences. In *Newton's scientific and philosophical legacy*, ed. Paul B. Scheurer and G. Debrock, 343–354. Dordrecht: Kluwer Academic Publishers.

Freeden, Michael, and Javier Fernandez-Sebastián. 2019. European liberal discourses: conceptual affinities and disparities. In *In search of European liberalisms: concepts, languages, ideologies*, ed. Michael Freeden, Javier Fernandez-Sebastián, and Jörn Leonhard, 1–35. New York: Berghahn.

Furfaro, Federica. 2016. *Recezione e traduzione della Pandettistica in Italia tra Otto e Novecento: Le note italiane al "Lehrbuch des Pandektenrechts" di B. Windscheid*. Torino: Giappichelli.

Ghiringelli, Robertino. 2012. Gian Domenico Romagnosi. In *Enciclopedia Italiana di Scienze, Lettere ed Arti. Ottava appendice: Il contributo italiano alla storia del pensiero—Economia*, 402–406. Roma: Treccani.

Gioli, Gabriella. 1989. *Il pensiero economico di Antonio Scialoja*. Pisa: Pacini.

———. 2006. Introduzione. In *Antonio Scialoja, Opere. Vol.1: I Principi della economia sociale esposti in ordine cronologico*, ed. Gabriella Gioli, xix–lxxviii. Milano: Franco Angeli.

Gioli, Gabriella, and Antonio Magliulo. 2007. La manualistica di Antonio Scialoja. In *L'economia divulgata: Stili e percorsi italiani (1840–1922)*, ed. Massimo M. Augello and Marco E.L. Guidi, 1–32. Milano: Franco Angeli.

Girardi, Giacomo. 2017–2018. *I beni degli esuli: I sequestri austriaci in Veneto tra controllo politico e prassi burocratica (1848–1861)* (Dissertation thesis). Milano: Università degli Studi di Milano. Dipartimento di Studi storici.

Goeth, Sarah. 2017. Fortpflanzung—Verpflanzung: Vegetabile Metaphern in den Wissenskulturen des 18. und 19. Jahrhunderts. In *Metaphorologien der Exploration und Dynamik (1800/1900)*, ed. Gunhild Berg, Martina King, and Reto Rössler, 61–82. Hamburg: Felix Meiner Verlag.

Goggans, Phillip. 2004. Political freedom and organic theories of states. *Journal of Value Inquiry* 38: 531–543.

Gregorio, Massimiliano. 2012. *Le dottrine costituzionali del partito politico: L'Italia liberale*. Firenze: Firenze University Press.

Guillo, Dominique, and Amy Jacobs. 2002. Biology-inspired sociology of the nineteenth century: a science of social 'organization'. *Revue française de sociologie* 43: 123–155.

Hegel, G.G. Federico. 1840. *Filosofia della storia, compilata dal dott. Edoardo Gans e tradotta dal Tedesco da G. B. Passerini*. Capolago: Tipografia e libreria elvetica.

———. 1848. *Filosofia del diritto, tradotta dall'originale tedesco da Antonio Turchiarulo*. Napoli: Stamperia e Cartiere del Fibreno.

Isabella, Maurizio. 2005. Riformismo settecentesco e Risorgimento: L'opera di Pietro Verri e il pensiero economico italiano della prima metà dell'Ottocento. *Il Pensiero Economico italiano* 13 (1): 31–50.

Isenburg, Teresa. 1977. Il dibattito su Malthus e sulla popolazione nell'Italia di fine '800. *Studi storici* 18 (3): 41–67.

Jacob, Margaret C. 1977. Newtonianism and the origins of the Enlightenment: a reassessment. *Eighteenth-Century Studies* 11 (1): 1–25.

Janssens, Marcel. 1963. Das Bild der Pflanze und der Organismusgedanke im Schrifttum des jungen Herder. *Jahrbuch des Wiener Goethe-Vereins* 67: 30–39.

———. 1964. L'image de la fleur dans les écrits du jeune Joh. Gottfr. Herder. *Revue belge de philologie et d'histoire* 42 (3): 913–925.

Kant, Immanuel. 2000. In *Critique of the power of judgement*, ed. Paul Guyer. Cambridge: Cambridge University Press.

Koerner, Lisbet. 1999. *Linnaeus: nature and nation*. Cambridge, MA: Harvard University Press.

Lanchester, Fulco. 2012. Romagnosi costituzionalista. *Giornale di Storia costituzionale* 23: 77–97.

Lentini, Orlando. 1980–1981. Organicismo e azione sociale da Ardigò a Pareto. *Quaderni di Sociologia* 30 (2): 192–215.

Leonhard, Jörn. 2004. From European liberalism to the language of liberalisms: the semantics of liberalism in European comparison. *Redescriptions. Yearbook of Political Thought and Conceptual History* 8: 17–51.

Luzzatto, G.L. 1966. Un salotto ebraico a Padova nel secolo scorso. *La Rassegna Mensile di Israel* 32 (1): 38–41.

MacDonald, J. Marc. 2017. Malthus and the philanthropists, 1764–1859: the cultural circulation of political economy, botany, and natural knowledge. *Social Sciences* 6 (4): 1–33.

Magliulo, Antonio. 2006. Introduzione. In *Antonio Scialoja, Opere, vol. 2: Trattato elementare di economia sociale*, ed. Antonio Magliulo, 5–6. Milano: Franco Angeli.

Majorana Calatabiano, Salvatore. 1847. *Ricchezza e miseria ossia nuovo trattato di economia politica*. Catania: Tip. del Real ospizio di beneficenza.

Marangon, Paolo. 2003–2004. Fogazzaro e l'evoluzionismo. *Atti e Memorie dell'Accademia Galileiana di Scienze, Lettere ed Arti, parte III: Memorie della Classe di Scienze Morali, Lettere ed Arti* 116: 53–66.

Mazzarella, Ferdinando. 2012. Darwinismo, storicismo, socialità: 'La nuova tendenza' di Giuseppe Vadalà-Papale. *Quaderni Fiorentini per la Storia del Pensiero Giuridico Moderno* 41: 583–626.

Mazzolini, Renato G. 1985. Adam Gottlob Schirach's experiments on bees. In *The light of nature*, ed. J. North and J.J. Roche, 67–82. Dordrecht: Martinus Nijhoff.

Meneghini, Andrea. 1851a. *Elementi di economia sociale ad uso del popolo*. Torino: Cugini Pompa e comp.

———. 1851b. Sulla cassa di risparmio. In *Scritti raccolti e pubblicati dalla Società d'Incoraggiamento per la provincia di Padova*, vol. 1, 290–301. Padova: Tip. Sicca.

———. 1859. Imposte ed estorsioni austriache nella Venezia. *Rivista contemporanea* 18: 318–356.

———. 1860. *Le finanze austriache*. Torino: Unione tipografico-editrice.

———. 1863. *Le imposte nella Venezia e nella Lombardia*. Torino: Unione tipografico-editrice.

———. 1865. *Sulla condizione finanziaria delle province italiane tuttora soggette all'Austria*. Torino: Stamperia dell'Unione tipografica editrice.

Meriggi, Marco. 1987. Il Regno Lombardo-Veneto. In *Storia d'Italia*, ed. Giuseppe Galasso, 18(2). Torino: UTET.

Messedaglia, Angelo. 1877. *La scienza statistica della popolazione: Prelezione al corso di Statistica presso la Regia Università di Roma il 12 dicembre 1877*. Torino and Roma and Firenze: Loescher.

Monteleone, Giulio. 1969. La carestia del 1816–1817 nelle province venete. *Archivio Veneto* 86: 23–86.

Moravia, Sergio. 1974. Introduzione. In *Giandomenico Romagnosi, Scritti filosofici, vol. I: Gnoseologia, psicologia, morale*, ed. Sergio Moravia, 2–52. Milano: Casa Editrice Ceschina.

Moscati, Laura. 1990. Savigny in Italia: Sulla fase iniziale della recezione. *Panorami* 2: 55–89.

———. 1997. L'interpretazione della *Geschichte* di Savigny nella scienza giuridica preunitaria. In *Nozione, formazione e interpretazione del diritto dall'età romana alle esperienze moderne: ricerche dedicate al Professor Filippo Gallo*, vol. III, 249–265. Napoli: Jovene editore.

Müller-Wille, Staffan. 2003. Nature as a marketplace: the political economy of Linnaean botany. *History of Political Economy* 35: 154–172.

Nassar, Dalia. 2016. Analogical reflection as a source for the science of life: Kant and the possibility of the biological sciences. *Studies in History and Philosophy of Science* 30: 1–10.

Novello, Elisabetta. 2020. *Giampaolo Tolomei: Giurista, economista, politico e illustre accademico dell'Università di Padova*. Padova: Cluep.

Osborne, Michael A. 2017. Parasitology, zoology, and society in France, ca. 1880–1920. In *Biological individuality: integrating scientific, philosophical, and historical perspectives*, ed. Scott Lidgard and Lynn Nyhart, 206–224. Chicago: The University of Chicago Press.

Padovan, Dario. 1998. Organicismo e neo-organicismo nelle scienze sociali fra le due guerra: Il contributo di Corrado Gini. *Sociologia* 1: 65–94.

Paillottet, Prosper. 1853. Eléments d'Économie social à l'usage du peuple, par M. Andrea Meneghini. *Journal des Économistes* 34: 464–468.

Palazzolo, Maria Iolanda. 1997. Geografia e dinamica degli insediamenti editoriali. In *Storia dell'Editoria nell'Italia contemporanea*, ed. Gabriele Turi, 11–54. Firenze: Giunti.

Palma, Luigi. 1877. *Corso di diritto costituzionale.* 3 vols. Firenze: Stabilimento Pellas.

Pertile, Antonio. 2011. Commemorazione del membro effettivo prof. Luigi Bellavite (1821–1885). In *Commemorazioni dei soci effettivi, 1843–2010*, ed. Michela Marangoni, vol. 1, 571–579. Venezia: Istituto Veneto di Scienze, Lettere ed Arti.

Petrarco, Silvana. 1996. *Numbers and nationhood: writing statistics in nineteenth-century Italy.* Cambridge: Cambridge University Press.

Petty, William. 1899 [1690]. *Political Arithmetick.* In *The economic writings of Sir William Petty*, ed. Charles Henry Hull, vol. 1, 233–313. Cambridge: Cambridge University Press.

Poggi, Francesco. 2005. Concezione organica della società e sviluppo economico in Giuseppe Toniolo. In *Giuseppe Toniolo: Il pensiero e l'opera*, ed. Romano Molesti, 187–198. Milano: Franco Angeli.

Poli, Baldassarre. 1827. *Saggio filosofico sopra la scuola de' moderni filosofi naturalisti coll'analisi dell'organologia, della craniologia, della fisiognomia, della psicologia comparata e con una teorica delle idee e dei sentimenti.* Milan: Francesco Sonzogno.

Reynolds, Andrew. 1999. What is historicism? *International Studies in the Philosophy of Science* 13 (3): 275–287.

Romagnosi, Gian Domenico. 1805. *Introduzione allo studio del diritto pubblico universale*, 2 vols. Parma: Dalla Stamperia Imperiae.

———. 1815. *Della costituzione di una monarchia nazionale rappresentativa*, 2 vols. Filadelfia (Lugano): s.n.

Romagnosi, Giandomenico. 1822. *Dell'insegnamento primitivo delle matematiche.* 2 vols., Milano: Giovanni Pirotta.

Romagnosi, Gian Domenico. 1832. *Dell'indole e dei fattori dell'incivilimento con esempio del suo risorgimento in Italia.* Milano: Società degli Editori degli Annali Universali delle Scienze e dell'Industria.

———. 1841–1848. *Opere,* riordinate ed illustrate da Alessandro De Giorgi, 8 volumes (15 books). Milano: Perelli e Mariani Editori; Padova: Tipi di Angelo Sicca.

Romani, Roberto. 1988. Romagnosi, Messedaglia, la 'Scuola lombardo-veneta': la costruzione di un sapere sociale. *Quaderni di storia dell'economia politica* 2: 105–145.

———. 1992. Policy issues and patriotism in Italian economic thought, 1815–1861. *Journal of European Economic History* 21 (2): 251–280.

Rosmini-Serbati, Antonio. 1836. *Il rinnovamento della filosofia in Italia (Opere, vol. 5).* Napoli: Caro Batelli e Comp.

———. 1837. La società ed il suo fine. In *Della sommaria cagione per la quale stanno o rovinano le umane società*. Milano: Tipografia e Libreria Pogliani.

———. 1839. *La società ed il suo fine*. Milano: Tipografia e Libreria Pogliani.

———. 1994. *Society and its purpose* (*Philosophy of politics*, vol. 2). Trans. Denis Cleary and Terence Watson. Durham: Rosmini House.

Rossi, Paolo. 1977. Introduzione. In *Antonio Fogazzaro, Ascensioni umane: Teoria dell'evoluzione e filosofia cristiana*, ed. Paolo Rossi, 7–44. Milano: Longanesi.

Rückert, Joachim, and Thomas Duve, eds. 2015. *Savigny international?* Frankfurt a. M.: Klostermann Verlag.

Sabbatucci, Giovanni. 1998. Trasformismo. In *Enciclopedia delle scienze sociali*. Roma: Treccani, http://www.treccani.it/enciclopedia/trasformismo_(Enciclopedia-delle-scienze-sociali)/. Accessed 20 October 2019.

Santinello, Giovanni. 1990. La presenza del pensiero di Kant a Padova e Baldassare Poli. In *Kant e la finalità della natura: A duecento anni dalla Critica del giudizio*, 189–214. Padova: Accademia Patavina di Scienze, Lettere ed Arti.

Savigny, Federigo Carlo di. 1839. *Il diritto del possesso: Trattato civile*, tradotto in Italiano dall'Abb. Pietro Conticini. Firenze: Tipografia Pezzati.

Savigny, Federigo Carlo di. 1853. *Prefazione alla sua opera Sistema del diritto romano attuale, tradotta dal dott. Leone Fortis e preposta alla versione italiana di quell'opera, fatta dal dott. Paride Zajotti*. Venezia: Co' tipi della Gazzetta Uffiziale.

Schabas, Margaret. 2005. *The natural origins of economics*. Chicago: The University of Chicago Press.

Schick, Edgar B. 1971. *Metaphorical organicism in Herder's early works: a study of the relation of Herder's literary idiom to his world-view*. The Hague and Paris: Mouton.

Scialoja, Antonio. 1840. *I Principi della economia sociale esposti in ordine ideologico*. Napoli: Tip. G. Palma.

———. 1846. *I Principi della economia sociale esposti in ordine ideologico*. Seconda edizione riveduta, corretta e aumentata. Torino: G. Pomba.

———. 1848. *Trattato elementare di economia sociale*. Napoli: G. Pomba.

Sheehan, Jonathan, and Dror Wahrman. 2015. *Invisible hands: self-organization and the eighteenth century*. Chicago: University of Chicago Press.

Simone, Giulia. 2012. *Il Guardasigilli del regime: L'itinerario politico e culturale di Alfredo Rocco*. Milano: Franco Angeli.

Simonis, Annette. 2015. Zur Metaphorik des Kulturtransfers und Kulturaustauschs um 1800: 'Verpflanzungen' bei Herder, Goethe, Schiller und Co. *Comparatio. Zeitschrift für Vergleichende Literaturwissenschaft* 7 (2): 237–261.

Sofka, James R. 1998. Metternich's theory of European order: a political agenda for 'perpetual peace'. *The Review of Politics* 60 (1): 115–149.

Stringari, Sandro, and Robert R. Wilson. 2000. Romagnosi and the discovery of electromagnetism. *Rendiconti Lincei* 11 (2): 115–136.

Tennemann, Wilhelm Gottlieb. 1832–1836. *Manuale della storia della filosofia*, tradotto da Francesco Longhena, con note e supplementi dei professori Giandomenico Romagnosi e Baldassare Poli. 3 volumes, Milano: Fontana.

Thienemann, August. 1909. Die Stufenfolge der Dinge, der Versuch eines natürlichen Systems der Naturkörper aus dem achtzehnten Jahrhundert. *Zoologische Annalen* 3: 185–274.

Thomas, Marion. 2017. Biological and social new orders: Charles Robin (1821–1885) and Comteanism in the Third Republic. *Historical Studies in the Natural Sciences* 47 (5): 653–677.

Trincanato, Pietro Giovanni. 2017. "Umiliare il ricco refrattario, proteggere il cittadino fedele": I sequestri austriaci a Venezia tra leggenda nera e prassi burocratica. *Mélanges de l'École française de Rome* 129 (2): 337–349.

Tusset, Gianfranco. 2009. The Italian contribution to early economic dynamics. *European Journal for the History of Economic Thought* 16 (2): 267–300.

———. 2016. Labour as culture: the Lombardo-Veneto School. In *Economic thought and history: an unresolved relationship*, ed. Monika Poettinger and Gianfranco Tusset, 177–191. London: Routledge.

———. 2018. *From Galileo to modern economics: the Italian origins of econophysics*. New York: Palgrave Macmillan.

Wolfe, Charles T., and Sylvie Kleiman-Lafon. 2021. Unsystematic vitality: From early modern bee swarms to contemporary swarm intelligence. In *Active materials*, ed. Peter Fratzl, Michael Friedman, Karin Krauthausen, and Wolfgang Schäffner. Berlin: De Gruyter.

Woodcock, George. 1993. *Pierre-Joseph Proudhon: a biography*. Montréal: Black Rose Books Ltd.

Woolf, Stuart. 1991. *The poor and how to relieve them: the restoration debate on poverty in Italy and Europe*. In *Society and politics in the age of the Risorgimento: essays in honour of Denis Mack Smith*, ed. John A. Davis and Paul Ginsborg, 49–69. Cambridge: Cambridge University Press.

Zamagni, Stefano. 2010. Catholic social thought, civil economy, and the spirit of capitalism. In *The true wealth of nations: Catholic social thought and economic life*, ed. Daniel K. Finn, 63–94. Oxford: Oxford University Press.

Zammito, John H. 2018. *The gestation of German biology: philosophy and physiology from Stahl to Schelling*. Chicago: The University of Chicago Press.

CHAPTER 7

Progress, Evolution, and Cellular Constitution

7.1 Patterns of Biological Organization

In the eighteenth century, mechanical metaphors of the body were common in biological as well as in socio-economic debates. During the first decades of the nineteenth century, we observe a shift toward organicist-biological views and metaphors, culminating in numerous analogies between the constitution of organisms and of the state in the late century. These analogies enjoyed such popularity that the distinction between social and biological entities was nearly erased during the heyday of naturalism (Nyhart 2017; Weindling 1981). With respect to the aims of this book, it is important to keep in mind that explicit analogies between the organization of (higher) living beings and a human state or society only emerged after 1848. The examples from the pre-1848 period, discussed in Chap. 6, instead reveal general, yet rather superficial acquaintance with contemporary knowledge of organismic constitution and function. Moreover, the simultaneous and similar use of political-anatomical metaphors must not seduce us into thinking that these reflect shared world-views transcending geographical and social boundaries or individual differences. On the contrary, closer examination of particular positions reveals a great diversity of political and scientific beliefs and basic assumptions. Antonello La Vergata has emphasized that metaphors are often

A. Dröscher, *Plants and Politics in Padua During the Age of Revolution, 1820–1848*, Palgrave Studies in the History of Science and Technology, https://doi.org/10.1007/978-3-030-85343-3_7

individual, multifunctional, and connect elements of many thought chains (La Vergata 1995). The question of (conscious or unconscious) analogies is even more complex. Many dimensions and implications of the analogies were at work simultaneously and differently for each scholar. Finally, yet importantly, as Renato Mazzolini (1988, pp. 96–112) has shown in the case of pathologist and politician Rudolf Virchow, analogies and metaphors do not necessarily work in both directions. Virchow's description of the constitution of organisms as states (of cells) (*Zellenstaat*) did not automatically imply that a state should be seen as an organism. On the contrary, for Virchow, the state certainly was not and would have never been an organism, at most a complex of organisms. Also Lynn Nyhart (2017, p. 627) argues for an asymmetry. By the 1870s, the state-as-organism metaphor structured the political discourse about the state, whereas scientists rarely drew upon political writings to justify their theoretical positions on organismal organization. Similarly, in the 1830s, the French botanist and politician François-Vincent Raspail considered the transformation of organisms and of societies side by side, yet drew a clear line between the life of animals and the life of humans (Vienne 2017, pp. 640–641).

Even if we limit our purview to the area of organism-state analogies, there was greater conceptual diversity than one may think. That is unsurprising. First of all, each author bestowed a different level of explanatory power to the bio-political analogy or comparison. Secondly, the level of acquaintance with the most recent findings of anatomical and physiological as well as social and societal inquiry differed. Scholars who championed the close affinity of political-legal models and biological facts tended to be more attentive to the results of recent biological research than the much greater number of authors who admitted only to their usefulness for general and vague inspiration. Yet, according to Weindling, even the development and use of the cell-state metaphor, around 1900, by protagonists of cell research, such as Oskar Hertwig, had poor biological grounds and completely dismissed any empirical microscopic data. Rather, "it was their ideals of the state which became incorporated into biological theory. [...] The theory of the cell state expressed the sense of social responsibility of the German professor to the nation, idealized as a *Kulturstaat*" (Weindling 1981, p. 116). Thirdly, even the most knowledgeable anatomists and physiologists of the time did not know exactly why the parts of an organisms hold and work together with such orderliness. The number of explanations that circulated cannot conceal that none provided a definite answer. The question of the whole and its parts continued to be passionately debated. As late as 1896, the eminent cytologist Edmund B. Wilson reminded biologists that

> There is at present no biological question of greater moment than the means by which the single cell-activities are co-ordinated, and the organic unity of the body maintained; for upon this questions hangs not only the problem of the transmission of acquired characters, and the nature of development, but our conception of life itself. (Wilson 1896, p. 41)

There was a wide range of possible stances between, on the one hand, models of loose aggregation of single parts voluntarily collaborating as long as it benefitted them, and, on the other hand, models of strict hierarchies of parts that mechanically obey commands or sacrifice themselves entirely for the sake of the unit or a superior good.

During the period under consideration in this book, cell theory was still in the making, and certainly had not yet found its way into the public socio-political debate. German political theorists of the first half of the century, and often even later authors, usually referred—consciously or unconsciously—to Kant's organism concept (Sect. 6.4 in Chap. 6). Yet, Kant's ideas about the nature of 'the parts' remained vague. Even when, in the 1830s and 1840s, cell theory had more concrete anatomical and experimental results to offer, the lack of evidence of how cells cooperate made answers to the central question of organic theory, that is, the nature of 'togetherness', necessarily speculative. The discussion about the degree of independence of the parts or their subordination to a higher entity, and about the nature of the tie that holds them together, stood on shaky empirical grounds. Models that sounded plausible to some scholars sounded far-fetched to others, and the reasons for favoring one model over another did not reside in scientific proof alone.

Despite all commonalities between biological and social organization, it is therefore unsurprising to find a preponderance of weak and strong analogies, rather than one or a few worked-out and dominant models. A closer look at the history of organism-state analogies confirms the diversity of conceptions. Herbert Spencer, for instance, mentioned the 'social organism' for the first time in 1843, but in referring to phrenology (Elwick 2003, pp. 48–54 and 58). In 1850, he proposed a vision of bodies as "a commonwealth of monads", thus emphasizing the deliberate choice of the individual parts. Consolidating his ideas ten years later in the essay *The Social Organism* (Spencer 1901 [1860]), he founded a social theory based on the conception of liberally compound, democratic, and anti-dirigiste organisms. In response, Thomas Huxley (1825–1895) championed an authoritarian view—alongside his call to governmental interventionism—in which each part was beholden to a dirigiste authority such as the will or the brain (Elwick 2003, pp. 54–60). The form such conceptions took did not depend on philosophical assumptions alone. The choice of experimental technique and research object could also lead to diverging results. Whereas salps, small planktic sea organisms that alternate between solitary and colonial habit, played a crucial part in Spencer's and in Huxley's conceptions, research on siphonophores fundamentally influenced the German discussion of the second half of the century. In 1853, for zoologist Carl Vogt, the structure and functioning of siphonophores and similar lower organisms, which live in colonies of closely collaborating units, resembled an anarchy; for his colleague Rudolf Leuckart, they lived together "like in a communist state" because "there are here no poor next to the rich, no hungry next to the sated—but also no lazy next to the industrious. Each contributes its own to the sustenance and health of the whole" (quoted from Nyhart 2017, p. 620). However, it was Leuckart's only political comment in his entire scientific corpus. Several decades later, Edmond Perrier, an expert in marine colonial invertebrates, also favored an associative view, calling living beings "Republics of plastids" (Thomas 2017, p. 676).

Cell research and cell theory added a new dimension to the discussion of biological organization. As soon as the cell was proposed as the elementary structural unit of living beings, the question of organization revolved around these nearly invisible entities, regardless of whether one agreed with the postulates of cell theory or not (Dröscher 2002). However, as we shall see in Sect. 7.2, conceptualizing the cellular level was not easy. There was no consensus among cell researchers on how to define individuality, nor on how to explain togetherness. For Rudolf Virchow, who leaned toward defending the value of individuals within compound structures, multicellular living bodies were organized like a republic of cells (Mazzolini 1988; Caianiello 2008), for the more authoritarian Theodor Schwann they were like military hierarchies (Vienne 2017, p. 649), and for Ernst Haeckel the form of bodily organization of vertebrates was a monarchy of cells (Reynolds 2008).

In the 1830s and 1840s, cell theory and the debate on biological constitution were still dominated by botanists. Due to the possession of definite walls, vegetal cells were more easily discerned as individual entities than the softer animal cells. Moreover, plants have a less complex internal structure. Interestingly, it appears that botanists were much less inclined to use social-biological metaphors than zoologists. Nevertheless, their conceptions of biological organization, given the speculative nature of these ideas, can give insights into their general theoretical framework. The analysis of Giuseppe Meneghini's (Fig. 7.1) scientific works therefore entails several tasks. As a pioneer of cytology, his writings illustrate the difficulties of botanists in coming to terms with the general question of organization; given that he was a member of the Paduan culture, it will be interesting to see to what extent his biological conceptions fitted into the socio-political currents of his time; and as brother of Andrea and an activist in Padua's reform movement, it is instructive to investigate to what extent his scientific views were noticed in the broader Paduan debate. Finally, historians of science mostly underestimate the role of religious beliefs in nineteenth-century biological debates, or interpret them as obstructive (except e.g., Vienne 2017). Meneghini's case shows that this was not necessarily the case.

Fig. 7.1 Giuseppe Meneghini (1811–1889), in 1857. Half-length portrait by Francesco Pierucci. By courtesy of Ministero della Cultura Italiano—Biblioteca universitaria di Pisa, B. co. F. V. 1. 23

7.2 Degrees of Perfection

In the early nineteenth century, many Italian naturalists contributed to the microscopic investigation of plant and animal tissue, yet most had not heard about the theoretical implications of the endeavor that revolutionized the biological sciences from the 1830s onwards. Meneghini by contrast, always in close contact with European colleagues, was soon part of the action and made cells an integral part of his studies and general biological thinking. His single most important scientific achievement in this regard came in 1837, with one of the first descriptions of cell division (Dröscher 2017; Meneghini 1837, pp. 342–343, 345, and 348–349; Meneghini 1838, p. 4 and 9), whereas his main contribution to the development of cell theory in Italy was theoretical and social in nature. Considerable high-level research in fine plant anatomy had been undertaken in Italy during the previous decades and centuries. In the 1830s and 1840s, many Italian botanists became embroiled in a veritable explosion of cryptogamic investigations, with centers of activity in Padua, Pavia, Genoa, and Naples. One reason for the primacy of Italian researchers was the availability of Giovanni Battista Amici's (1786–1863) achromatic microscopes, which enabled excellent detailed investigations of microscopic features (Nimis and Hawksworth 1994, p. 14). Meneghini received one of these instruments in 1842 from Archduke Leopold II of Tuscany (1797–1870) as a reward for donating his collection of algae to the Erbario centrale in Florence. However, he had started such observations already in his youth, when his father Agostino bought him a microscope.

Despite the wide interest and technical advancement, few Italian naturalists ventured into broader philosophical debates on organization, individuality, and cooperation, in contrast to their German, British, Belgian, and French counterparts, who engaged in very much livelier theoretical discussions (Lidgard and Nyhart 2017). In particular, during the early years of cell theory, speculation played a fundamental role. The 'cellular revolution' of the doctrine of bodily constitution was in fact not due to technical advancements alone. It also required a conceptual revolution, in particular a developmental approach to anatomy and a rethinking of the relationships between parts and wholes. The latter thrived on a huge grey area between observational data and their possible interpretations. Most

Italian researchers were uncomfortable with speculation. Moreover, in Italy, political-social allusions were potentially dangerous. Centuries of indirect and direct censorship by governments and the Church, self-censorship, as well as the ideal of positive empirical science had turned Italian scholars into masters at hiding their personal opinions behind seemingly neutral descriptions of facts.[1] Even an open-minded scholar like Meneghini rarely laid bare his general vision of organic organization. Nevertheless, his views are indirectly discernable in his writings. In this context, his medical doctoral thesis *De axe cephalo-spinali* is helpful. It was published in 1834 in Latin and translated into Italian in Naples nine years later by the physician and revolutionary Angelo Camillo De Meis (1817–1891) (G. Meneghini 1834; G. Meneghini 1843a).

Meneghini's dissertation considered the anatomy and physiology of the central nervous system of vertebrates, including man. His choice of the topic was not accidental. As we have seen in Chap. 5, his master and mentor Giuseppe Bernardi had based his vision of the history of humankind on his conception of man's mental powers. Meneghini may have also heard or read the essay of Baldassarre Poli, professor of theoretical philosophy in Padua between 1837 and 1852, who invited his students to investigate the history of philosophy combined with a naturalistic-organological approach, yet not as an abuse of analogy, but as an empirical and positive science (Poli 1827, p. 43; Poli 1837). Moreover, neuroanatomy and neurohistology had a long tradition in Italy and, mainly in Leopoldo Marc'Antonio Caldani (1725–1813), an outstanding representative in Padua. Meneghini's points of reference, however, were mainly foreign. He based his anatomical investigations on recent currents that had revolutionized the field of brain research. Setting himself explicitly in the tradition of Albrecht von Haller (1707–1777), Félix Vicq d'Azyr (1746–1794), Antoine Étienne Renaud Augustin Serres (1786–1868), and Friedrich Tiedemann (1781–1861), he imbued his studies with a unitary, physiological, comparative, and developmental framework (G. Meneghini 1834, p. 7).

Meneghini adopted from Vicq d'Azyr not only a new type of neuroanatomy, but also a new way of conceiving anatomy in general. The French physician had been the main innovator in studies of the brain in the late eighteenth century (Mandressi 2018; Parent 2007). He saw himself as heir of Albrecht von Haller and thus further developed the use of experimental techniques and the merging of anatomy and physiology to establish a science of functional anatomy. Years before George Cuvier (1769–1832), he pioneered the transformation of traditional comparative anatomy into a science of organization with a focus on common patterns and 'essential forms'. In the introduction of his *Traité d'Anatomie et de Physiologie* (1786), Vicq d'Azyr stated

> Nature thus appears to follow one general type or model, not only within the structure of the diverse animals, as I have already said, but also within that of their different organs. And one does not know what to admire more, the abundance with which the forms seem varied or the constancy of the sort of uniformity that an attentive eye discovers within the immense extent of its productions. (Vicq d'Azyr 1786, 1, p. 12)[2]

This project included human anatomy. Regarding the hotly debated question of whether living nature, in all its diversity, displays a common basic plan in all creatures or rather divides into several neatly separate groups, Vicq d'Azyr advocated a unitary view. With the same holistic idea in mind he also approached the study of single organisms, conceived of as organized and living bodies (*corps organisés et vivants*) and morphological units. This therefore required a focus on the correlation (*compensation*) of bodily parts rather than separate studies of single organs. In this perspective, diversity did not imply distinctness but specific functional adaptation.

Several decades later, Friedrich Tiedemann introduced a developmental dimension in this area of research (Kruta 2018), then also pursued by Antoine Étienne Renaud Augustin Serres and Johann Friedrich Meckel (1781–1833). At the beginning of the nineteenth century, the belief that living nature follows the general pattern of a *scala naturae* of increasing perfection was still strong. Yet, it had become more and more difficult to establish linear sequences of organisms of growing complexity among the myriad of different species. The consideration of the earliest stages of development opened up new perspectives, because adult organisms could be viewed as products of individual and adaptive variations on basic plans that were visible in their most simple form only when they first appeared in embryos. Tiedemann as well as Serres applied this general approach also to studies of the brain and published two influential treatises in 1816 and in 1824–1827 respectively. In the introduction of his *Anatomie und Bildungsgeschichte des Gehirns*, Tiedemann offered the following synopsis: "the formation of the brain in the [human] embryo and the foetus passes during the months of pregnancy through the major formative stages, whereupon the brain of the animals seems to be stunted throughout their life" (Tiedemann 1816, p. vi).[3] Hence, he advocated a hierarchic recapitulation theory, a view that became increasingly popular in the nineteenth century and that also came to constitute a fundamental element of Meneghini's thinking in his anatomical, botanical, and even geological work. According to Tiedemann, all animals followed a same basic pattern, yet 'inferior' ones arrested their embryonic development at earlier stages, whereas humans, as the highest form of development and complexity, passed through them all. Meneghini incorporated this approach into his doctoral thesis, convinced that "one must follow the evolution of the embryos of each species" (G. Meneghini 1834, p. 7).[4]

A second remarkable feature of Tiedemann's teachings, which influenced the young Meneghini, was his egalitarianism. In spite of his hierarchic conception of nature, and in contrast to a great majority of contemporaries, Tiedemann drew no essential distinction between the varieties of the human species. In his (English) essay *On the Brain of the Negro, Compared With That of the European and the Orang-Outang* (1836), he maintained that his extensive comparative and quantitative studies of human skulls had shown considerable variation regarding size, weight, and internal structure of the brains of European and African humans and apes. Yet, this variability was found in all races, whereas the ranges within the different races, and between men and women, overlapped and were, therefore, fundamentally the same. Moreover, Tiedemann recognized a correlation with body size and age (Gould 1999). Based on these results, he made the clear statement that "the intellectual faculties of the Negroes do not in general seem to be inferior to those of the European and other races" (Tiedemann 1836, p. 524).

Even in his later paleontological work, Meneghini never touched the discussion about human races. We can, however, trace arguments concerning human equality and the dignity of 'lower' organisms in his botanical writings (see below). He might have also learned about these currents through Stefano Gallini. In his dissertation Meneghini cited Gallini's *Discorso sulla superiorità dell'uomo* (1831). Gallini had met Vicq-Azyr, Louis-Jean Marie Daubenton (1716–1800), William Hunter (1718–1783), Joseph Priestley (1733–1804), and many others before becoming, in 1806, professor of Comparative anatomy and physiology, and then of Fine anatomy and physiology in Padua. Gallini however insisted, like other French pioneers of cell theory such as Henri Dutrochet and François-Vincent Raspail, on the need to apply chemistry and physics in the investigation of fine anatomy and physiology, a call that Meneghini did not heed. In fact, during Meneghini's university studies a profound switch from the ideas of the French to those of the German-speaking world had taken place. Whereas in the 1790s many Paduan university professors, in particular those of the scientific disciplines, were, like Gallini, proponents of French ideas, after the 1820s, almost all professors in medical disciplines in Padua had studied in Vienna (Bonuzzi 1999), and the broad arrival and translation of Austrian and German-speaking science had started (Chap. 6). Among these was, in 1824, De Visiani's translation of Nicolò Giuseppe de Jacquin's botanical handbook into Italian (Jacquin 1824), and the translations of agronomist Luigi Configliachi (Sect. 9.1). Meneghini's dynamic and organic conception of life indeed had more in common with the views propounded by German transcendental morphologists (Richards 1992, p. 38).

Meneghini became deeply committed to holistic and developmental ideas, and they remained his guiding principles throughout his scientific career. His research was embedded in debates about hierarchies of organization, teleological transformism, development, and the basic uniformity of life, connecting individuals and species, past and present. In Padua, as early as the 1710s, ideas of a *scala naturae* had been authoritatively advocated by the influential naturalist Antonio Vallisneri, for instance in his *Academic lecture around the order of the progression and the connection which all created things have in common* (Vallisneri 1721). For Vallisneri, even the simplest creature played a fundamental role in the harmony of nature. Yet, whereas Vallisneri was inspired by Leibniz's metaphysical optimism and inevitable progressivism (Generali 2020), Meneghini considered, similar to the classification later proposed by Richard Owen (1804–1892), the level of perfection as corresponding to the level of integration of the parts. Regarding the brain, he opined, that "the nervous system, hence, is nothing else than a manifestation of the most perfect organization, and the most noble animality" (G. Meneghini 1834, p. 15).[5] By analogy to the position of the brain among the various tissues, Meneghini allocated a position to man by virtue of possessing the highest degree of perfection in the hierarchy of species. Man was not equipped with new exclusive organs, but rather "distinguishes himself by the perfection of the development and complexity of the parts to which a long series of other animals has led insensibly and gradually" (ibid., p. 6).[6] Hence, the development of individuals and species was based on the same principles. Both were expressions of one *typus*, common to all vertebrates, both moved from homogeneity to heterogeneity, and both exhibited a continuous progression toward perfection. Comparative embryology was the key to understanding the common structural plan, because the gradations of perfection corresponded to the degree of progression along the developmental path. In his view, "when the inferior beings halt their development, they represent the embryonic state of the superior being" (ibid., p. 6).[7] Therefore, man does not essentially differ from the other vertebrates, but represents a higher level of development. Man's embryonic stages *comprehend* the permanent stages of all inferior beings (ibid., p. 7).

Meneghini's discourse on the collaboration of the parts reveals other aspects of his general conception of living organization. He viewed the nervous system as the most complex and therefore the most evolved of all tissues, yet nowhere did he maintain that it 'governed' the other organs or, as Thomas Huxley (1872, p. 72) envisioned in his dirigiste model in 1873, that it "rules the individual components with a rod of iron [...] the brain, like other despots [...] is above law". Meneghini rather saw the function of the brain as connecting the parts and thus guaranteeing that everything worked smoothly. The brain was a part among other parts. It had acquired its specific form through developmental steps, exactly like the other parts of the organism. Monarchists like Cuvier and later Haeckel attributed to the brain a central directing power. Meneghini instead conceived of its task as intercommunication, maintaining that "Every operation of Nature is tied together with a necessary and indissoluble link" (G. Meneghini 1838, p. 46).[8] His conception was thus closer to the social philosophy of his brother Andrea and to the organic theories of the state (Sect. 6.3 in Chap. 6) that regarded the state (the whole body) to be the supreme entity; whereas the monarch (the brain) was a central and prominent part, but still only an organ whose *raison d'être* derived from acting for the wellbeing of the whole body (Dohrn-van Rossum and Böckenförde 1978, p. 598–600).

When we delve into the particulars of his concept of perfection, we find that Meneghini considered the senses and movement as the two principal elements without which no life was possible (G. Meneghini 1834, pp. 269–270).[9] The greater the perfection of an animal, the more these two elements developed independently, became more complicated, and diversified. Yet, the measure of 'organic progress' was the degree of communication between these elements. Hence, it was a view of a harmonious interplay between parts, an organic form of associationism not monarchical subordination. The same held true for the organization of the nervous system itself. Comparing it in different animal classes, Meneghini focused on the mutual relationship and cooperation of single parts. He regarded structures that exhibited tighter anatomical connection and higher levels of communication as more functional and perfect (G. Meneghini 1834, p. 20). We can find a similar bottom-up understanding of organization in his defense of Charles Gaudichaud-Beaupré's (1789–1854) theory of the merithallus of 1841 (Gaudichaud 1841). In contrast to most of his Italian colleagues, Meneghini supported the Frenchman's view that single internodes, together with the annexed leaves, form repetitive structural units. Even though Gaudichaud was an opponent of cell theory (Tirard 2016), Meneghini shared his developmental approach to organography and emphasized that this meant starting from below, from the single organs, not from the point of view of the entire and adult plant. Even more important, similar to the deductions that Herbert Spencer drew from his study of the repeated segments of earth-worms, insects, and other animals about two decades later (Spencer 1865–67, II, p. 91; on this, see Elwick 2003, pp. 57–58), for Meneghini Gaudichaud's theory pointed to a considerable degree of independence of the single parts of a plant. Conceiving the single organs or merithalli as individuals capable of autonomous growth and variation furthermore meant that all organs "cooperate in the formation of the common body that participates in the organization and in the life of all" (G. Meneghini 1844b, p. 26; G. Meneghini 1844c).

However, Meneghini did not view living bodies to be wholly self-sufficient systems and his conception of life was not entirely materialistic. All interactions were admittedly mechanical, but "physiological research on cerebral functions proves that there exists something non-material within us, a particle of the divine aura (*divinae scilicet aurae particulam*), which elevates us to the highest end" (G. Meneghini 1834, p. 274).[10] This assertion, in the final lines of his doctoral thesis, was one of the very few declarations of his religious beliefs in the entirety of his scientific work. Moderate in his scientific as well as his political views, Meneghini did not put forward a fully secularized science. Although his scientific argumentation was in large parts materialistic and mechanistic, it did restrict itself to the invocation of natural laws, but also relied on a superior guiding principle. He made no further statement whether this 'non-material something' was the soul or something else. Nor did he entertain any hypothesis or embark on any investigation to reveal the interplay between the body and the particle of divine aura.

This indicates that he tended to separate the religious from the scientific sphere, and it suggests that religion continued to be a fundamental aspect of his political as well as scientific thought, even if this was not explicitly stated. The defense of religious values and the opposition to materialistic views in scientific publications were common in Padua, as may be shown by Achille De Zigno's naturalistic work. De Zigno was a descendant of a rich Italian-Irish family that had settled, first, in Venice in 1777, and then in Padua (Lonigo 2013a; Omboni 1897; Pearson 2005). His father bought the villa *La Certosa* at Vigodarzere, a town north of Padua, where his Irish wife, Lady Mary Creagh Maguire, created an English landscape garden (Fig. 1.1, no. 12). She was also the person who awakened and sustained Achille's interest in plants and bestowed him an Amici microscope (Lonigo 2013b, p. 43). The politically conservative De Zigno family maintained numerous bonds with noble families in Veneto and in Austria, and in their salon hosted many notable political and cultural figures from Austria, Britain, and Ireland. After years of travels, in 1833, Achille started his career as administrator of his father's possessions, yet continued to carry on botanical, in particular cryptogamic, studies (De Zigno 1833). From 1838, he started to assume public offices, and from 1846 to 1856, he was *podestà* (mayor) of Padua.

Science was never more than a gentleman's pursuit for him, but he took his studies seriously, was always well informed about current European science and soon acquired an international reputation for his paleontological studies. In his early years, he was receptive to the Paduan cryptogamic enterprise. In a series of microscopic studies, carried out between 1833 and 1835, he set out to refute the idea of spontaneous generation then still prevalent in Italy. His aim, to demonstrate the cellular origin of algae, was therefore not motivated by observational data alone but was for him a plausible alternative to spontaneous generation. For the Catholic De Zigno, explaining a vital process as fundamental as generation as "dependent on physical-chemical laws alone" was an "absurd proposition of poorly exact observers" invaded by the spirit of "superficial philosophers who support the doctrines of materialism" (De Zigno 1839, p. 7 and 14).[11] A few years later, owing to serious problems with his eye-sight, he abandoned microscopic research and switched to geology and paleontology. According to Pietro Corsi, he approached these fields with the same combination of openness toward new theories and religiously motivated hesitation to embrace such theories fully (Corsi 1991, p. 629).

In the wider European context, such an attitude toward science and religion was by no means the exception. On the contrary, there are many contemporary examples of intimate connections between biological and political progressivist and constitutional thinking and the inclusion of religious beliefs into scientific concepts. Atheistic currents were growing during the first half of the nineteenth century, but they were still far from mainstream, even among scientists. In the German landscape of popular science and nature writing, for instance, the radical materialists were known best, but they were actually a minority (Daum 2002, p. 135; Vienne 2017). Vico and Rosmini may have been Giuseppe Meneghini's models, and, like Herder and Hegel, they had offered secular, but not anti-religious visions of science and progress. As was typical in their day, they explicitly attempted to reconcile Enlightenment and scientific philosophy with religion. Yet, this did not imply that science had to be apolitical. On the contrary, Herder was openly committed to liberalism, republicanism, and democracy. Both Meneghini brothers were religious, and Giuseppe probably wanted to forestall the reproach of materialism. We cannot say if he knew that the Catholic Church had placed Raspail's *Nouveau Systéme de Chimie organique* (1833) onto the *Index Librorum Prohibitorum* in July 1834 (Anon. 1841, p. 313; on this see Vienne 2017, p. 637), but he certainly knew that Giuseppe Bernardi was often accused of materialism (Parazzi 1871, p. 813).

A comparison with visions of biological and political constitution proposed shortly thereafter by other researchers reveals similarities as well as differences. Compared to the conceptions of Schwann and Raspail, Meneghini's position appears both vaguer and more moderate. It tallies neither with Herbert Spencer's liberal-individualistic vision of spontaneously interacting parts that come together to form compound and anti-dirigiste organisms, nor with Thomas Huxley's authoritarian view, in which each part is beholden to a dirigiste authority such as the will or brain. It amounted neither to Virchowian praise of individualism nor to Haeckelian centralized subordination. Meneghini's conception was different. He conceived of living beings as assemblies of parts with equal dignity cooperating to create higher structures. Yet, the compound structure was neither a monarchy nor an anarchy. Rather, the whole was guided by a (not further defined) superior and divine aura. In this, Giuseppe Meneghini echoed the convictions of many moderates in Padua. Republicanism, as a term as well as a concept, recalled Mazzini's revolutionary movement as much as the Republic of Venice, and both provoked considerable discomfort among the moderates, whereas the idea of the Pope as the supreme guide of the future state was largely welcomed.

7.3 "The March of Nature Is Always Progressive"

If materialistic ideas threatened to provoke opposition and rebuke, advocating a general evolutionary design, even if this included human beings, seems to have caused few troubles. In fact, in parallel to the historicist-progressivist currents in Padua's political circles, we find a remarkable presence of pre-Darwinian evolutionary ideas during the 1830s and 1840s. The most noteworthy example was Giuseppe Meneghini's dynamic concept of organic organization, whereas De Visiani was a conservative Linnaean and opposed any form of transformism. The situation in Padua confirms recent statements by Pietro Corsi and other historians that a multi-faceted European debate on the transformation of life forms did not begin with Darwin but already commenced at the beginning of the nineteenth century (Corsi 2005, p. 68; Spahn 2015).

Historical accounts of early evolutionary ideas in Italy are still rare (e.g., Pancaldi 1991, esp. pp. 3–38; Forgione 2018). Yet in Veneto, transformist ideas were common. Eighteenth-century scientific discussion about the genesis of the Earth was particularly lively, and plied by eminent and widely known figures like Giovanni Arduino (1714–1795) and Abbot Alberto Fortis (1714–1803). During his stay in Paris, Fortis wrote the two volumes of *Mémoires pour server à l'histoire naturelle de l'Italie* (1802). He defended an adaptationist transformist view of life, according to which the form of species, including man, changed through adaptation to changing environments (Ciancio 1995; Ciancio 2013). Another reknowned transformist some years later was Gian Battista Brocchi (1772–1826). Brocchi had studied in Padua, but then went to Brescia and Milan. His theory about the transformation of Earth moved between Cuvier, Lamarck, and Schelling (Dominici and Eldredge 2010; Pancaldi 1991, pp. 3–39). Based on his studies of local molluscan fossils, Brocchi developed a transmutationist theory that became popular in Britain and was also one of Charles Darwin's starting points when he set out to empirically test the process of transmutation. Brocchi conceived of species as distinct entities that, much like individuals, were born, flourished, and then died out—an idea that resembles Vico's cyclic philosophy of history. Species were, however, connected with 'allied species' through successions of births and deaths forming lineages of ancestry and descent.

In 1796, Brocchi, founder of the Botanical Garden in Brescia, also authored a small treatise on floriculture and gardening for a wedding festivity of the Paduan Papafava family (Brocchi 1796). Moreover, in his famous portrait, Brocchi's sits amidst a landscape park. Indeed, the English garden style may have lent credence to transformist currents more than the static French-style gardens. The romantic desire to become one with nature and the experience of wildly and exuberantly growing life in the landscape parks made evolutionary ideas as well as the recognition of development and metamorphosis as central features of life intuitively more plausible and acceptable.

Giuseppe Meneghini's thinking about historical change was influenced more by German romantic *Naturphilosophie*, than by the philosophies of Vico and Rosmini, who had served as models for Bernardi and his brother Andrea. Unsurprisingly, he chose a quotation from Étienne Geoffroy Saint-Hilaire (1772–1844) rather than from Lamarck for the epigraph of his dissertation thesis: "As historians of how things are, our role is limited to saying that things are as they are" (G. Meneghini 1834, internal cover sheet).[12] The quote was from Geoffroy Saint-Hilaire's *Discours préliminaire* of his *Cours de l'histoire naturelle des mammifères* (Geoffroy Saint-Hilaire 1829, p. 33). In this work, which appeared one year before the famous Cuvier-Geoffroy debate, Geoffroy defended his principle of the unity of organic composition as resulting from accurate observation and having the status of Newtonian law, both results of the Creator's will. More interesting for Meneghini was that Geoffroy was Lamarck's colleague and defender, but he focused on embryological development and comparative anatomy, and that his transcendental transmutationist vision was much closer to that of German morphologists like Goethe (1749–1832).[13] Goethe saw organisms not as aggregates of distinct characteristics, but as organic wholes (see, e.g., Zemplén 2017). The unity-in-plurality of vegetal forms had been one of his major concerns. He translated his intimate conviction of the existence of a fundamental law in nature into research on affinities that ought to do justice to the abundance as well as to the diversity of forms. Goethe's key to understanding nature was metamorphosis. He asserted that "Form is something mobile, something becoming, vanishing. The science of form is the science of transformation. The doctrine of metamorphosis is the key to all signs of nature" (Goethe 1987, p. 349).[14] His visit to Padua's Botanical Garden in September 1786, where he was confronted with a great variety of plants, had been of seminal importance for this concept. His botanical works have raised considerable attention by recent scholarship (e.g., Bies 2015; Holmes 2017; Kelley 2012; Robin 2011).

Meneghini was an outspoken admirer of 'the immortal Göthe' (G. Meneghini 1836, p. 2). In his anatomical and physiological studies, he took up Herder's and Goethe's view that diversity was only apparent because all variations are linked through an underlying and unifying plan. For the Paduan botanist, "the variety of forms and functions [refer] to simple modifications of one single constant type" (ibid., p. 1).[15] The study of the developmental process was indispensable for understanding the pattern of the plant kingdom. Herder, as well as Goethe, advocated a 'genetic' approach (Herder 1784–91), which meant comparatively following the progressive and linear development of single parts from the simple to the complex (Barr Nisbet 1967; Barr Nisbet 1970; De Souza 2018; Spahn 2015, pp. 675–676). In his 1844 popular essay *Metamorphosis of the plants*, Meneghini similarly explained: "one recognizes the evident uniformity of organic composition and a progressive serial coordination that gradually leads from the simplest to the most complicate [...]. This progressive metamorphosis that restarts at any moment and halts at every step yet never recedes" (G. Meneghini 1844d, p. 237).[16] He followed this up with an unambiguously transformist statement, declaring that "the extant animal as well as vegetable forms originate from the antique ones through successive metamorphoses, and, like these, are themselves just transitory and mutable in the progression of time" (G. Meneghini 1844d, p. 237).[17]

If this line of thought was followed through to its genetic consequence, then all life originated in "a first germ that has generated all the vegetable marvels through successive and uncountable modifications" (G. Meneghini 1844d, p. 238).[18] Meneghini thus can be characterized as a pre-Darwinian evolutionist, but for him significance of this view was restricted to philosophical speculation and as such it was "not applicable to the brief moment which forms the object of science" (G. Meneghini 1844d, p. 240).[19] The metamorphological hypothesis seemed plausible and provided a useful framework to handle variety in plant organization and to deduce general laws, yet Meneghini was aware that it required historical reasoning, and historical reasoning was—he underlined—not part of the scientific research into immediate causes. Consequently, during his time in Padua, he made no contribution to ideas about species development. Hence, Meneghini advocated and disseminated evolutionary thinking, but as far as his own scientific investigations were concerned, he preferred topics that were accessible by empirical means.

Nevertheless, the historicist-developmental conception deeply undergirded his botanical work. When Meneghini turned his attention from brain research to botany—a realm of species without brains—his approach to vital phenomena continued to be organicist, developmental, and focused on the interrelations of the parts. His framework was probably further influenced by his stay in Vienna in 1838. In the Habsburg capital, he was in close contact with the botanist Stephan Endlicher (1804–1849), a supporter of Lorenz Oken and later involved in Vienna's 1848 uprising (Riedl-Dorn 2019). For Oken as well as for Schelling, the purpose of natural history was not to distinguish and catalogue species, but to grasp of the common and unifying principles. Some of the most prominent tools for revealing the unity beneath the great diversity that came to the fore in the early nineteenth century were history and development. Endlicher and, even more forcefully, his successor Franz Unger merged Oken's pre-Darwinian evolutionary ideas with cytology (Dröscher 2016; Gliboff 1998; Klemun 2009), an approach that also became fundamental for Meneghini.

In 1842, Unger met Meneghini again when he attended the Fourth Congress of Italian Scientists in Padua. On that occasion, Meneghini's organicist-romantic leanings became evident. He gave a talk about a type of abnormal development—namely, a case of chloranthy or phyllody with antolysis—in the pistil of the Ranunculaceae *Delphinium amoenum* (today better known as *Delphinium laxiflorum*). His explanation of this phenomenon returned to typical romantic assumptions. He asserted that this malformation was explainable by means of the "theory of the successive formation of the elements which constitute all buds" and the "law of organic balancing", and declared that "the partition [is] always symptom of exuberant vegetative energy", and hence of intrinsic organic forces (G. Meneghini 1843c, p. 268).[20] Unlike French naturalists like Geoffroy St.-Hilaire and Raspail, who allowed for the effect of external influences in their views of progressive transformation and put much weight on chemical analyses, Meneghini assumed a slow and gradual innate drive. This is even more astonishing as his brother Andrea ceded considerable importance to the role of external factors like education in the development of the (human) elements of societies (Sect. 6.4 in Chap. 6).

Giuseppe did not deny that culturing and a change of the external conditions could induce noteworthy modifications, but he maintained that these did not affect the normal disposition that characterized the type (G. Meneghini 1844d, p. 238). He felt it was more important to emphasize development as a continuous succession of generations. In a series of essays addressed to the agronomists of the Veneto provinces, he explained that in the same way as a twig developed into a trunk, the twig had developed out of a bud. And the principal branches and the trunk

> originated from a bud enclosed in the seed in the form of an embryo, we notice that the most modest herb, like the most gigantic tree, equally repeats its formation through a long series of successive generations of buds, one deriving from the other, and one developing after the other. This is a consideration of great relevance, because it enables us to consider from the current state of a plant its entire past history. (G. Meneghini 1847, p. 11)[21]

Be this as it may, Meneghini undertook no further investigations into the exact mechanisms underlying transformation or metamorphosis. There is no evidence that he subjected the phenomenon to any systematic experimental inquiry. Like for most romanticists, the 'law of development' was an ideal concept for him, not a real working hypothesis. Moreover, possibly he preferred not to venture too deeply into a question that touched on the grounds of religious belief.

Between 1836 and 1848, Giuseppe investigated the internal organization of plants and algae. Along with his historical, hierarchical, and progressivist vision, and in the wake of Endlicher and Unger's approach, a further element became fundamental: the cellular approach. The first time Meneghini referred to cells was in 1836, that is, a few years before Schleiden and Schwann's formulation of cell theory, but at the same time he did not mention cells in a prominent manner. Following Goethe's idea, in order to trace the great variety of plant forms and functions back to one single type, he compared the stem structure of fifteen different genera of monocotyledons (G. Meneghini 1836, p. 1). When he received kind words about his work from Mohl, Treviranus, de Candolle, and other eminent and like-minded botanists (Canavari 1889, p. 13), he began to focus on mosses (G. Meneghini 1837) and finally algae. As we have seen in Sect. 5.1 in Chap. 5, Meneghini had been attracted to these 'inferior' plants since his youth and considered them God's small marvels. First and foremost, however, he believed that the study of algae offered the best means to approach the main biological questions of his time, in particular the theories of panspermia, spontaneous generation, organization, and the metamorphosis of species. For Meneghini, all these phenomena manifested more conspicuously in these simplest of organisms, yet were not essentially different from those of higher organisms, plants as well as animals (G. Meneghini 1838, p. 47).

In algae, all of these inquiries led to cells. Meneghini was not as fine an experimentalist like Lazzaro Spallanzani (1729–1799), Bonaventura Corti (1729–1813), Felice Fontana (1730–1805), or others who had studied the vital phenomena in algae before him. Meneghini's approach was more botanical-floristic, but he saw anatomy and physiology as fundamental for revealing the affinities in and the unity of living nature. His first phytological publication, *Conspectus algologiae Euganeae*, summarized the following year in the essay *Notes on the organography and physiology of algae*, was one of the first to provide a description of cell division (G. Meneghini 1837, pp. 342–343, 345, and 348–349). The throttling of infusoria and segments of algae and the appearance of diaphragms followed by division into two portions (or four, in the case of pollen) had been observed since the second half of the eighteenth century (Ratcliff 2016). However, the phenomenon had not received much attention. The few scholars who, like Bonaventura Corti, had paid attention to the process, did not consider it a research topic of its own, but rather an argument in the broader epigenesis-preformationism-debate or concerning the question about the border between life and non-life (Dröscher 2022; Monti 2010, pp. xii–xvi). By the late 1830s, the theoretical framework had changed considerably and the division was identified as a cellular process and considered in its developmental dimension. Yet Meneghini, like his European botanist peers, did not regard division as the only type of cell reproduction (Dröscher 2017). In the cyanobacteria *Nostoc* and *Microcystis* and in the chlorophyte *Haematococcus*, he noticed an 'evolution', that is, an endogenous formation of cells inside other cells. In *Phytoconis* (today *Lichenomphalia*), he observed the *ex novo* generation of cells in mucus, and in the green alga *Chlorococcum* and the filamentous diatom *Oscillaria*, he described division, a throttling of one cell into two (G. Meneghini 1838, p. 4 and 9).

Meneghini was not a cell researcher in the strict sense. He was not interested in the cell itself, but rather developed his cellular perspective because it enabled him to conceive of a unified concept of the metamorphosis of plant species, and to trace the process of organic progress back to its first stages. Returning to his thoughts on animal evolution, as laid out in his dissertation, for him, "all plants, howsoever many there may be, can be reduced to a small vegetable vesicle, and the various phases of its development may encompass in a few expressions the 60,000 species counted by botanists" (G. Meneghini 1838, p. 44).[22] In 1840 in a German botanical journal, he rejected Christian Gottfried Ehrenberg's (1795–1876) morphological criterion for classifying the species of the green algae *Desmidiaceae* and instead proposed to use the shape of its cells (G. Meneghini 1840). His cellular approach thus conflated anatomy, taxonomy, and individual and species development. In 1846, in his work on diatoms, translated into English in 1853 by the Ray Society, Meneghini explained how his cellular recapitulation theory could unveil species relationships:

> Science is now put in possession of a most important truth,—that within the superior organic type, as well in the vegetable as in the animal kingdom, there is included, so to speak, in a summary manner, the history of the lower, which present in a permanent form their various intermediate states: that the same histological and morphological facts which appear manifest in the more simple organisms, are repeated in the more complicated; that the primitive organic structure is very similar in the two kingdoms; in short, that in the first instance every plant, every animal, and every tissue in the one or the other, proceeds solely from cells. And since the primitive state, which in superior beings is only transitory, remains permanent in the inferior, we have thus, as well in plants as in animals, very simple beings, reduced indeed to the simplicity of a single cell. (G. Meneghini 1853, 352–353)

Meneghini thus transported Tiedemann's developmental doctrine to the plant kingdom and adapted it to the cellular dimension. As set out in Chap. 6, Padua's intellectual milieu in the 1830s and 1840s encouraged visions of transformation and perfectioning. Conversely, scientific studies influenced political views. In the context of censorship and hidden political messages (Sect. 9.3 in Chap. 9), Meneghini's insistence on statements such as "From being to being, from the less to the more complicated, the march of nature is always progressive" (G. Meneghini 1838, p. 48)[23] did not express a strictly scientific creed alone. Rather, it voiced the ideals and enriched the rhetoric of the emerging reformist movement. Even more so, his opinion that

> Certainly, these beings do not differ among themselves, neither in essence, nor in nature, only the degree of development distinguishes one from the other: in reality there do not exist but individuals and all the systematic divisions and subdivisions are creations of man, not of nature. (G. Meneghini 1838, p. 48)[24]

For Meneghini, organic nature was but one, all natural phenomena were subject to the same laws, and all life forms were but different developmental gradations. This assertion from 1838 reflected a common attitude in his time and foreshadowed the calls for social cooperation among Padua's moderate circles (Chap. 6) and was the ideological basis for the two plebiscites, held during the short revolutionary period (Chap. 10). It also reverberated in the declarations of advocates of self-help like those of his brother Andrea (Sect. 9.2 in Chap. 9) as well as those of liberal conservative politicians of the second half of the century, such as deputy Angelo Papadopoli (1843–1919). Papadopoli, member of one of the richest landowning families of Venice, stated in 1871: "today the legal divisions between classes do not exist anymore." Yet his key to social rise was not education, but wealth. He justified wealth and the possibility of social rise as a survival of the fittest, because wealth "has become a heritage of those who know how to gain it and to conserve it … misery, as a consequence, avoiding to use more severe words, is often a fault!" (Papadopoli 1871, pp. 12–13).[25]

In 1989, Adrian Desmond (1989, pp. 2–4) provocatively proposed that in the 1830s many in Britain found the notion of evolution unacceptable, not because of scientific disagreement, but rather for its disturbing social and political meaning. The French demagogues of the July Revolution had incorporated Lamarck's godless self-developing and 'ascending-from-below' evolutionary theory into their radical and egalitarian rhetoric. Therefore, the Anglican elite was anxious to keep the bouts of violence and the uprisings that were taking place on the European continent as well as the secularization and republican movements that were growing in strength away from their country. For them, evolutionary ideas were French and hence hostile. To meet the democratic threat, conservative physicians promoted a biology based on the Platonic ideal forms of the German *Naturphilosophie* (Desmond 1989, p. 13). Nonetheless, Lamarck's ideas (or what was sold as such) circulated among British radical physicians and in illegal penny prints of the insurrectionary working class.

As we have seen earlier, many of Meneghini's conceptions too originated from romantic and idealistic philosophies. Yet, there is no evidence that he favored these conceptions due to pressure from governmental authorities; even less that they bore conservative connotations. Rather, the idea of slow and steady progress agreed with the ideology of Padua's elite. Like their British counterparts, moderate liberals in Padua did not adopt the term 'evolution', but for other reasons.[26] Meneghini spoke of 'metamorphosis', 'progressive metamorphosis', or 'progression', whereas for him 'evolution' was, as we have seen above, an embryological term, denoting endogenous cell formation. In fact, sometimes he used it, like his friend Zanardini (see below), as synonym for individual development (Zanardini 1842, p. 110). De Zigno, following the tradition of eighteenth-century Veneto geologists, adopted the term 'transformism' (De Zigno 1839, p. 25). Transformism and progression were also key terms in moderate political circles before and even after the advent of Darwinism (De Mattei 1940), and remained a peculiar aspect of the North Italian political culture in the second half of the century (Camurri 1992, p. 7; Lanaro 1984). In 1880, the influential Italian senator Fedele Lampertico from the nearby city of Vicenza defended suffrage in parliament as conforming to the natural law and to the historical progress of societies (Pombeni 1992). His statement that politics must never halt or turn back, but gradually and continuously progress following natural evolution, strongly resembles Giuseppe Meneghini's words of some forty years earlier. Likewise, Lampertico's invocation of a 'social physiology' (a term coined by Romagnosi) and an 'organic equilibrium' shows the enduring influence of such thought patterns among Venetian progressive liberals throughout the nineteenth century. "Just as in a human organism the vital forces need to be equilibrated", Lampertico argued, "in the social body not one of these [forces] must drop the others" (quoted from Pombeni 1992, p. 297).

Another prominent figure who shared this intimate credo was Marco Minghetti (1818–1886), an influential politician from Bologna with strong ties to moderates in Veneto. In 1878, he outlined in a famous speech what he considered to be the core of the future moderate program: "We apply the process of nature to the political society, and we want to realize in the moral subjects the same principle that today dominates the physical ones, i.e. the law of slow evolution" (quoted from Camurri 1992, pp. 39–40). Minghetti and Veneto's liberals were in fact crucial for preparing the basis for Depretis' 'transformist' government (Sect. 6.5 in Chap. 6). However, it is also evident that in both the Veneto and the Italian political context of the late nineteenth century, scientific as well as socio-political discussion of evolution continued to have non-Darwinian meanings. Anything but a quarrel or fight for survival, the Italian *trasformismo* was a slow, progressive, and lawful process.

7.4 Giuseppe's Mission

However transformist and holistic Meneghini's concept of life was, it came up against one boundary that defied transgression, and that was the line between plants and animals during their lifetime (for the animal-plant border, see GGibson 2015). He rejected the then still popular opinion, maintained for instance by Franz Unger in his book *Die Pflanze im Momente der Thierwerdung* (The plant at the moment of becoming an animal) (Unger 1843), that especially inferior organisms can be plants in one phase of their life and animals in another. Interestingly, Meneghini did so on cytological and chemical grounds. As a result of his work on diatoms, he argued that, on applying traditional criteria, both kingdoms indeed exhibited the same phenomena. Yet if "[we] will now compare together such elementary or primordial cells taken from beings respecting whose animal or vegetable nature there can be no doubt [and subject them] to the action of chemical reagents, they manifest different results; hence we are led to conclude that there exists a difference in their chemical composition" (G. Meneghini 1853, pp. 13–14). The difference between animal and plant cells was therefore factual, material, and pre-existent.

The volume on diatoms was Meneghini's final scientific contribution to cell research. He was one of the very few Italian naturalists who truly grasped the import of the cell for the revolution in the life sciences that had only begun and would soon extend to almost all fields of biological and medical inquiry. Unfortunately for him and for the development of cytology in Italy, he failed to instill his conviction in a broader circle of like-minded colleagues, even less to establish a cytological school or tradition in Padua. The beginnings of his mission in support of cell theory had been promising. He considerably influenced especially the Paduan naturalists, a small and heterogeneous group, but active, expert, and well integrated into the European scientific community.

Along with Achille De Zigno's above-mentioned 1839 paper on the cellular origin of algae and mosses, other cryptogamists had begun to supplant Linnaean views with ideas of progression, and the traditional concept of structure with a cellular one. One of these was Vittore Trevisan, a member of the wealthy Venetian aristocracy and later politically active as De Zigno's advisor. He graduated with a degree in philosophy in 1842 and held several administrative and military positions, but his passion was for cryptogams. During his student years, count Nicola da Romano gifted him the herbarium of Abbot Girolamo Romano (1765–1841). When

Trevisan died, the herbarium had grown to over one million samples (Nimis and Hawksworth 1994, p. 17). He published 141 scientific papers and became an internationally renowned lichenologist. Yet, even in the late 1840s, he continued to accept an ascending linear order of living beings (Trevisan 1845, p. 472). Meneghini's influence is evident when Trevisan proposed a new taxonomy of inferior algae based on cellular criteria at the Venice congress of 1847. Some years before, Meneghini had proposed similar standards for the family Desmidiaceae (G. Meneghini 1840) that had been adopted by the English botanist John Ralfs (1807–1890) as "the best work on this family which I have seen" (Ralfs 1845). Trevisan dedicated his essay to Robert Brown, Heinrich Link, and Ludolph Christian Treviranus, who all had attended his talk in Venice, but his terminology and methodology followed Meneghini, who in fact advocated its publication (Anon. 1847, p. 11 and 47; Trevisan 1848).

Meneghini's deepest impact was on his brotherly friend Giovanni Zanardini (De Toni and Levi 1888). The Venetian physician indeed developed similar ideas of progressive perfection, cellular constitution, and divine providence. In the early 1830s, he began to study the algae of the Adriatic Sea with Meneghini, and in 1840, dedicated to his friend a new cyanobacterium species, *Oscillaria meneghiniana* Zanardini (later called *Spirulina meneghiniana* Zanardini) (Zanardini 1840, p. 199; Zanardini 1842, p. 248 and plate III, fig. 3). In 1839, Zanardini set out to "investigate the essence of the various modifications produced over time by the vital movement on the primitive cell" (quoted from G. Meneghini 1878–1879, p. 394). In his monograph on Adriatic algae, he devoted much space to general questions about structure and development. In 1842, he described the anastomoses in the genus *Mougeotia* as a consociation of individuals (*plurium individuorum consociationem*) (Zanardini 1842, p. 115) and opposed François Raspail's view that confervae (filamentous green algae) originated from 'fruits'. He instead demonstrated that these algae always reproduced through cell division (Ivi, p. 121.). He emphasized that this had far-reaching consequences for the concept of plant structure because, with a cellular type of reproduction, the alga "retains with each subdivision the properties of the mother cell" (Ivi, p. 114–115).

Cell theory had in fact given new impulses to the old question why complex organisms hold together as a functional unit. Vitalists answered this question by drawing on some sort of overarching force. Gottfried Reinhold Treviranus (1776–1837), for instance, ascribed the togetherness of the parts of an organism to the activity of a *Bildungsprincip*, or formative principle (Treviranus 1814, p. 639). Zanardini, instead, accepted the fundamentally organicist view that it was due to a cellular genealogical bond. Moreover, cells also shaped Zanardini's understanding of natural affinities. He opined that all individuals originated from a single cell, which represents an entire individual at its lowest level. On undergoing 'vital movement' (*motu vitali*) and divisions, the originally individual cells constituted all higher and more complex forms (Zanardini 1842, pp. 120–121). Finally, in line with Meneghini's view, Zanardini's concept of life was not materialist; rather, the wisdom of the provident divine Mind (*provida divinae Mentis sapientia*) operated within the body (ivi, p. 119).

In addition to leading Padua's circle of cryptogamists, Meneghini sought to assert his cytological expertise on the national stage. He found an ally in his younger colleague Filippo Parlatore (1816–1877) from Florence who maintained similar ideas (Parlatore 1844, pp. 44–45). Yet Parlatore's principal attention was directed not to cells but to Goethe's idea of morphology and the promotion of a Humboldtian vision of plant geography at the congresses (Visconti 2020). Like Parlatore, Meneghini was an active participant at nearly all congresses of the Italian scientists (Sect. 2.4 in Chap. 2). He missed no opportunity to call his colleagues' attention to the importance of theorizing about cells and their role in vital phenomena (Dröscher 1996, pp. 35–38). Hosting attendees like Amici, Guglielmo Gasparrini (1804–1866), von Mohl, Link, Dutrochet, Brown, Morren, Mirbel, Unger, Fée, and Treviranus, the audience of the botanical section was particularly attentive and skilled in arguments concerning cells. At the congress in Padua in 1842, Meneghini put the question "if what we currently know about the development of inferior algae harmonizes with the observations of Schleiden and other recent studies about the formation of vegetable tissue in higher plants" at the top of the agenda of future congresses (Anon. 1843, pp. 337–338). At the Venice congress, he informed the audience about the most recent advancements in cell research, emphasizing their importance for all botanists (Anon. 1847, p. 11).

But despite all tireless efforts, Meneghini did not succeed in instituting the cell or a cellular approach as a major research topic in Padua. Nor can one find traces of his specific concepts of the organization of algae and plants in contemporary Paduan or Venetian socio-political discussion. The reasons for this 'failure' are varied. The most decisive one is certainly the post-1848 repression (Sect. 10.3 in Chap. 10). Meneghini's still uncertain professional situation surely did not help to establish him as a key influencer of the general political debate, either. As a propaedeutic lecturer in natural sciences for future surgeons, he had many students but no disciples. Moreover, he was still in his thirties and thus had not had time to found a school. He probably discussed about bodily organization and individuality with his friends at the Caffé Pedrocchi, yet the cell-body debate was still only in an early and ill-defined stage. Indeed, most of his thoughts on the matter are scattered amidst various pieces of writing. The sketchiness of organicist conceptions is confirmed by the absence of explicit bio-political analogies; these only became popular during the post-1848 debate, in Padua as well as in the rest of Europe. Still, the parallelism between generic biological and socio-political ideas is striking, but probably due to the general intellectual atmosphere of those years and to values shared by circles that included both naturalists and political economists.

Finally, a third reason is the general belief that politics and science should be kept separate, even if in many cases the scholars were deeply involved in both fields. This view, which also characterized the relationship between science and religion, did not impede naturalistic argumentations, yet the analogies remained generic. The political organicist analogies that were proposed actually showed a rather poor scientific standard. On the other hand, scientifically skilled scholars like Giuseppe Meneghini consciously refrained from philosophical speculation concerning biological issues, for instance the monophyletic origin of life, and surely would have refused to seriously envision a new societal order based on his observational data.

Notes

1. Neil Tarrant (2014), among others, challenges this view. Even though historiographic accounts of a 'clash' between the Catholic Church and modern science surely need revision and diversification, it is equally evident that the Church exerted a considerable authority on the intellectual circles, and ending up on the *Index librorum prohibitorum* had severe consequences even in the early nineteenth century. On the more subtle methods of exerting censorship in eighteenth-century Italy, see Delpiano (2018).
2. Vicq d'Azyr 1786, 1, p. 12: "La Nature paroît donc suivre un type ou modele général, non-seulement dans la structure des divers animaux, comme je l'ai déjà dit, mais encore dans celle de leurs différents organes; et l'on ne sait ce que l'on doit le plus admirer, ou de l'abondance avec laquelle

ses formes paroissent variées, ou de la constance et de l'espece d'uniformité qu'un œil attentif découvre dans l'immense étendue de ses productions" (my translation).

3. Tiedemann 1816, p. vi: "… dass die Bildung des Hirns im Embryo und Fötus während den verschiedenen Monaten der Schwangerschaft die Haupt-Bildungsstufen durchläuft, worauf das Hirn der Thiere das ganze Leben hindurch gehemmt erscheint."
4. G. Meneghini, 1834, p. 7: "… in singulis speciebus embryonales evolutiones persequet decet."
5. G. Meneghini, 1834, p. 15: "Systema itaque nervosum nihil aliud est, nisi perfectioris organizationis, nobiliorisque animalitatis manifestatio."
6. G. Meneghini, 1834, p. 6: "… maximumque perfectionis gradum in homine adipiscitur, qui non jam novis organis exclusivis insignitur, at evolutionis implicationisque praestat perfectione, sibimetipsi tantum propria, ad quam sensim ac predetentim longa aliorum omnium animantium series perducit."
7. Ibid. : "… quod quum inferior inter se evolutionem suam suspendant, […], embryonalem statum entis superioris debeant simulare."
8. G. Meneghini 1838, p. 46: "Le operazioni tutte della natura sono insieme collegate di nesso necessario ed indissolubile."
9. G. Meneghini 1834, pp. 269–270: "Vitae animalis phenomena a duobus magnis elementis repetenda sunt, sine quibus vita existere non potest. Quum in infimis tum in altissimis animalitatis gradibus, necessariae ejusdem conditiones sitae sunt in sensu et motu."
10. G. Meneghini 1834, p. 274: "Praeter igitur tot ac tanta testimonia eloquentissima et inconcussa, quae nos intus adloquuntur, etiam physiologica investigatio in cerebralis functiones nobis persuadet, aliquid non materiale in nobis existere, divinae scilicet aurae particulam, quae nos ad altissimum finem extollit, suavissima spe solatur, atque optimo jure efficit, ut sublimem nostram naturam praedicemus."
11. De Zigno 1833, p. 7 and 14: "dipendente dalle sole fisico-chimiche leggi"; "Assurda proposizione di poco esatti osservatori" […] "filosofi superficiali, che si fecero a sostenere le dottrine del materialismo."

    Another influential advocate of vitalism in Padua was the professor of theoretical medicine Giacomo Andrea Giacomini (Federspil and Martini 1991).
12. G. Meneghini 1834, internal cover sheet: "Historiens de ce qui est, notre rôle se borne à dire, que les choses sont ainsi."
13. However, Geoffroy argued for a direct effect of environmental factors in inducing variations of the body plan of an organism or even producing a new species. The effect was particularly strong at the embryological stage when it perturbed the developmental process.

14. Goethe 1987, p. 349: "Die Gestalt ist ein Bewegliches, ein Werdendes, ein Vergehendes. Gestaltenlehre ist Verwandlungslehre. Die Lehre der Metamorphose ist der Schlüssel zu allen Zeichen der Natur."
15. G. Meneghini 1836, p. 1: "la varietà delle loro forme e delle loro funzioni a semplici modificazioni di un solo tipo costante."
16. G. Meneghini 1844d, p. 237: "si riconosce evidente l'uniformità di organica composizione ed un progressivo coordinamento in serie che dalle più semplici guida passo a passo fino alle più complicate [...] Questa metamorfosi progressiva, che ad ogni istante ricomincia e ad ogni passo s'arresta ma non recede giammai."
17. G. Meneghini 1844d, p. 237: "che le forme attuali si degli animali che delle piante siano provenute da successive metamorfosi delle antiche, ed al pari di quelle siano esse pure transitorie e mutabili col progresso del tempo."
18. G. Meneghini 1844d, p. 238: "un primo germe che colle successive ed innumerevoli modificazioni generò tutte quante sono le meraviglie vegetali."
19. G. Meneghini 1844d, p. 240: "Essa non è quindi applicabile al breve istante che forma l'oggetto della scienza."
20. G. Meneghini 1843c, p. 268: "la teoria della formazione successiva degli elementi costituenti ogni gemma"; "essendo la partizione sempre sintomo di esuberante energia vegetative."
21. G. Meneghini 1847, p. 11: "... provenne dalla gemma inchiusa sotto forma di embrione nel seme, rileviamo che la più umile erbetta al pari dell'albero più gigantesco egualmente ripetono la formazione loro da una serie di generazioni successive di gemme successivamente le une dalle altre originate, e successivamente le une sulle altre sviluppate. Considerazione di grande rilievo, perché ci pone in grado di giudicare dallo stato attuale della pianta di tutta la sua storia passata."
22. G. Meneghini 1838, p. 48: "Tutte le piante, quante esse sono, si ridurrebbero per noi ad una vescichetta vegetale, e le varie fasi del di lei sviluppo tutte inghiottirebbero in poche espressioni le sessantamila specie annoverate dai botanici."

    Some years before, Raspail had made a similar declaration (Vienne 2017, p. 635), but included the animal kingdom. It goes "Donnez-moi une cellule dans le sein de laquelle puissant s'élaborer à l'infini, et s'infiltrer à mon gré d'autres cellules, et je vous rendrai toutes les formes du monde organisé" (Raspail 1827, p. 306).
23. G. Meneghini 1838, p. 48: "Di essere in essere, dal meno complicato a quello che lo è maggiormente sempre progressiva è la marcia della natura."
24. G. Meneghini 1838, p. 48: "Certamente quegli esseri non differiscono fra loro, né in essenza, ne in natura, solo il grado di sviluppo l'uno dall'altro li

distingue: in realtà non esistono che individui e tutte le divisioni e suddivisioni sistematiche sono creazioni dell'uomo e non di natura."

25. Papadopoli 1871, pp. 12–13: "le divisioni legali fra classe e classe non esistono più, la ricchezza, in massima, è diventata il retaggio di chi sa acquistarla e conservarla ... la miseria, per conseguenza, a non voler usare parole più severe, è spesso una colpa!"
26. Lamarck himself never used the term 'evolution' either. He spoke about 'marche de la nature' (march of nature).

References

Anon. 1841. *Index Librorum Prohibitorum Sanctissimi Domini Nostri Gregorii XVI Pontificis maximi, jussu editus.* Romae: Ex Typographia Reverendae Camerae Apostolicae.

———. 1843. *Atti della Quarta Riunione degli scienziati tenuta in Padova nel settembre del 1842.* Padova: Co' Tipi del Seminario.

———. 1847. *Diario del Nono Congresso degli scienziati convocati in Venezia nel settembre MDCCCXLVII.* Venezia: Co' Tipi di Giovanni Cecchini.

Barr Nisbet, Hugh. 1967. Herder, Goethe, and the natural 'type'. *Publications of the English Goethe Society* 37 (1): 83–119.

———. 1970. *Herder and scientific thought.* Cambridge: The Modern Humanities Research Association.

Bies, Michael. 2015. Staging knowledge of plants. Goethe's elegy 'The Metamorphosis of Plants'. In *Performing Knowledge, 1750–1850*, ed. Mary Helen Dupree and Sean B. Franzel, 247–267. Berlin: de Gruyter.

Bonuzzi, Luciano. 1999. La medicina padovana fra '800 e '900 (ascesa ed evoluzione del costituzionalismo). *Annali di Storia delle Università italiane* 3: 171–179.

Brocchi, Giovanni Battista. 1796. *Trattato delle piante odorifere e di bella vista da coltivarsi ne' giardini.* Bassano: Stamperia Penada.

Caianiello, Silvia. 2008. La federazione delle parti: Sul concetto di individuo in Rudolf Virchow. *Medicina nei secoli* 20 (1): 43–89.

Camurri, Renato. 1992. Introduzione. In *La scienza moderata: Fedele Lampertico e l'Italia liberale*, ed. Renato Camurri, 1–54. Milano: Franco Angeli.

Canavari, Mario. 1889. *Commemorazione di Giuseppe Meneghini fatta nell'aula magna dell'Università Pisana ai XXIV Marzo MDCCCLXXXIX.* Pisa: Tipografia T. Nistri.

Ciancio, Luca. 1995. *Autopsie della terra: Illuminismo e geologia in Alberto Fortis (1741–1803).* Firenze: Olschki.

———. 2013. I segni del tempo: Teorie e storie della Terra. In *Il contributo italiano alla storia del pensiero*, ed. Antonio Clericuzio and Saverio Ricci, 332–343. Roma: Istituto dell'Enciclopedia Italiana.

Corsi, Pietro. 1991. De Zigno, Achille. In *Dizionario Biografico degli Italiani*, vol. 39, 627–630. Roma: Treccani.

———. 2005. Before Darwin: Transformist concepts in European natural history. *Journal for the History of Biology* 38: 67–83.

Daum, Andreas W. 2002. Science, politics, and religion: Humboldtian thinking and the transformations of civil society in Germany, 1830–1870. *Osiris* 17: 107–140.

De Mattei, Rodolfo. 1940. *Dal trasformismo al socialismo*. Firenze: Sansoni.

De Souza, Nigel. 2018. Herder's theory of organic forces and its Kantian origins. In *Kant and his German Contemporaries. Vol. 2: Aesthetics, History, Politics, and Religion*, ed. Daniel O. Dahlstrom, 109–127. Cambridge: Cambridge University Press.

De Toni, Giovanni Battista, and David Levi. 1888. *L'algarum Zanardini*. Venezia: Tip.-Lit. M. Fontana.

De Zigno, Achille. 1833. *Plantae crytogamae in provincia Patavina hucusque observatae*. Patavii: Typis Seminarii.

———. 1839. *Sopra alcuni corpi organici che si osservano nelle infusioni: Cenni*. Padova: Tip. Cartallier e Sicca.

Delpiano, Patrizia. 2018. *Church and censorship in eighteenth-century Italy: Governing reading in the Age of Enlightenment*. New York: Routledge.

Desmond, Adrian. 1989. *The politics of evolution: Morphology, medicine, and reform in radical London*. Chicago: The University of Chicago Press.

Dohrn-van Rossum, Gerhard, and Ernst-Wolfgang Böckenförde. 1978. Organ, Organismus, Organisation, politischer Körper. In *Geschichtliche Grundbegriffe: Lexikon zur politisch-sozialen Sprache in Deutschland*, ed. Otto Brunner, Werner Conze, and Reinhart Koselleck, vol. 4, 519–622. Stuttgart: Klett-Cotta.

Dominici, Stefano, and Niles Eldredge. 2010. Brocchi, Darwin, and transmutation: Phylogenetics and paleonology at the dawn of evolutionary biology. *Evolution: Education and Outreach* 3: 576–584.

Dröscher, Ariane. 1996. *Die Zellbiologie in Italien im 19. Jahrhundert*. Halle: Deutsche Akademie der Naturforscher Leopoldina.

———. 2002. Edmund B. Wilson's *The Cell* and cell-theory between 1896 and 1925. *History and Philosophy of the Life Sciences* 24: 357–389.

———. 2016. 'Lassen Sie mich die Pflanzenzelle als geschäftigen Spagiriker betrachten': Franz Ungers Beiträge zur Zellbiologie seiner Zeit. In *Einheit und Vielfalt: Franz Ungers (1800-1870) Konzepte der Naturforschung im internationalen Kontext*, ed. Marianne Klemun, 169–194. Wien: Vienna University Press.

———. 2017. Regola o caso speciale? Franz Unger (1800–1870) e la scoperta della divisione cellulare. *Mefisto* 1 (1): 125–144.

———. 2022, in press. Inspiring imagination – embarrassing analogies: coping with the causes of cytoplasmic streaming. *Intellectual History Review* 32.

Elwick, James. 2003. Herbert Spencer and the disunity of the social organism. *History of Science* 41: 35–72.

Federspil, Giovanni, and Chiara Martini. 1991. Giacomo Andrea Giacomini: A Padua clinician between metaphysics and science. *History and Philosophy of the Life Sciences* 13 (1): 73–95.

Forgione, Fabio. 2018. *Il potere dell'evoluzione: Il dibattito sulla variabilità delle specie nella Torino dell'Ottocento.* Milano: Franco Angeli.

Gaudichaud, Charles. 1841. *Recherches sur l'organographie, la physiologie et l'organogénie des vegetaux.* Paris: Fortin, Masson at Co.

Generali, Dario. 2020. Dall'uniformità della natura alla teoria del contagio vivo in Antonio Vallisneri. *ISPF-LAB* 17. http://www.ispf.cnr.it/system/files/pandemia/pdf/001_022_D_Generali_contagio_vivo.pdf. Accessed 25 August 2020.

Geoffroy St.-Hilaire, Étienne. 1829. *Cours de l'histoire naturelle des mammifères.* Paris: Pichon et Didier.

Gibson, Susannah. 2015. *Animal, vegetable, mineral? How Eighteenth century science disrupted the natural order.* Oxford: Oxford University Press.

Gliboff, Sander. 1998. Evolution, revolution, and reform in Vienna: Franz Unger's ideas on descent and their post-1848 reception. *Journal of the History of Biology* 31 (2): 179–209.

Goethe, Johann Wolfgang. 1987. Fragmente zur vergleichenden Anatomie. In *Johann Wolfgang Goethe. Sämtliche Werke, Briefe, Tagebücher und Gespräche*, 45 vols., 1. Abteilung – Band 24: *Goethes Schriften zur Morphologie*, ed. Dorothea Kuhn. Frankfurt a. M.: Deutscher Klassiker Verlag.

Gould, Stephen Jay. 1999. The great physiologist of Heidelberg. *Natural History* 108: 26–36.

Herder, Johann Gottfried von. 1784–91. *Ideen zur Philosophie der Geschichte der Menschheit.* 4 vol. Riga – Leipzig: Johann Friedrich Hartknoch. Engl. edition: Herder, J. G. 1803. Outlines of a Philosophy of the History of Man (Trans. T. Churchill). London: J. Johnson (Luke Hansard).

Holmes, Tove. 2017. "Beweglich und bildsam": Goethe, plants, and literature. *Literatur für Leser* 40 (2): 91–105.

Huxley, Thomas H. 1872. *More critics on Darwin and administrative nihilism.* New York: D. Appleton and Company.

de Jacquin, Nicolò Giuseppe. 1824. *Introduzione allo studio dei vegetabili, tradotta, illustrata ed accresciuta da Roberto De Visiani.* Padova: Tipografia della Minerva.

Kelley, Theresa M. 2012. *Clandestine marriage: Botany and romantic culture.* Baltimore: Johns Hopkins University Press.

Klemun, Marianne. 2009. Franz Unger and Sebastian Brunner on evolution and the visualization of earth history: A debate between liberal and conservative catholics. *Geological Society Special Publications* 310: 259–267.

Kruta, Vladislav. 2018. Tiedemann, Friedrich. In *Complete Dictionary of Scientific Biography*. Encyclopedia.com. http://www.encyclopedia.com/science/dictionaries-thesauruses-pictures-and-press-releases/tiedemann-friedrich. Accessed 7 March 2018.

Lanaro, Silvio. 1984. Genealogia di un modello. In Storia d'Italia: Le regioni dall'Unità a oggi. Il Veneto, ed. Silvio Lanaro, 5–96. Torino: UTET.

La Vergata, Antonello. 1995. Herbert Spencer: Biology, sociology, and cosmic evolution. In *Biology as society, society as biology: Metaphors*, ed. Sabine Maasen, Everett Mendelsohn, and Peter Weingart, 193–229. Dordrecht: Springer.

Lidgard, Scott, and Lynn K. Nyhart, eds. 2017. *Biological individuality: Integrating scientific, philosophical, and historical perspectives*. Chicago: The University of Chicago Press.

Lonigo, Alberto. 2013a. Achille De Zigno: La famiglia, l'impegno pubblico e gli studi di uno scienziato padovano dell'800. *Storia veneta* 20: 2–19.

———. 2013b. Il Barone Achille De Zigno: La fammiglia, il contest familiar ed I rapport col mondo scientifico. In *Achille De Zigno: Impegno civico e attività scientifica*, ed. Guido Roghi, Mariagabriella Fornasiero, and Alberto Lonigo, 35–62. Padova: Padova University Press.

Mandressi, Rafael. 2018. *Félix Vicq d'Azyr: L'Anatomie, l'état, la médecine. Une carrière scientifique entre Ancien régime et Révolution*. Paris: Bibliothèque interuniversitaire de médecine. http://www.biusante.parisdescartes.fr/histoire/medica/vicq.php. Accessed 1 March 2018.

Mazzolini, Renato G. 1988. *Politisch-biologische Analogien im Frühwerk Rudolf Virchows*. Marburg: Basilisken Presse.

Meneghini, Joseph [Giuseppe]. 1834. *De axe cephalo-spinali dissertation inauguralis*. Patavii: Ex officina sociorum titulo Minerva.

Meneghini, Giuseppe. 1836. *Ricerche sulla struttura del caule nelle piante monocotiledoni*. Padova: Tipi della Minerva.

Meneghini, Joseph [Giuseppe]. 1837. *Conspectus algologiae euganeae, germanicis naturalium rerum scrutatoribus Pragae anno 1837 convenientibus*. Pataviae: Typis Minervae.

Meneghini, Giuseppe. 1837. De Bryopsidum fructificatione. *Botanische Zeitung* 20 (2): 721–726.

———. 1838. *Cenni sulla organografia e fisiologia delle alghe*. Padova: Tipi della Minerva.

Meneghini, Joseph [Giuseppe]. 1840. Synopsis Desmidiearum hucusque cognitarum. *Linnea: Ein Journal für die Botanik in ihrem ganzen Umfange* 14: 201–240.

Meneghini, Giuseppe. 1843a. *Intorno all'asse cerebro-spinale. Memoria tradotta dal latino da A. C. De Meis per cura dello studio privato del prof. Pietro Ramaglia*. Napoli: Barnaba cons.

———. 1843c. Relazione. In *Atti della Quarta Riunione degli scienziati tenuta in Padova nel settembre del 1842*, 266–271. Padova: Co' Tipi del Seminario.
———. 1844b. Sulla teoria de' Meritalli di Gaudichaud. *Giornale Botanico Italiano* 1: 17–26.
———. 1844c. Lecture. In *Atti della quinta unione degli scienziati italiani tenuta in Lucca nel settembre del MDCCCXLIII*, 360–362. Lucca: Tipografia Giusti.
———. 1844d. Sulla metamorfosi delle piante. *Giornale Euganeo* 1: 236–240.
———. 1847. Teorica degl'innesti. Articolo secondo. *Il Tornaconto* 2 (14 gennaio 1847): 11–13.
———. 1853. On the animal nature of the Diatomeae, with an organographical revision of the genera established by Kützing. In *Botanical and physiological memoirs*, ed. Arthur Henfrey, 343–513. London: Ray Society.
———. 1878–1879. Commemorazione di Giovanni Zanardini (1804–1878). *Atti del R. Istituto Veneto di Scienze, Lettere ed Arti* 37: 391–400.
Monti, Maria Teresa. 2010. Logica in azione e utili studi. In *Bonaventura Corti, Il giornale degli animaluzzi*, ed. Maria Teresa Monti, vii–cxxii. Firenze: Olschki.
Nimis, Pier Luigi, and David L. Hawksworth. 1994. The lichenological activity of Vittore Trevisan earl of San Leon (1818–1897). *Opera Naturalistica Classica* 2: 13–27.
Nyhart, Lynn. 2017. The political organism: Carl Vogt on animals and states in the 1840s and '50s. *Historical Studies in the Natural Sciences* 47 (5): 602–628.
Omboni, Giovanni. 1897. *Commemorazione del Barone Achille De Zigno*. Venezia: Stab. Tipografia-litografia C. Ferrari.
Pancaldi, Giuliano. 1991. *Darwin in Italy: Science across cultural frontiers*. Bloomington: Indiana University Press.
Papadopoli, Angelo. 1871. *Della necessità di un nuovo indirizzo nella pubblica beneficenza in Venezia*. Venezia: Marco Visentini.
Parazzi, L. 1871. L'abate Giuseppe Bernardi da Noventa di Piave. *La Gioventù: Rivista dell'Istruzione pubblica in Italia* X/2 (Quaderni XXIII–XXIV): 812–815.
Parent, André. 2007. Félix Vicq d'Azyr: Anatomy, medicine and revolution. *Canadian Journal of Neurological Sciences* 34: 30–37.
Parlatore, Filippo. 1844. Sullo spirito nelle scienze naturali nel secolo passato e nel presente. *Giornale Botanico Italiano* 1: 31–49.
Pearson, Hugh Lance. 2005. Baron Achille De Zigno: An Italian palaeobotanist of the 19th century. *Geological Society, London, special publications* 241: 86–94.
Poli, Baldassarre. 1827. *Saggio filosofico sopra la scuola de' moderni filosofi naturalisti coll'analisi dell'organologia, della craniologia, della fisiognomia, della psicologia comparata e con una teorica delle idee e dei sentimenti*. Milan: Francesco Sonzogno.
———. 1837. *Elementi di Filosofia teoretica e morale. Vol. 1: Psicologia empirica*. Padova: Coi Tipi del Seminario.
Pombeni, Paolo. 1992. Fedele Lampertico e la scienza politica dei moderati: Note di lettura. In *La scienza moderata: Fedele Lampertico e l'Italia liberale*, ed. Renato Camurri, 287–307. Milano: Franco Angeli.

Ralfs, John. 1845. On the British Desmidieae. *Annals and Magazin of Natural History* 15: 149–160.

Raspail, François. 1827. Recherches chimiques et physiologiques destinées à expliquer non-seulement la structure et le développement de la feuille, du tronc, ainsi que des organes qui n'en sont qu'une transformation, mais encore la structure et le développement des tissus animaux. *Mémoires de la Société d'Histoire naturelle de Paris* 3: 17–88 and 209–313.

Ratcliff, Marc. 2016. *Genèse d'une découverte: La division des infusoires (1765–1766)*. Paris: Publications Scientifiques du Muséum National d'Histoire naturelle.

Reynolds, Andrew. 2008. Ernst Haeckel and the theory of the cell state: Remarks on the history of a bio-political metaphor. *History of Science* 46: 123–152.

Richards, Robert J. 1992. *The meaning of evolution: The morphological construction and ideological reconstruction of Darwin's theory*. Chicago/London: The University of Chicago Press.

Riedl-Dorn, Christa. 2019. *Ein uomo universale des 19. Jahrhunderts und sein wissenschaftliches Netzwerk: Stephan Ladislaus Endlicher und seine Korrespondenz mit Wissenschaftlern seiner Zeit*. Göttingen: V&R Unipress.

Robin, Nicolas. 2011. Heritage of the romantic philosophy in post-Linnaean botany: Reichenbach's reception of Goethe's metamorphosis of plants as a methodological and philosophical framework. *Journal of the History of Biology* 22 (2): 283–304.

Spahn, Christian. 2015. *Evolution*. In *Oxford handbook of German philosophy of the 19th century*, ed. Michael Forster and Kristin Gjesdal, 674–694. Oxford: Oxford University Press.

Spencer, Herbert. 1865–67. *The principles of biology*. 2 vols. London/Edinburgh: Williams and Norgate.

———. 1901. The social organism (1860). In *Essays*, ed. Herbert Spencer, vol. 1, 265–307. London: Williams and Norgate.

Tarrant, Neil. 2014. Censoring science in sixteenth-century Italy: Recent (and not so recent) research. *History of Science* 52 (1): 1–27.

Thomas, Marion. 2017. Biological and social new orders: Charles Robin (1821–1885) and Comteanism in the Third Republic. *Historical Studies in the Natural Sciences* 47 (5): 653–677.

Tiedemann, Friedrich. 1816. *Anatomie und Bildungsgeschichte des Gehirns im Fötus der Menschen, nebst einer vergleichenden Darstellung des Hirnbaues in den Thieren*. Nürnberg: Steinische Buchhandlung.

———. 1836. On the brain of the Negro, compared with that of the European and the Orang-Outang. *Philosophical Transactions of the Royal Society of London* 126: 497–527.

Tirard, Stéphane. 2016. Les botanistes français et la cellule dans les années 1840. *Bulletin d'histoire et d'épistémologie des sciences de la vie* 23 (2): 121–134.

Treviranus, Gottfried Reinhold. 1814. *Biologie, oder Philosophie der lebenden Natur für Naturforscher und Aerzte*. 4. Band. Göttingen: Johann Friedrich Röwer.
Trevisan, Vittore. 1845. Relazione. In *Atti della sesta Riunione degli scienziati italiani tenuta in Milano nel settembre del MDCCCXLIV*, 471–472. Milano: Coi Tipi di Luigi di Giaomo Pirola.
Trevisan, Vittore B. A. 1848. Saggio di una monografia delle alghe coccotalle. Padova: Coi Tipi del Seminario.
Unger, Franz. 1843. *Die Pflanze im Momente der Thierwerdung*. Wien: Beck's Universitäts-Buchhandlung.
Vallisneri, Antonio. 1721. Lezione accademica intorno all'ordine della progressione, e della connessione, che hanno insieme tutte le cose create. In *Istoria della generazione dell'uomo e degli animali se sia da' vermicelli spermatici, o dalle uova*, 421–438. Venezia: Gio. Gabbriel Hertz.
Vicq d'Azyr, Félix. 1786. *Traité d'anatomie et de physiologie avec des planches coloriées représentant au naturel les divers organes de l'Homme et des Animaux*. Paris: François Didot l'aîné.
Vienne, Florence. 2017. Worlds conflicting: The cell theories of François-Vincent Raspail and Theodor Schwann. *Historical Studies in the Natural Sciences* 47 (5): 629–652.
Visconti, Agnese. 2020. Tracce di Humboldt nella sezione di Botanica dei congressi di Padova e Venezia. In *Scienziati italiani a congresso nel Veneto asburgico (1842, 1847)*, vol. I, eds. Valeria Mogavero and Maria Pia Casalena. *Venetica: Rivista di storia contemporanea* 58 (1): 95–105.
Weindling, Paul. 1981. Theories of the cell state in imperial Germany. In *Biology, medicine and society 1840–1940*, ed. Charles Webster, 99–155. Cambridge: Cambridge University Press.
Wilson, Edmund B. 1896. *The Cell in development and inheritance*. New York: The Macmillan Company.
Zanardini, Giovanni. 1840. Sopra le alghe del mare Adriatico. Lettera seconda. *Biblioteca Italiana ossia Giornale di Letteratura Scienze ed Arti* 99: 195–229.
Zanardini, Joannis [Giovanni]. 1842. Synopsis Algarum in mari Adriatico hucusque collectarum cui accedunt monographia Siphonearum nec non generales de algarum vita et structura disquisitiones. *Memorie della Reale Accademia delle Scienze di Torino* 2/IV: 105–256.
Zemplén, Gábor Áron. 2017. Structure and advancement in Goethe's morphology. In *Marking time: Romanticism and evolution*, ed. Joel Faflak, 147–172. Toronto: Toronto University Press.

CHAPTER 8

The Sweeping Power of Horticulture

8.1 Flowers in Paduan Culture

Despite his interest in cryptogams, Giuseppe Meneghini confessed that flowers were the sources of the tenderest emotions: "The little flowers, tenaciously clinging on the earth raises its incenses to heaven, almost as if it talks to man about the love of homeland and the worship to God" (G. Meneghini 1844–1846, 1, p. 314).[1] Similar aesthetical-religious-patriotic sentiments were shared by botanophiles at many places, yet in Veneto they were particularly strong. Interest in plants and flowers was not restricted to gardening, but an essential part of its cultural history. During the early modern period, vegetal imagery and allegory featured prominently in Veneto's literature. Humanists and writers like Pierio Valerian (1477–1558), born Giovanni Pietro dalle Fosse in Belluno, Pietro Bembo (1470–1547), and Sperone Speroni (1500–1588) reinvented themselves as men of letters outside the courts, founding academies, assuming ecclesiastic positions, rising to university chairs, and thus refashioning traditional ways of life. The garden figured in their works as a key venue of courtly life, a 'pleasant place' (*locus amoenus*) far from the hustle and bustle of affairs in Venice, and explicitly continued the Decameronian imagery of sites of shelter and protection (Vianello 1993, pp. 41–43). The plant world also played a leading role in pastoral drama and pastoral

A. Dröscher, *Plants and Politics in Padua During the Age of Revolution, 1820–1848*, Palgrave Studies in the History of Science and Technology, https://doi.org/10.1007/978-3-030-85343-3_8

poetry, one of the most characteristic literary genres of the early modern period (Jeffery 1924; Sampson 2006). It emerged in Northern Italy in the mid-sixteenth century, although the very first pastoral drama, *Favola di Orfeo*, appeared as early as 1472 in Mantua, and resumed the Hellenistic and Virgilian tradition of praising romanticized nature and the idyll of simple country life. The success of Torquato Tasso's *Aminta*, first performed in 1573 during a garden party at the court of Ferrara, started a European fashion, especially in Elizabethan England. The fact that the pastoral genre continued to enjoy success in Italy, in particular in Paduan cultural circles, even during the early seventeenth century when the fashion had waned in most of Europe, demonstrates the deep and persistent influence of imagery depicting pristine nature and classical Greek motives. A peculiarity of the North Italian form was the conflation of poetry, morals, and (anti-magical) therapeutic herbal-pharmaceutical knowledge, as exemplified in Tasso's *Aminta* as well as in Giovanni Battista Guarini's *Pastor Fido*, which premiered twelve years later (Schneider 2016).

Flowers and ornamental plants played a crucial role during some of Padua's most important political events. In 1778, the vast elliptic square *Prato della Valle* (Figure 2.1, A), usually the venue of festivals and horseraces, was transformed into a flowering garden to celebrate the arrival of Padua's new *podestà* (mayor) Domenico Michiel (or Micheli) (Sorgato 1845, p. 15). In this period indeed, as confirmed by the naturalist Giovanni Battista Brocchi (1796, pp. 5–12), botany and in particular the study of 'sweet flowers' and 'gentle herbs' had acquired the greatest popularity among all the studies of nature. Therefore, Brocchi, who promoted the transformation of the study of fossils into a science, did not scorn sitting for his portray dreamy in his landscape garden, incanting hyacinths, anemones, and violets, and publishing a small treatise on his personal expertise in floriculture. Demanding scientific standards for the study of nature and expressing one's romantic passion for living nature did not exclude each other.

Only one year later, the French-born writer and agronomist Pier Luigi Mabil read his *Essay on the Nature of Modern Gardens* (1798) at the Paduan Academy of Science. This essay, published two decades later (Mabil 1817), and his treatise *Theory of Garden Art* (Mabil 1801) contributed greatly to the diffusion of landscape gardens in Veneto. Importantly, the North Italian type of landscape gardens permitted flowers, strictly prohibited in the early English designs. Paduan writers and garden owners like Mabil continued to praise the smell and beauty of flowers, and to treat trees as if they were jewels. Mabil's treatise, dedicated to Giovanni de Lazara, whose circle included Agostino Meneghini, contained indeed a long chapter on trees (36 pages) and a short one on flowers (3 pages).

The role of flowers and trees in Padua's landscape gardens is illustrated by the Treves garden, which combined the new romantic fashion with the ancient if artisanal function of gardens as repositories of rare species, experimental plots for breeding and cultivation, and depots for intensive scholarly exchange, monetizable for social and political purposes. A famous historical example for such a combination of intents is the garden of Versailles (Hyde 2005). In addition to these aspects, the botanical interests of the Treves brothers were surely more than just a temporary fashion or a formal social convention. Their aunt Enrichetta supported the plan to create a garden with scientific aspirations, later praised by De Visiani for its collection of exotic palms (De Visiani 1840, p. 6). A similar project had been realized in 1805, under the auspices of Gian Battista Brocchi, in Bassano by Alberto Parolini, another of Enrichetta Treves's frequent guests. The transformation of a garden into a scientific plant collection required specific infrastructure, that is, glasshouses, a heating system, a subterranean cistern, and, in the case of the Treves garden, a bridge for bringing water to the parts on the other side of the channel (Massaro 2018; Massaro 2019). The connection between the Treves garden and the nearby academic Botanical Garden (Fig. 2.1, no. 4) was strong, and many rare plant species were present in both gardens (Ronconi 1853). Notwithstanding their scientific interests, the garden was for the Treves brothers also an intimate expression of their self-conception. We can surmise what nature in flower signified for the banker Giacomo Treves from the letters of his former instructor Don Zaccaria Cappello (1762–1837). These were replete with botanical metaphors. One of these equated the image of a learned man with

> a garden abounding in flowers, well placed on their stems and regulated by the hands of an industrious gardener, with cleared and cleaned flowerbeds, with all angles expressing clarity and decency, [...] lines of well placed fruit trees [...] showing fruits of every season [...], and which forms the satisfaction of its grower. (Quoted from Massaro 2014–2015, pp. 75–78)

Like the collection of exotic palms of the Treves garden and the collection of rare conifers in Alessandro Papafava's garden, exotic or particularly beautiful trees were the highlight of many gardens in Padua, presented to visitors much like treasures. In the sixteenth century, Lorenzo Priuli (1489–1559), son of the 83rd Doge of Venice, displayed in his Paduan garden (Fig. 2.1, no. 3) the *Scamonea di Aleppo*, the cardamom *Amomo*, exotic medicinal plants imported from Syria and mentioned by Luigi Anguillara, Conrad Gesner (1516–1565), and later Joseph Pitton de Tournefort (1656–1708). In the eighteenth century, Giovanni Marsili (1727–1795) described seventeen Paduan gardens and their noteworthy plants and visitors (Marsili 1840, pp. 12–13; see also De Toni 1887). The Paduan tradition of (living) non-arboreal plant collecting too continued throughout the nineteenth century. Given the lack of public exhibits, private collections fulfilled an important museal function in the local culture of that time. Many figures made a considerable effort to also present scientifically valuable specimens. For instance, professor of Civil law, Alessandro Racchetti, and professor of Obstetrics, Rudolf Lamprecht (1781–1860), owned plant collections that Giuseppe Meneghini deemed noteworthy, Giuseppe Cristina cultivated camellia species, Alessandro Papafava conifers, and the physician Domenico Martinati (1774–1855) renowned collections of fungi and cacti (De Toni 1909, p. 14).

Another custom that reveals the high esteem in which plants continued to be held by Padua's *notabili* in the early decades of the nineteenth century was the common giving of botanical poetry and dissertations to spouses and their families, as illustrated in Sect. 3.2 in Chap. 3. The Abbot Girolamo Romano, teacher at the Seminary of Rovigo and librarian of the Academy of Science in Padua (Saccardo 1869, pp. 79–83), was particularly active in this regard. His botanical expertise helped him win the friendship of many eminent figures in Padua. One example of this are his flower recommendations to embellish the Cromer-Meneghini garden (Romano 1823a; Romano 1831). Romano also published other works for the Meneghini family, namely, an essay on the cultivation of *Iris* on occasion of Anna's wedding and a list of phanerogams of the Euganean hills for Andrea's wedding. Two previous editions of this list had been gifts for the wedding of count Giordano Emo Capodilista (1800–1843) and Lucia Maldura (1805–1892) in 1823, and for the wedding of Giovanni Cittadella and Laura Maldura (d. 1829) in 1828 (Romano 1823b; Romano 1828). His most successful book was the translation, in 1819–1820, of the second edition of Georges Louis Marie Dumont de Courset's (1746–1824) practical manual *Le Botaniste cultivateur* (1811–1814), to which Romano added a short list of Veneto gardens, among them that of Agostino Meneghini (Dumont de Courset 1819, pp. 55–57).

8.2 The Festival of Flowers

In addition to the sentimental, esthetical, and representational role of gardens, floriculture also became an increasingly profitable industry during the nineteenth century. Plants, in all their variety and use, had always been of utmost commercial importance in Veneto. Padua's Botanical Garden had indeed been founded for economic reasons. In 1545, the governors of the *Serenissima* agreed to create a botanical garden to promote the commerce in plants and seeds for the production of medical remedies, lotions, and dyes, for woodworking, and many other uses that became increasingly important in the sixteenth century. The introduction of exotic species from overseas prompted attempts to cultivate them in Europe. Venetian apothecaries and traders had made several attempts to install a big botanical garden in their city, but Venice's soil was too wet and saturated with chlorides. Sabrina Minuzzi (2016, pp. 92–94) maintains that just the absence of such a central and authoritative botanical institution in Venice stimulated many to install small private 'everday-*horti*'. Yet, for the creation of a big botanical garden, Padua was chosen, and retained its monopoly for centuries.

During the 1840s, the cultivation of flowers and ornamental plants took an economic turn, decisively so, and the Paduan population amplified this development. This is reflected by frequent articles on floriculture that appeared in Andrea Meneghini's journal *Il Tornaconto*, which in fact bore the subtitle *Journal for agriculture, horticulture, industry, commerce and municipal economy for the Venetian provinces* (Sect. 9.3 in Chap. 9). Over the seventeen months of its existence, thirty-two articles on the Gardening Society and on the cultivation of flowers and ornamental plants appeared in the journal, most of them written by Angelo Giacomelli (1816–1907) from Treviso, who ran a European plant trade (see below). In January 1848, he added a four-paged catalogue supplement of dahlias and other flowering plants (Anon. 1848).

The emerging Venetian floriculture industry received a major boost from the activities of De Visiani's Gardening Society. Its principal aim was the promotion of gardens and urban greening (Sect. 2.3 in Chap. 2), and for most the principal reason for membership in the *Società promotrice di giardinaggio* was pragmatic, that is, to exchange information and flower samples, as well as to exhibit, advertise, and sell one's flowers and eventually to win distinctions and prizes. On the occasion of the Society's festival of flowers, many exhibited a great number and a great variety of plants, some more than 200 species (Anon. 1846b; Anon. 1847a; Anon. 1854). The floral exhibitions brought together horticulturists and park owners from all of Veneto. To name just some of the most famous non-Paduan garden owners that attended the festivals, there was Teresa Papadopoli Mosconi, who commissioned the scenographer and architect Francesco Bagnara (1784–1866) for her huge romantic Papadopoli garden in Venice; Drusilla Loschi Dal Verme (1811–1885), commissioner of the romantic Jappelli garden near Vicenza, today known as Parco Villa Zileri; Giulia Piovene Porto, owner of the garden of Villa Piovene Porto; and Girolamo Polcastro (1763–1839), owner of the park today known as villa Wollemborg, about 25 kilometer north of Padua (Fig. 1.1, no.14). Representatives of Vienna's garden culture also participated: Ernest Engelbert von Arenberg (1777–1857) and his second wife Sophie Karoline Marie Prinzessin von Auersperg (1811–1901), who owned, among other things, a famous *Gartenpalais*.

Padua's first floral occasion, the first 'Festival of Flowers', took place as early as the eleventh century, imitating similar French pagan events (Sberti 1818, pp. 36–37; Sgaravatti Montesi 1966). It was a joyous day for the entire city. Houses and squares were adorned with flowers and garlands. The streets filled with musicians, actors, jongleurs, and jesters. About one hundred years later, on 23 June 1164, some insurgents took advantage of the crowd and confusion due to the festival to start a rebellion, to besiege and conquer the castle of emperor Barbarossa's deputy, count Pagano della Torre, to throw him out of his castle, and to install the 'free Municipality of Padua'. During subsequent festivals, a *Game of the Castle of Love* took place to celebrate this historical success. According to Paduan legend, during the upheaval of 1164 Paduan people also liberated the beautiful damsel Speronella, held captive in count Pagano's castle. And several Venetian cities participated in a game in memory of this heroic rescue, during which young men tried to climb up a wooden tower on a cart. Whoever conquered the top of the tower would attempt to conquer the ladies who were waiting there and dropping flowers and fruit, before another contender took his place (Sorgato 1845, pp. 11–12). Yet, the fun turned into a second dangerous combination of flowers and politics a few years later, in 1214, when Venetian youngsters did not respect the rules of the game and dropped gold coins instead of flowers. This provocation caused a fight and profound resentment between the men from Venice and those from Padua and Treviso, which eventually led to an armed conflict between the three cities (Anon. 1852).

Fig. 8.1 Railway ticket of an attendee of the Congress of Italian scientists in Venice in 1847 to travel to the Festival of Flowers in Padua. By courtesy of Ministero della Cultura—Biblioteca Nazionale Marciana, Per 1152.A.9

Whether any of these stories correspond to historical facts is uncertain. Nor is it clear whether Roberto De Visiani was conscious of the anti-German and tumultuous history of the event, when, in 1845, he asked the Austrian authorities for permission to revive the Festival of Flowers to celebrate the third centennial of the Botanical Garden of Padua (Fig. 8.1.). But he received immediate approval from Vienna, still under the rulership of the flower emperor Francis I (Sect. 2.3 in Chap. 2). Similar events already took place in other European and North Italian cities. In the Botanical Garden of Modena, Giovanni De Brignoli di Brunhoff (1774–1857) organized annual floral expositions from 1843, and two years after that, the Mutual Aid Institution for gardeners in Milan arranged a 'Flower Fair' in the public gardens (Anon. 1846a, pp. 69–70). In 1844, Camillo Benso di Cavour (1810–1861) argued for the inclusion of floriculture at the Industrial Exposition in Turin (Anon. 1844, pp. 10–11).

De Visiani nurtured an "almost morbid love for plants" (Curti and Menegalle 1996, p. 79). His opening speech as professor of Botany in Padua was a romantic hymn to the "radiant mantle of verdure and flowers coating and covering the squalid nudity of earth! This splendid garment, where the suave temper of green seems to be competing with the vague azure of the skies" (De Visiani 1837, p. 5).[2] This enthusiasm found favor with many of his peers, also attested to by the broad positive response to the Gardening Society. The Festival of Flowers found an even greater and varied audience and transformed the Botanical Garden, a few years after its almost complete devastation, into a favorite place of Padua's elite. The first festival took place in June 1846 with the participation of seventeen competitors and the exposition of 857 plants (Anon. 1846b; Pasquali 1846, pp. 193–194). The society absorbed the costs for the transport of the plants, which came from all Veneto provinces, and bought 192 of them: 133 camellias, 28 azaleas, 6 gardenias, 14 Daphne, 4 rhododendrons, 2 Olea, 3 Fuchsia, 1 magnolia, and 1 Salvia. The advertisement of nine different prizes further boosted participation. About 8000 people visited the event, attracted by music, nocturnal illumination, and fireworks (Sgaravatti Montesi 1966, p. 78).

The great success of the first festival prompted a further event during the following year. This time, it took place in September in order to promote autumn flowers. It was opened by the viceroy, archduke Rainer Joseph of Austria, himself a passionate botanist and gardener, and members of his family. Meanwhile the number of associates of the Gardening Society had grown from 60 to 223, providing the association with more funds for prizes and other incentives. The second exposition saw the participation of twenty-six competitors and 1752 plants. This time, seventeen prizes were awarded (Anon. 1847a; Ronconi 1847). The date was also chosen because it coincided with the Ninth Congress of the Italian scientists, which took place in Venice. According to a report by Guglielmo Stefani, during the afternoon of 21 September, the attendees of the congress were transferred by train to Padua (Fig. 8.1), where "a long line of carriages and omnibuses, assigned for the short ride, waited" to bring them to the exposition and festivities (Stefani 1847, p 330).[3] In the evening, the whole group moved to the Caffé Pedrocchi to toast "among remote brothers, who feel the need to talk with each other affectionately about common joys and common hopes", calling out "Cheers to Padua!" and "Cheers to the Congress!" At ten o'clock in the evening, accompanied by Bengal fires and music bands, coaches took them to the station, where the train for the return to Venice was waiting.

The festival's success is in good part attributable to Padua's elite, their representational needs, and appetite for ornaments. Among the members of the Gardening Society were some of the most notable Veneto and Austrian families (Sect. 2.3 in Chap. 2), and in his report de Visiani did not omit to mention all the innovations in their gardens: new flower species, new types of glass houses and tepidaria, and eight new positions for gardeners (De Visiani 1847, pp. 6–10; Sgaravatti Montesi 1966, p. 79). In addition to the festivals, the Gardening Society also promoted flower shows in private gardens. Yet, two further important and closely interrelated trends surfaced in these festivals, namely the changing social composition of the participants and the growing economic dimension of floriculture. Analyzing the list of participants at the festival of flowers, we indeed notice an interesting social transformation and, closely linked with this, a shift from the romantic engagement with plants and flowers to an economic one. Floricultural activity ceased to be an exclusive privilege of the nobility. In 1846, only three out of seventeen competitors at the festival were noblemen (Anon. 1846b). The growing demand for garden as well as cut flowers indicates that flowers began to have strong esthetical, social, and symbolic appeal for the lower middle class, too. In 1847, De Visiani proudly reported that the love for plants and flowers also extended to the houses of the lower classes (De Visiani 1847, p. 10). Consequently, the market of potential purchasers grew steadily and this provided new opportunities for floricultural entrepreneurial activities. Alongside amateur floriculturists like the aforementioned abbot Romano and Pier Luigi Mabil, new figures stepped out of the shadow of anonymity in the late 1840s.

In a similar vein to the entrepreneurial activities of von Hügel in Vienna (Sect. 2.3 in Chap. 2), gardeners in Veneto sensed the economic possibilities. Most active was the previously mentioned Angelo Giacomelli. After a career in politics, he founded a garden in Portello di Treviso in 1842 (today Barriera Garibaldi). Under the aegis of his gardener Massimiliano Zaubek, it soon became famous for its collection of Cacteae (Saccardo 1895, p. 526). Giacomelli travelled extensively to visit European gardens, learning new techniques and trading in plants and seeds. He regularly published articles and catalogues in Andrea Meneghini's journal *Il Tornaconto*, further evidence of the economic importance of the flower market. Apart from this love for gardens and flowers and his entrepreneurial skills, Giacomelli shared other interests with Padua's reformist circles. He published articles on agricultural advancements, maintained Catholic-liberal ideas, and was active in the revolutionary movement. He was one of the group, tragically remembered as the Belfiore martyrs, whose members were hanged in Mantua between 1852 and 1853 during the Austrian repression. Giacomelli's father succeeded in saving Angelo's life. In 1852, he married Maria Rosmini, a relative of the philosopher Antonio Rosmini. After a financial crash in 1880, he was forced to sell all his belongings (Proietti e Biadene 2000).

Aside from the rise of bourgeois owners of smaller and less famous gardens, the 1840s also saw gardeners gaining wider official recognition. Even today their role is still underappreciated, yet recently started to attract some attention by historians (Cazzato 2009; Macellari 2019; Maniero and Macellari 2005; Wulf 2008). The first steps toward public recognition of the profession were taken in Milan in 1838 with the foundation of the *Pio Istituto dei Giardinieri di Milano*, a mutual aid institution, and with the celebration of San Foca, the patron of gardeners (and sailors), from 1840 in Turin, though soon adopted in many other cities. Floral expositions greatly contributed to the public visibility of gardeners, too. The organizers of Padua's Festival of Flowers indicated for each exhibited collection the names of both the garden owners and their gardeners. Interestingly, almost all prizes were awarded to the gardeners, not to the owners.

Several gardeners went into business for themselves. Carlo Maupoil (1777–1856), born in Paris and author of *Il buon giardiniere* (1826; *The good gardener*) (Maupoil 1826) and numerous works on experimental agriculture, founded in 1822 near Venice one of the most important horticultural-agricultural dynasties in Northern Italy (Manfrin 2016, pp. 79–84; Saccardo 1895–1901, p. 107). Another competitor at the Padua exposition was Giuseppe Maria Ruchinger (1809–1879), gardener of the Botanical Garden of San Giobbe in Venice. In 1866, he purchased the garden and transformed it into a plant nursery. His brother Giuseppe (1805–1856) had been assistant at the Botanical Garden in Padua from 1832–1834, and then became professor of Pharmacology in Prague (Saccardo 1895–1901, I, pp. 142–143 and II, p. 94; R. Vianello 1996; R. Vianello and Giormani 2018). In the mid-nineteenth century, several well-equipped companies for the production and commerce in plants and flowers set up shops around Padua: Gribaldo, Rizzi, Sgaravatti, and Van den Borre (Macellari 2018; Macellari 2020; Ronco 2002, p. 57). Probably the most important of these was the company founded by Angelo Sgaravatti (1798–1865) in 1820 at Saonara, also home of the Cittadella Vigodarzere garden. Sgaravatti started as gardener of the noble Morosini and Farsetti families. He soon became an expert, set up his own company and became rich enough to buy the villas and gardens of his impoverished ex-patrons. Thanks to clever marketing strategies, his enterprise expanded, including into other European markets where Sgaravatti sold his products to royal houses and ministries. In the early twentieth century, the production in Saonara alone covered 260 hectares and employed 338 workers (Belussi and Sedita 2008; Celetti 2013; Gardano 1992–93).

The Festival of Flowers thus reflected the multifaceted role of ornamental plants in Padua and the socio-economic transformations that took place in Veneto during the 1840s. The festival was a major social event, an opportunity for the *notabili* to display the marvels of their gardens, an occasion of encounter of people from different geographical, social, and gender backgrounds, a delight as well as a pleasure for the population, and a showcase for emerging economically profitable businesses. Giuseppe Meneghini was involved as vice-president of the Society and in 1847 as a member of the commission for the flower competition (Anon. 1847b).[4] However, the Society as well as the festival were the pet projects of De Visiani. There can be no doubt that the festivals of 1846 and 1847 represented the highpoint of floriculture in Padua. After the 1848 revolution, the festival had two more editions, one in June 1854 and the other in May 1868, but then, owing to political controversies, were discontinued with no successor event until the late twentieth century.

Notes

1. G. Meneghini 1844–1846, 1, p. 314: "Il fiorellino, che abbarbicato tenacemente alla terra innalza i suoi incensi al cielo, sembra quasi parlare all'uomo l'amore di patria e il culto a Dio."
2. De Visiani 1837, p. 5: "questo ridente ammanto di verzura e di fiori, che tutta avvolge e ricopre la squallida nudità della terra! questo splendido vestimento, in cui la soave tempra dei verdi gareggiar sembra col vago azzurro de' cieli."
3. Stefani 1847, p. 330: "… il breve tragitto alle assegnate carrozze ed agli omnibus che in lunga fila disposti stavano pronti."
4. The other members were De Visiani, Ronconi, Configliachi, Alberto Parolini, Giuseppe de Salvi, Vittore Trevisan, and Giuseppe Cristina. Given the participation of attendees of the scientific congress in Venice, De Visiani also invited von Hügel, Robert Brown, and the Scottish archeologist Alexander Henderson.

References

Anon. 1844. *Catalogo dei prodotti dell'industria de' R. Stati annessi alla pubblica esposizione dell'anno 1844 nelle sale del castello del Valentino*. Torino: Tipografia Chirio e Mina.

———. 1846a. *La Scuola di Minerva: Strenna genealogico-storica per l'anno 1846*. Milano: Tipografia Pirotta e C.

———. 1846b. *Prima esposizione di piante e fiori, tenuta dalla Società promotrice del giardinaggio nell'I. R. Orto botanico di Padova nei giorni 7 ed 8 giugno 1846*. Padova: Tipi del Seminario.

———. 1847a. *Seconda esposizione di piante e fiori tenuta dalla Società promotrice del giardinaggio nell'I. R. Orto botanico di Padova*. Padova: Crescini.

———. 1847b. Processo verbale della Commissione incaricata di aggiudicare i premi alle piante e collezioni più meritevoli fra quelle prodotte al concorso nella Esposizione tenuta nell'I. R. Orto Botanico di Padova dalla Società promotrice del giardinaggio. *Il Tornaconto* 39 (30 September 1847): 323–325.

———. 1848. Catalogo delle Dahlie ed altre piante vendibili presso Massimiliano Zaubek giardiniere di Casa Giacomelli in Treviso. *Il Tornaconto* 3 (20 January 1848): supplement.

———. 1852. *Storia veneta, espressa in 150 tavole inventate e disegnate da Giuseppe Gatteri sulla scorta delle cronache e delle storie più riputate e secondo i vari costumi del tempo incise da Antonio Viviani ed illustrate da Francesco Zanotto*. Venezia: A. Viviani.

———. 1854. *Terza esposizione di piante e fiori tenuta dalla Società promotrice del giardinaggio nell'I. R. Orto botanico di Padova*. Padova: L. Penada.

Belussi, Fiorenza, and Silvia Rita Sedita. 2008. The symbiotic division of labour between heterogeneous districts in the Dutch and Italian horticultural industry. *Urban Studies* 45 (13): 2715–2734.

Brocchi, Giovanni Battista. 1796. *Trattato delle piante odorifere e di bella vista da coltivarsi ne' giardini*. Bassano: Stamperia Penada.

Cazzato, Vincenzo, ed. 2009. *Atlante del giardino italiano, 1750–1940: Dizionario biografico di architetti, giardinieri, botanici, committenti, letterati e altri protagonisti*. 2 vols. Roma: Istituto Poligrafico dello Stato.

Celetti, David. 2013. *Vivai Sgaravatti*. Padova: Il Poligrafo.

Curti, Luigino, and Fernanda Menegalle. 1996. Roberto de Visiani. In *Professori di materie scientifiche all'Università di Padova nell'Ottocento*, ed. Sandra Casellato and Luisa Pigatto, 77–85. Padova: Edizioni Lint.

De Toni, Giovanni Battista. 1887. *Intorno ad alcuni alberi e frutici ragguardevoli esistenti nei giardini di Padova. Padova*. Tipografia Gio. Batt. Randi.

———. 1909. Una lettera inedita del botanico padovano Giuseppe Meneghini. *Bollettino del Museo civico di Padova* 12: 13–16.

De Visiani, Roberto. 1837. *Della utilità ed amenità delle piante. Discorso inaugurale letto il 20 aprile nella grande aula dell'Università di Padova*. Padova: Minerva.

———. 1840. *Illustrazione delle piante nuove o rare dell'Orto Botanico di Padova. Memoria I*. Padova: Tipi di Angelo Sicca.

———. 1847. *Parole indirizzate il giorno 15 di Dicembre 1846 all'adunanza generale della Società promotrice del Giardinaggio in Padova*. Padova: Tipi del Seminario.

Dumont de Courset, Georges Louis Marie. 1819. *Il botanico coltivatore, recato in italiano dall'ab. Girolamo Romano*. Padova: Tipografia e Fonderia della Minerva.

Gardano, Stefano. 1992–93. *La famiglia Sgaravatti e il florovivaismo da reddito in Saonara (tesi di laurea)*. Padova: Università degli Studi di Padova, Facoltà di Scienze politiche.

Hyde, Elizabeth. 2005. *Cultivated power: Flowers, culture, and politics in the reign of Louis XIV*. Philadelphia: University of Pennsylvania Press.

Jeffery, Violet M. 1924. Italian and English pastoral drama of the Renaissance: Sources of the 'Complaint of the Satyres against the Nymphes'. *The Modern Language Review* 19 (1): 56–62.

Mabil, Pier Luigi. 1801. *Teoria dell'arte de' giardini*. Bassano: Remondini.

———. 1817. *Saggio sopra l'indole dei giardini moderni*. Verona: Mainardi.

Macellari, Elena. 2018. Il delirio della Botanica: Vecchi cataloghi e stabilimenti d'Orticoltura. https://wsimag.com/it/scienza-e-tecnologia/36962-il-delirio-della-botanica. Accessed 15 February 2020.

———. 2019. Grandi giardiniere italiane. https://wsimag.com/it/architettura-e-design/51689-grandi-giardiniere-italiane. Accessed 15 February 2020.

———. 2020. La storia avvincente della floricoltura. https://wsimag.com/it/scienza-e-tecnologia/37518-la-storia-avvincente-della-floricoltura. Accessed 15 February 2020.

Manfrin, Mauro. 2016. Tra Mira e Dolo nel 1866: il plebiscito delle donne, i comizi agrari e la sperimentazione agricola: Appunti di storia atipica. *L'Esde: Fascicoli di Studi e di Cultura* 11: 69–92.

Maniero, Federico, and Elena Macellari. 2005. *Giardinieri ed esposizioni in Italia (1800–1915)*. Perugia: Ali&No Editrice.

Marsili, Giovanni. 1840. *Notizie inedite dei patrizii veneti dotti nella cognizione delle piante e dei loro orti botanici più rinomati*. Padova: Tip. Cartallier e Sicca.

Massaro, Martina. 2014–2015. *Giacomo Treves dei Bonfili collezionista e mecenate (1788–1885): La raccolta di un filantropo patriota*. Dissertation thesis, Università IUAV di Venezia, Venezia.

———. 2018. Visualizing the Treves botanical garden in Padua: from documentary research to laser survey and 3D modeling. In *Visualizing Venice: Mapping and modeling time and change in a city*, ed. Kristin L. Huffman, Andrea Giordano, and Caroline Bruzelius. Chap. 7. Abingdon/New York: Routledge.

———. 2019. *Palazzo Treves dei Bonfili e il suo giardino*. Padova: Il Poligrafo.

Maupoil, Carlo. 1826. *Il buon giardiniere*. 2 vols. Venezia: Gaspari S. Felice.

Meneghini, Giuseppe. 1844–1846. Lezioni di Botanica Popolare. *Giornale Euganeo* 1: 314–320 and 394–402; 2(1): 197–207 and 407–412; 2(2): 31–39, 266–273, and 301–311; 3(1): 251–259; 3(2): 194–208, 432–450, and 497–519.

Minuzzi, Sabrina. 2016. *Sul filo dei segreti: Farmacopea, libri e pratiche terapeutiche a Venezia in età moderna*. Milano: Edizioni Unicopli.

Pasquali, A. 1846. Solennità della Società promotrice del Giardinaggio in Padova, e in particolare della festa dei fiori seguita in quell'IR Orto Botanico. Il Vaglio XI/25 (20 giugno 1846): 193–194.

Proietti, Andrea, and Giovanni Biadene. 2000. Angelo Giacomelli. In *Dizionario Biografico degli Italiani* 54, http://www.treccani.it/enciclopedia/angelo-giacomelli_(Dizionario-Biografico)/. Accessed 15 January 2020.

Romano, Girolamo. 1823a. *Le iridi cultivate per le auspicatissime nozze Gaudio-Meneghini*. Padova: pei tipi della Minerva.

———. 1823b. *Le piante fanerogame Euganee: per le nobilissime nozze Emo-Capodilista e Maldura*. Padova: pei tipi della Minerva.

———. 1828. *Le piante fanerogame Euganee: per le nobilissime nozze Cittadella-Maldura*. Padova: tip. del Seminario.

———. 1831. *Le piante fanerogame Euganee: per le auspicatissime nozze Meneghini-Fabris*. Padova: tip. del Seminario.

Ronco, Roberto. 2002. *La filiera florovivaistica nel Veneto*. Udine: Industria Grafica Editrice Multigraf.

Ronconi, Giovanni Battista. 1847. Seconda esposizione di piante, fiori e frutti fatta in Padova dalla Società promotrice del giardinaggio. *Il Tornaconto* 1(40) (7 October 1847): 333–334 and 1(41) (14 October 1847): 341–343.

———. 1853. Appendice, curiosità botaniche, il Giardino Treves, a Padova. *Gazzetta uffiziale di Venezia* 191: 761–762.

Saccardo, Pier Andrea. 1869. *Della storia e letteratura della Flora veneta. Sommario*. Milano: Valentiner e Mues Librai-Editori.

———. 1895. Contribuzioni alla storia della botanica italiana. *Malpighia* 8: 476–539.

———. 1895–1901. *La botanica in Italia: Materiali per la storia di questa scienza*. Vol. 2 vols. Venezia: Tipografia Carlo Ferrari.

Sampson, Lisa. 2006. *Pastoral drama in early modern Italy: The making of a new genre*. London: Routledge.

Sberti, Anton Bonaventura. 1818. *Degli spettacoli e delle feste che si facevano in Padova*. 2nd ed. Padova: Adolfo Cesare.

Schneider, Federico. 2016. *Pastoral drama and healing in early modern Italy*. London/New York: Routledge.

Sgaravatti Montesi, Marisa. 1966. *Giardini a Padova e manifestazioni floreali*. Prefazione di C. Cappelletti. Padova: Associazione Pro Padova.
Sorgato, Gaetano. 1845. *Memoria sugli spettacoli e sulle feste di Padova*. Padova: Tipi del Seminario.
Stefani, Guglielmo. 1847. Un po' di cronaca padovana: La Festa dei Fiori. *Il Caffè Pedrocchi* 2(39) (26 September 1847): 329–330.
Vianello, Valerio. 1993. *Il "giardino" delle parole: Itinerari di scrittura e modelli letterari nel dialogo cinquecentesco*. Roma: Jouvence.
Vianello, Riccardo. 1996. L'Orto botanico di San Giobbe a Venezia. *Atti e memorie dell'Accademia italiana di storia della farmacia* 3 (2): 129–138.
Vianello, Riccardo, and Virgilio Giormani. 2018. *Orti botanici e parchi pubblici. Un frammento di storia di Venezia nel XIX secolo*. Venezia: Istituto Veneto di Scienze Lettere ed Arti.
Wulf, Andrea. 2008. *The brother gardeners: Botany, empire and the birth of an obsession*. Cornerstone: William Heinemann.

CHAPTER 9

Cultivating Land and People

In the 1840s, Veneto had about two million inhabitants, the great majority of whom made their living from agriculture. No analysis of Padua's cultural settings can gloss over this matter of fact. The following pages will not even attempt to provide a social history of a topic as extensive and complex as agriculture or agronomics in Italy, since the late 1980s subject of numerous insightful treatises (e.g., Bevilacqua 1989–1991; Biagioli and Pazzagli 2004), nor in Veneto or the Padua area. Rather they will concentrate on some specific items I consider significant for illustrating the interaction between politics and plant science and the role of the Meneghini brothers in this interplay.

During the Enlightenment, the meaning of the term 'to cultivate' was not restricted to the growing of crops. More generally, it meant 'to improve', and 'cultivation' was assimilated to 'culture'. The analogy is antique actually. The Roman poet Horace (65–8 d. C.) already compared rude men with fields, which needed cultivation to bear fruit. The metaphors of Jesus Christ as a sower as well as the 'seeds of knowledge' and the German *Kindergarten* also stem from this connection. The cultivation of land was conceived of as a similar civilizing process as the cultivation of people. Progressivists saw themselves as farmers who do their utmost to create the best possible conditions for their fields, to improve techniques, and to select species so that the single plants (or men) could do their part that the harvest

A. Dröscher, *Plants and Politics in Padua During the Age of Revolution, 1820–1848*, Palgrave Studies in the History of Science and Technology, https://doi.org/10.1007/978-3-030-85343-3_9

would progressively improve. During the Enlightenment, this imagery became particularly intriguing. Jean-Jacques Rousseau, for instance, made an analogy between the cultivation of the mind and the cultivation of agricultural and horticultural lands; however, he distinguished between passive female cultivation and active male education (George 2005). Other Enlightenment savants regarded the uniform ranks of crops as a symbol of order and containment, and therefore helpful in controlling nature and disciplining farmers. Herder, too, used analogies of human and plant cultivation (Goeth 2017, p. 70–71). As we have seen in Sect. 6.2 in Chap. 6, Romagnosi's *incivilimento* was a kind of cultivation, too, even though he also conceded an important degree of agency to the individual elements and their interactions. My point here is that the partial overlap of the agricultural and the social meaning of cultivation strongly influenced social theory in Veneto, as manifest in Andrea Cittadella Vigodarzere's peculiar romantic agro-vegetal worldview that will be the topic of Sect. 9.2.

Philosophy is not the only area of overlap between plant science and politics. They also came together in more practical ways. The activities of the *Istituto Veneto* (Sect. 2.4 in Chap. 2) and of the Paduan associations (Sects. 2.2 and 2.3 in Chap. 2) reflected the enduring dominance of horticulture and agriculture in the thinking and operation of the Paduan intellectual elites. Between 1840 and 1847 the *Istituto* awarded nine prizes for scientific merit. Three of these were given in agriculture, three in economics, and only three in a basic science, namely physics (Gullino 1996, p. 44–48). Likewise, we can also observe in agricultural matters a vigorous attempt by Paduan notables to use botanical knowledge to assume a leading role in the modernization of the basic economic pillar of the entire region. The foundation of the Agricultural Garden in Padua as well as the *Society for the Encouragement of Agriculture and Industry in Padua* and the direction of the journal *Il Tornaconto* (The Profit) (Sect. 9.3) provide illuminating case studies for investigating this question. Yet, while the Agricultural Society and the journal were typical offshoots of the 1840s, the story of the Agrobotanical Garden illustrates how an eighteenth-century vanguard institution was not able to meet the new needs of mid-nineteenth century agronomy.

Both Meneghini brothers played a crucial role in the interplay between science and agricultural ideologies and policies. Andrea was a prime mover behind the *Society for the Encouragement* and director of the journal *Il Tornaconto*, which engaged with economic and agricultural questions. He was active on the political level, in particular organization, information, and networking, whereas his brother Giuseppe promoted the adoption and dissemination of a scientific approach to agriculture. In this context, debates around the potato cultivation and potato failure in Veneto in the first half of the nineteenth century, treated in Sect. 9.3, are particularly insightful. A flashpoint of international exchange, growing nationalism, resistance to innovation, and existential food crises, the potato debate eventually fostered the increasing involvement of botanists that studied the causes of the crop failures.

9.1 The Agrobotanical Garden of Padua

Not all projects to create scientific institutions devoted to applied botany in Padua have come to fruition. Success depended on agreement, at any one time, with the respective governing authorities in Venice, Paris or Vienna. Agroforestry, for instance, did not find endorsement, even though there had been evident necessity. In the northern parts of Veneto, forests played a decisive absorbing role in damming the river waters percolating down from the Alps. Besides, throughout the modern age, Venice deployed considerable (though insufficient) administrative and legislative activity to satisfy its incessant need for wood, in particular larch, fir, oak, and beech trees, for the shipyards and arsenals (Lazzarini 2014–2018). Napoleon had been aware of the dangers of excessive logging. Yet, the need for firewood, especially in times of war, had created an acute shortage of wood. He therefore launched public reforestation programs, to plant and maintain rows of trees along the roads of the Reign (including Italy), and to create an intricate network of functionaries, nurseries, and scientific botanical institutions, which experimented with and exchanged new tree seeds (Ambrosoli 1998, p. 711–712). His reforestation program achieved many of its goals and the fundamental one of reestablishing a secure wood supply (Sect. 4.1 in Chap. 4). Yet, howsoever important trees were for the economy of Veneto, forestry was of minor interest for the Paduan province that only had wooded areas in the Euganean hills (Celetti 2008, p. 68–85). Whether it was for this reason or another, Padua did not succeed in establishing scientific dominance in this field, because the Austrian government denied the requisite funds for the creation of chairs of forest science at the universities of Padua and Pavia, preferring to train the keepers for the Venetian woods at their own academy in Mariabrunn (Lazzarini 2013, p. 10–11). Some attempts to provide silvicultural advice in Padua were made in the context of the Festival of Flowers, which cleverly took advantage of the Paduan tradition in tree collecting (Sect 8.1 in Chap. 8) and incorporated exotic trees into the exhibition as well as into many Paduan gardens, thus promoting the introduction of useful species for timber or fruit cultivation (Anon. 1847a). Similar attempts of promoting silviculture in public and private parks were made in other European countries (e.g., Tenzer 2017).

The case of agronomy was different. From the mid-eighteenth century, enlightened minds across Europe saw an urgent need for systematic approaches and radical reform and interventions in agriculture. The first dedicated academy was founded in 1736 in Dublin. Many other cities followed soon thereafter. The Republic of Venice, aware of the decreasing flow of wealth and goods into its harbors, turned its attention to the ill-kept agriculture of its hinterland. In 1765, following a series of devastating rural epidemics and serious food shortages, the *Serenissima Signoria*, the government of the Republic of Venice, created the first chair for experimental agriculture in Padua and appointed Pietro Arduino (1728–1805). In his first report to the Senate in Venice, Arduino recommended the establishment of academies in all major cities of the territory, a project that was promptly implemented. Soon a network of agricultural academies covered the entire republic. In 1768, the academies of Treviso and Verona commenced work, the following year it was the turn of Belluno, Conegliano, and Vicenza, and in 1770 the academy of Feltre opened its doors. Minor academies also opened in Bassano, Bergamo, Brescia, Capodistria, Cefalonia, Crema, Oderzo, Rovigo, Salò, Spalato, Trau, Udine, Zara, and other cities, yet many remained practically inactive due to the lack of adequate funding (Luzzatto 1928; Molesti 2006, p. 157–186). Padua's agricultural academy opened in 1769, but merged ten years later with the *Accademia dei Ricovrati* to form the *Accademia di Scienze Lettere ed Arti*, thus including agriculture in the city's general scientific, cultural, and social life.

The task of the academies was not to come up with grand social or political visions, but to propose and disseminate concrete and practical improvements. This included the systematic collection of data and compilation of reports, discussion of possible legislative, administrative, and infrastructural interventions, evaluation of new machines, methodologies, and techniques to increase the yields from cultivations, the study of plant pathology, proposals for new species for introduction into the monotone crop repertoire of the Venetian farmers, and the like. In 1773, for instance, during a food shortage, the Venetian government turned to Padua's

Academy and requested that its associates carry out eight experiments. The Academy's means were those of classical academies: scientific research, discussion and dissemination of knowledge through prize competitions and publication of proceedings, books, and journal articles.

In contrast with the educational functions of many Nordic, in particular Scandinavian agricultural gardens (Schnitter 2011), and in contrast with the moral education through gardening promoted by John Amos Comenius (1592–1670) in Bohemia and Rousseau in France, Padua's agricultural institution distinguished itself through its scientific outlook and its close ties with botanical research. In fact, the newly founded discipline of agronomy was not housed in the faculty of law, as it happened at other Italian universities, but among the natural sciences in the faculty of philosophy.[1] Still another indicator of the close ties between agriculture and botany in Padua is that Pietro Arduino, before becoming Italy's first university professor of agriculture, had directed (even though he only retained the position of keeper) Padua's Botanical Garden. His new chair was called *Cathedra ad agriculturam experimentalem*, thus emphasizing the charge of undertaking experimental research. In order to do so, the *Serenissima* instituted, as early as 1766, the first university agricultural garden, the 5.3-hectare *Campi della pubblica scuola di agricoltura* [Fields of the public school of agriculture] (Zanetti 1992; Zanetti 1996) (Fig. 2.1, no.2 in Chap. 2). The agricultural garden of Milan was created in 1781, the one of Pavia in 1807. Pietro as well as his son and successor Luigi Arduino (1750–1833) preferred a botanical and experimental rather than a merely didactical approach (L. Arduino 1807; P. Arduino 1766). Yet, seeing themselves as teachers of an applied science, they held their lectures in Italian, not in Latin, in order to address farmers and land owners.

In 1829, the direction of the garden passed into the hands of the Barnabite Abbot Luigi Configliachi (1787–1864) (Busatto 2015; Casellato 1996; Casetta 2012). Configliachi had taught in Milan and Mantua. In 1819, he was called onto the chair of Rural economy in Padua, and he also lectured on general natural history. This choice was not entirely fortunate. In his early years in Padua, Configliachi translated a series of German-speaking scientific and technical textbooks, among them *Ludolph Christian* Treviranus's volume on plant anatomy (Treviranus 1806; Treviranus 1822), and Leopold Trautmann's (1766–1825) successful agronomy handbook (Trautmann 1810–1811; Trautmann 1820–1821). It therefore appears likely that agricultural teaching in Padua was built on romantic scientific principles. Around 1810, when the originals of both books appeared, Treviranus and Trautmann had been among the pioneers of microscopic-anatomical and chemical approaches in their respective fields. However, in the 1820s and, even more so, in the 1830s, when Configliachi used them for his teaching, recourse to dynamic forces acting in and through matter was already outdated.

Configliachi never published a proper book or scientific paper. One of his very few editorial outputs was a small volume on agriculture in Padua, written for the wedding of the nephew of the missionary Jesuit Abbot Gaetano Zara (1819–1853) (Configliachi 1831, p. 5–38). In this short treatise of 1831, Configliachi provided some practical but rather generic advice on how to plant trees, in particular fruit trees. When he discussed mulberry trees, he abandoned the usual style of eighteenth-century natural history and became rather indignant about the ignorance of Paduan farmers. He reproached them for trimming the poor trees 'like barbarians'. He complained that farmers of the neighboring provinces, instructed by their owners, cultivated their stock correctly and therefore did not lack product, whereas "in the Paduan territories all these outputs are miserable, due to a cloud of prodigious errors deriving from deep-rooted ignorance" (ibid. p. 21).

Configliachi collected over 300 agricultural machines, yet these served for teaching not for research. It is unlikely that he carried out any systematic experience or experimented in the *Orto agrario.* This was not his personal choice alone. Some years earlier, the president of the Executive Board of the Cisalpine Republic, Pietro Moscati (1739–1824), had threatened Configliachi's predecessor Arduino with transfer to a different post if he did not stop his research and dedicated his time exclusively to teaching. Therefore, Pietro Casetta argues, Configliachi followed Austrian directives. Casetta (2012, p. 17) furthermore emphasizes that Configliachi acted as a 'priest agronomist', that is, not as a scientist searching innovative solutions but as a fatherly cleric who educated the rural population and refined existing techniques. Wherever the blame may fall, in the late 1830s and 1840s, when times had changed and a more scientific approach was possible and required, Configliachi did not change his attitude. He was a visible presence in Padua's circles and associations (Table 2.1 in Chap. 2), but did not exert a demonstrable innovative influence. He directed most of his energy toward philanthropy, founding an educational institute for blind children, the *Asilo caritatevole per la sociale istruzione dei ciechi.*

The history of Padua's Agrobotanical Garden in the 1840s is therefore most certainly a story of a missed opportunity, for it failed to contribute to the modernization of agronomy in Veneto in the same way as other European institutions. More reformist and far-sighted agronomists like Gherardo Freschi (1805–1883) from Friuli and Francesco Luigi Botter (1818–1878) from Treviso, both linked to Padua, could have ushered in effective collaboration between botany and agriculture, but they moved to other cities and implemented successful agronomical reforms elsewhere. Gherardo Freschi graduated from Padua in law and in natural sciences. He soon stood out due to his European experience and reformist spirit. Francesco Luigi Botter was, in the 1840s, promoted as successor of his master Luigi Configliachi, but accepted an invitation of the Papal States to found an important agronomical school in Ferrara. The example of potato cultivation in Padua shows how Giuseppe Meneghini tried to step into the breach (Sect 9.4). However, neither the Austrian governors nor the local elite seem to have taken measures to modernize the agronomical teaching and research in Padua.

9.2 The Land Is a Garden: Romantic Cultivation

In this section, I argue that the influential circle of big Paduan landowners, headed by Andrea Cittadella Vigodarzere (Fig. 9.1), promoted a particular 'agro-botanical policy' to underpin their economic, social, and political power. I call it 'agro-botanical policy', because, firstly, to a good extent it was based on an agricultural philosophy. Agriculture was fundamental for Veneto's economy, but for the wealthy class it was even more than a branch of production. It was central to their social ideology. The group of nineteenth-century intellectuals of the region mostly coincided with the group of landowners, and the latter considered agronomy a social and educational discipline, rather than a mere technical-economic one (Fumian 2004, p. 226–227). Secondly, I do so because to some extent the policy was based on a botanical philosophy. Plant-based crop production dominated Veneto's agriculture, whereas animal husbandry was almost non-existent. This had historical reasons. For a long time, the Republic of Venice had relied on imports for its meat consumption, which meant that the agriculture of its hinterland almost exclusively focused on crops (Berengo 1963, p. 12–15; Cavalli 1851, 1, p. 21–23). This mindset became deep-rooted among landowners and peasants and was later, when imports became too expensive, extremely difficult to overcome. The second botanical dimension concerns the Paduan conception of romantic gardens which also provides insights into the social vision of their owners.

Fig. 9.1 Andrea Cittadella Vigodarzere (1804–1870) in 1846. Xylograph by Francesco Ratti (1819–1895). From Scolari (1846, p. 5). By courtesy of Biblioteca delle Collezioni d'Arte e di Storia, San Giorgio in Poggiale, coll. TC R6

Andrea Cittadella Vigodarzere exemplifies the prevailing Veneto model of social and economic domination of the rural population. He is the main focus of this section. He was one of the richest landowners in Veneto and generally known for cultural patronage and considerable involvement in civil philanthropy. He presided over the local Orphanage of S. Maria delle Grazie, the *Casa di Ricovero per anziani* (Asylum for the old), and the Charity Commission. Later, he was among the most fervent advocates for the creation of kindergartens. His public protagonism found a congenial expression in his participation and promotion of the emerging Paduan associations and, in 1848, meant that he became embrangled in political affairs. He acted as secretary of the Section of Literature of the Academy of Science (1838–1847), as president of the Reading Cabinet (1839–1845), as president of the *Istituto Veneto* (1845–1847), as general president of the Science Congress in Padua (1842), and as president of the (politically most delicate) agronomic section of the Science Congress in Venice (1847). The Agricultural Society over which he presided in the initial years of its history played a pivotal role in his political strategy. Nevertheless, models other than Cittadella Vigodarzere's existed. Andrea Meneghini actively supported Cittadella Vigodarzere's initiatives, but his vision and actions were subtly different.

Any aspiration for social and economic progress in Veneto, in rural as well as urban contexts, had to factor in the agricultural dimension. Indeed, there were numerous well-intentioned reform ideas and projects, but proposals formulated in the academies often failed to reach the rural population. It was hoped that the foundation of agricultural gardens would compensate for this shortcoming, yet, as we have seen in the previous section, in the 1830s and 1840s, Configliachi was not too successful in this regard. Another instrument for innovation was the creation of agricultural societies. In Northern Italy, the first such societies date back to the closing decades of the eighteenth century (Molesti 2006). In 1768, the *Serenissima* installed the *Accademia di Agricoltura di Verona*, which equipped itself with a botanical garden, a library, and a museum of natural history, and soon became the principal cultural institution of the city (Ferrari 2012). Eight years later, Empress Maria Theresa created the *Società Patriotica di Milano*, around which assembled a network of noble landowners, naturalists, botanical experts, and parish priests who investigated the introduction and usefulness of new plant species, like olive trees, potatoes, Chinese radish, and Siberian malt, in their private gardens and estates (Visconti 2013). Napoleon also left an imprint. He mounted massive interventions to encourage agricultural progress in France and its client state, the Kingdom of Italy (Ambrosoli 1998, p. 716–717). His administrative reorganization of Northern Italy included the creation of a centralized network to connect its various botanical and agronomical institutions—academies, botanical gardens, private nurseries, and farmers, but also local mayors—and link them directly to Paris. The Botanical Garden of Turin became a center in the dissemination of seeds from Paris to Italy, and the intense circulation of information, orders, and seeds considerably changed the Italian flora and urban landscapes (ibid., p. 732–733).

The situation in Padua was different. The city played a rather marginal role in Napoleon's plan, and during the Restauration, the Austrians did not resume projects like Maria Theresa's. Reformist ideas circulated in Padua's Francophile salons. Antonio Vigodarzere and Agostino Meneghini had grown up in that world, and their efforts to drain the uncultivated areas of their estates and to plant trees in their gardens can be considered part of the Napoleonic legacy. Other enlightened landowners actually set out visions of social reform, but these ideas remained abstract words printed in books. The academies suggested concrete improvements, too, yet resistance to innovation in Veneto's rural society was one of the main obstacles to putting them into practice. Therefore, the reform wave of the late eighteenth century amounted to little more than a flash in the pan. Venerable academic discussions were scarcely effective if they did not get beyond city walls to reach landowners and farmers. Even worse, urban aristocratic landowners were often involved in harsh legal disputes with their *affittuari* (renters) and had mostly lost contact with the peasants. The failed introduction of the potato in the initial decades of the nineteenth century (Sect. 9.4) can be seen as a clear manifestation of the alienation between urban and rural populations, between landowners and peasants, and between learned scholars and illiterates. Most peasants deeply distrusted advice coming from above or from the academies and only relied on their own experience, whereas most of the Venetian nobility disdained the peasants for being rude, illiterate and immoral, and feared the 'power of the masses' (Bernardello 1997; Dandolo 1817, p. 246–252).

The project to create an agrarian society, the first of its type in Veneto, was launched in 1841 as an initiative of Padua's Chamber of Commerce. Eighty-one persons subscribed the official manifesto, among them, of course, Andrea Cittadella Vigodarzere and Andrea Meneghini (Solitro 1930, p. 268–269). One year later, the marquis Pietro Selvatico advertised the project at the Fourth Congress of Italian scientists in Padua. He complained about the low level of instruction among peasants and owners, "the inertia of the torpids", and the scarcity of country parsons, who might instruct the young "with the heart of a father" (Selvatico 1842, p. 230). The norms he proposed as guiding principles for the agrarian society reflected his roots in Enlightenment philosophies. He saw the Agricultural Society as an instrument for disciplining the people as much as to improve agriculture.

Selvatico's call for the establishment of similar societies in all Italian provinces and his emphasis that these should be "municipal in the application but Italian in their moral action" (ibid. p. 231)[2] did not go unnoticed by the Austrian government, which subsequently erected a series of political and administrative obstacles. The *Society for the Encouragement of Agriculture*, or *Società d'incoraggiamento per l'agricoltura*, only became active three and a half years later, in January 1846. The proceedings and the budget were initially published in the local journals, *Giornale Euganeo* and *Il Tornaconto*. Its first president was Andrea Cittadella Vigodarzere.

For him, this task was the most heartfelt, because the Society of Encouragement was to become the instrument of his utmost desire, the reform of agriculture.

In concert with his moderate peers, Cittadella Vigodarzere was aware of the urgent need for modernization and, like them, placed his trust in education, that is, better schools, popular science books, technical agricultural advice, and moral teaching by the parishes. Similar to Milan's *Società Patriotica*, Padua's Society was intended to serve in equal parts as an academy, advertising prizes (Solitro 1930, p. 106–111), producing publications, and offering a place where "minds of high thought are preparing this sacred bread of popular instruction" (Selvatico 1842, p. 230),[3] and as a network of country parsons. Parishes were the principal vectors in the dissemination of agricultural knowledge in the provinces. Luigi Configliachi was a typical, albeit better educated than most, example of this personage (Sect. 5.1 in Chap. 5). The network in Veneto was particularly dense. Especially in the rural areas, priests were not only the religious and moral authorities, but also responsible for the registry office and the administration of philanthropic institutions. The country parson was at the same time a civil servant and a public employee responsible for the general wellbeing of his community (Lanaro 1984, p. 24–28). If any innovation wanted to succeed among the farmers, traditionally averse to change, especially if proposed by urban academics, it needed to be embraced by him.

The Society for Encouragement typifies Padua's associationism (Sect. 2.2 in Chap. 2). It saw itself as an open and socially dynamic civil congregation explicitly devoted to concrete practical purposes, yet, indirectly, it created an organ of increasing political influence. Indeed, members of the Society often invoked the spirit of collaboration. Cittadella Vigodarzere (1843, p. xxvi, xxxvi, and xlvi) praised the 'union of will' and, in the same breath, man's historical progress from slavery to civil equality. In 1844, count Ferdinando Cavalli (1810–1888), together with Selvatico and Angelo Orlandi founding commissioner of the Society, and in 1850–1856 its president hailed "this marvellous event of mutual aid and exchange of services that is the basis of civil life" (Cavalli 1853, p. 6).[4] However, the ideal of collaboration among equals did not extend to everybody. Rather, it was restricted to the upper class of the *possidenti* and to the intellectual elite (often one and the same), who indeed cooperated, exchanged, and negotiated strategies of joint action, yet exerted a hierarchic control over the lower strata of society.

Cavalli was the great-great-grandson of Paolo Renier (1710–1789), the penultimate Doge of Venice. He had studied law in Padua and had been assistant to Alessandro Racchetti, professor of Criminal procedure and a disciple of Gian Domenico Romagnosi. When, in 1824, he inherited, like Cittadella Vigodarzere, a great fortune, he dedicated his life to politics and valued systematic studies of political economy. He shared Cittadella Vigodarzere's liberal-conservative vision of agricultural progress and social responsibility for the land and the peasants. Yet, under the auspices of Cavalli and Cittadella Vigodarzere, the Society did not actually admit the lower classes, even less so 'villagers'. They saw the Society as an honorable association of wealthy and learned persons. Whilst the first draft of the statutes, discussed and signed during the congress in Padua, had foreseen membership fees of three Austrian Lire (Anon. 1842b), the actual amount of the 196 subscriptions paid in its first year was 25 Austrian Lire (Anon. 1847f). This was more than eight times and thus represented an effective means of selection. On the other hand, it was in line with the fee of the Gardening Society and the annual subscription to *Il Tornaconto* (24 Lire per year), and lower than that of the Reading cabinet, which in 1844 was 15 Lire for enrolment and a 6 Lire of bimonthly usage fee.[5]

Cittadella Vigodarzere's project went beyond mere promotion of economic incentives. He saw the Society as an instrument for the realization of his general vision. Terenzio Maccabelli (2000, p. 301–302) has convincingly argued that the Society's aim was not to overthrow traditional structures but to redefine the identity of the landowner into a persona that ought to bring together innovative technology, social engagement, and political economy. In comments that reveal the Society's fundamental *raison d'être*, Cittadella Vigodarzere attacked 'the inattentive master' as well as quasi-industrial exploitation. He declared that "we consider false, unjust and harmful the opinion of those who want to afflict the order and social quiescence to become the helotism of the minor classes, and want to perpetually condemn the villagers to be little more than agrarian instrumental material" (Cittadella Vigodarzere 1851, 1, p. vii–viii).[6] His hero was his uncle Antonio, from whom he had inherited his immense wealth. In his obituary, he heaped adulation on him as "a landowner, aiming to better fields and farmers; who arouses fertility in the former, as he spreads benevolence, instruction, and aid to the latter [...] Oh! This owner is a sort of image of Divine Providence" (Cittadella Vigodarzere 1856, 1, p. 21).[7]

As president of the *Istituto Veneto*, Cittadella Vigodarzere defended the utility of machines in the manufacturing industry (Cittadella Vigodarzere

1846), yet as president of the Venice Congress of Italian scientists, he omitted to invite industrialists to the congress, and only after the much awkward insistence by many influential attendees did he organize an exhibition—the first of its kind in Italy—for the machines and industrial products already present at the congress (Soppelsa 2001, p. 264–265). In reality, only agriculture made his heart leap. As was typical for Northern Italian political economists, he criticized socialism as well as British capitalism, warning of the social damages it caused (Marino 1974, p. 108–109 and 124–125). His intimate belief was that: "The greed for other bigger and quicker, yet less secure profits should not divert us, essentially agrarian people, from cultivating our fields. Since the prestigious wealth deriving from the boldness of cumbersome manufactures and the risks of wide commerce frequently suffers the vicissitude of immediate and miserable disarray" (Cittadella Vigodarzere 1856, 1, p. xiii-xiv).[8] He was both aware of the need for modernization and indulged in romantic nostalgia. It sounded rather archaic when he declared that the real mastery of agriculture, "the most useful of all arts", required a country life, "a long experience burnt in by the rays of sun, macerated by the rains, aged in the midst of fogs, frosts, and winds" (ibid. p. xi-xii)[9] His creed was a mixture of positivist faith in the truth and social usefulness of science and the conviction that religion is the highest of all truths.

For Cittadella Vigodarzere, agriculture was the foundation of all civilization. In this vein, Cavalli, who in 1846 carried out a systematic survey of the 'very sad' condition of the Paduan countryside (Cavalli 1851, p. 23), reasoned, "Agriculture is a school of moral philosophy, because it makes you feel every day with your hands the variety of concerns necessary for a good life" (Cavalli 1853, p. 13).[10] Men like Cittadella Vigodarzere and Cavalli looked at nature through the eyes of a farmer, with the aim of perfecting nature and rendering it useful and efficient for the civilizing process. "Land is the supreme laboratory of wealth" opinioned Cavalli at the end of his survey of 1846, "but abandoned land is completely sterile or barely fruitful. In order to develop the immense apparatus of its treasures, it requires the fertilizing hand of man, who develops and even increases its productive force" (Cavalli 1851, p. 173).[11] In this same spirit, Cavalli and Cittadella Vigodarzere aimed to perfect man. Enthused by their enlightened reformist convictions and by the aim of economic and moral progress, they dreamed of cultivating both land and men through supervision, care, and conservation.

This was essentially the same philosophy as that which motivated their romantic gardening endeavors. If Tuscan reformers and landowners like Cosimo Ridolfi (1794–1865) and his contemporaries dreamed about installing agriculture as a form of manufacture (Biagioli 2004), Cittadella Vigodarzere's ideal was the garden. Jappelli's landscape parks were not simply spaces of vegetation (Chap. 4), but can be understood as sites of existential research. Jappelli, on behalf of his commissioners, sought to influence the visitors, and to manipulate and naturalize their senses (Autiero 2006, p. 67–68). His vegetal scenographies presented images of a unified life and of the harmonious interplay between the parts (the individuals) and the whole (the world). People like Cittadella Vigodarzere were attracted to the English garden style precisely for this reason (Fig. 9.2). His park reflected these holistic views and idyllic ideas of countryside. His chosen vegetal imagery was not one of exuberant and ever-changing living nature, but of a cyclic and gradually (if ever) developing one. By analogy, his romantic vision of agriculture reflected his desire for stability in a changing world, where industrialization, socialism, and republicanism threatened to bring profound change to social and economic settings. Even if English gardens rejected the absolute control and artificial deformity of the French parks, they were by no means sites of unbound wilderness or disorder. Indeed, some liberty was conceded to nature, nonetheless, in the romantic parks, gardeners mustered all their skills to create the illusion of a natural nature, yet selected and fashioned all characteristics to agree with the owner's vision. Similarly, romantic landowners like Cittadella Vigodarzere cared for their peasants—their 'human cultures'—by creating an idyllic socio-cultural asset that was meant to remain under control and unaltered through time.

Niccolò Tommaseo's lengthy obituary of Andrea Cittadella Vigodarzere strikingly resembles the piece Andrea had compiled for his uncle Antonio some twenty years earlier. Tommaseo praised his classical culture and poetic skill, but even more his religiosity, modesty, and respect:

> Rather than the land, he intended to improve the condition of his peasants: and, by making right a duty, he knew everyone name by name and treated them with that dignified simplicity, with those affectionately gentlemanlike manners for which he became worthy of being called *decorum of nobility*, *pearl of nobility*. [...] rightly it has been noticed that wealth, used in this way, can resemble a living *fountain*, never drying, distributed in *runlets* which all,

Fig. 9.2 The garden of Cittadella Vigodarzere in Saonara near Padua. Lithograph by G. B. Cecchini. From Anon. (1842a, p. 529). By courtesy of Biblioteca Capitolare di Padova, Coll. 700.AA5.7

if they had no master, would be thin trickles of water, oozing away uselessly in the sand. (Tommaseo 1870, p. 24; his italics)[12]

The agricultural philosophy of Cittadella Vigodarzere and his like-minded peers, although apparently a pre-1848 modernizing reform movement, was rooted in the ideas of the preceding century. Adrian Lyttleton (1979, p. 107) holds that the moderate reformers of the eighteenth century wanted "to transform the mentality of the nobility and [...] to convert their fellow nobles from an order of feudal office-holders educated in jurisprudence into a class of enlightened landlords educated in economics". The example of Cittadella Vigodarzere shows that in Veneto this held true well into the first half of the nineteenth century. The Paduan aristocrat was not alone. On the contrary, according to Silvio Lanaro, the model applied across Veneto. The code of behavior of Veneto's landed upper classes included a gentlemanlike relationship of the landowners toward their peasants. The *possidente* was the peasant's port of call in all administrative and political concerns. In return, the landlord looked after the peasant's wellbeing, performing a multifunctional role as benefactor; head of philanthropic, credit, and educational institutions; and representative in various local, regional (and later national) political and administrative positions (Lanaro 1984, p. 64–66; Preto 1982). Cittadella Vigodarzere, like Cavalli, Selvatico, and others, personifies a hierarchic paternalistic model that remained a peculiar feature of Veneto society throughout the century (Fumian 1984, in part. p. 99–109). Gianfranco Tusset (2000, p. 97–98) emphasizes that this type of paternalism was not the same as philanthropy, but a sophisticated form of mutual social relationship. It was based on order without constraint and wealth through cooperation. It was a tacit accord between the two sides, an exchange of a guarantee of social protection for loyalty and labor, and of comprehensive moral responsibility for obedience.

This paternalistic ideology of the nineteenth century is thus a child of eighteenth-century Italian moral philosophy. The reestablishment of moral philosophy had been one of the principal aims of Italian Enlightenment philosophers. Scholars like Ludovico Antonio Muratori (1672–1750), Francesco Maria Zanotti (1692–1777), Antonio Genovesi (1713–1769), and Pietro Verri (1728–1797) conceived of ethics as a 'natural morality', independent of divine revelation (Cremaschi 2020, p. 3). More concretely, the ideal of the benefactor in Italy relied on the prevailing concept of *pubblica felicità*, or public happiness, a concept that had no equivalent in other national Enlightenments. It did not require the fulfilment of all pleasures and desires, but a general peace of mind. The goal or even moral duty of a prince or good ruler, according to these enlightened philosophers, was to act wisely and to seek public happiness to guarantee progress, peace, and peace of mind for citizens, so enabling them to pursue their individual happiness in cooperation with the others. In Selvatico's words, the soil had to be prepared so that "the poor seedling will soon grow up to become a robust tree" (Selvatico 1842, p. 233).[13] Probably

Andrea Cittadella Vigodarzere, as well as his uncle Antonio, saw themselves as such princes, assuring for the peace of mind of 'their' people.

This intertwining of the concepts of plants, cultivation, agriculture, and societal organization becomes evident in Cittadella Vigodarzere's inaugural speech at the Congress of Italian scientists in Padua. He began with the imagery of a "botanical garden [...] that resembles an empire composed of many and different people [...] a happy and graceful empire that, if it fears the wraths of heaven, will never be subverted by revolts or troubled by wars; it grows each day under the loving eye of the naturalist, who is its monarch" (Cittadella Vigodarzere 1843, p. xl).[14] The allusions to the multi-ethnic Habsburg Empire and its botanophile monarchs are ostentatious, yet the imagery of gardening also reflects Cittadella Vigodarzere's intimate worldview. Unsurprisingly, the only quasi-contemporary scientific-naturalistic authority he cited was the celebrated French anatomist and paleontologist Georges Cuvier, a professed monarchist and anti-transformist (ibid. p. xxvii). For Cuvier, the world was in constant flux, but remained unaltered in its essence because all change was cyclic and local, and ultimately cancelled each other out. As we have seen in the chapter on historicism (Chap. 6), the celebration and recourse to past authorities like Cuvier were not necessarily a sign of antiquatedness, but expressed a general predilection of the time to reconnect and create a continuity between a glorious past and a hoped-for golden future.

Veneto's paternalism constituted a particularly successful social model. Local sporadic upheaval was frequent in periods of famine, especially in the years 1816–1817 and 1846–1847, yet in Veneto, such unrests did not become as violent as in other parts of Italy. Unsurprisingly, the first labor associations, founded some years later, also showed paternalistic structures. In 1867, a group of 'meritorious citizens' founded a Worker Society (*Società Operaia*) in Monselice, a small city near the southern slopes of the Euganean Hills (Fig. 1.1 in Chap. 1). Only nineteen out of the 215 founding members were actually workers (eleven stonecutters and eight factory workers). The other members were landowners, professionals, and artisans (Bergamasca 2004). The fact that most members of Veneto's intellectual elite were also *possidenti* explains why Cittadella Vigodarzere's vision was echoed in the writings of many exponents of the 'Lombardo-Veneto school' of political economy during the second half of the century. In Veneto, any socio-economic ideology—enlightened physiocracy, liberalism, bourgeois revolution, and even fascism—had to dovetail with these deep-rooted paternalistic structures. Even Andrea Meneghini found this out when, in 1866, many of his progressive-liberal companions, in spite of their being popular heroes in exile, lost the post-unification elections to conservative-liberal *possidenti* like Cavalli (Mogavero 2017; Zangrando 2017, p. 31–32).

Andrea Meneghini's concept of social organization was indeed slightly discordant. He shared many of Cittadella Vigodarzere's basic convictions, for instance, the Veneto version of liberalism, associationism, the right to property, and anti-socialism (Chap. 6). Cittadella Vigodarzere and Meneghini were both prime instigators in the Society for Encouragement during its initial years, the former as president and visionary, and the latter as its manager. Meneghini provided for the contacts, compiled the reports, drafted the statutes, and reported to the Congress of Italian scientists. However, he did not simply follow or execute Cittadella Vigodarzere's ideas. We cannot know how the Meneghini brothers would have run the estates of their father Agostino, since he lost all of his lands. Therefore, this analysis of their social theory relies on their theoretical reflections. Giuseppe's scientific writings show a greater focus on simple parts (organs, cells) than on the entire organism, and a predilection for the 'lower' (algae, diatomeae) over the 'higher' organisms (Sect. 7.2 in Chap. 7). Moreover, although he did not articulate a fully 'bottom-up' vision, he directed more attention toward the collaboration of parts than toward the directing force of higher structures like the brain. For Giuseppe, the brain had a coordinating, not a paternalistic function. Andrea's view, too, was less conservative, less hierarchic, and more progressivist. In his writings and actions, he focused on the people. He did not consider his wealth as 'a living fountain' for the others, as Cittadella Vigodarzere did (see above)—a system that would ultimately require that all water flow toward the fountain before it can be distributed. Andrea actually spent most of his (small) fortune helping others, a habit that meant that he and his family continuously lived in dire financial straits toward the end of his life (Sect. 10.3 in Chap. 10). He promoted the *Cassa di risparmio* as an opportunity for social ascent (Sect. 6.3 in Chap. 6), supported, probably as notary, the *Istituto medico-chirurgico-farmaceutico di mutuo soccorso in Padova*, founded in 1847 (Anon. 1848), and promoted the project of creating the Society of Mutual Aid among landowners (Anon. 1847j). In 1848, he organized a popular suffrage for Padua's revolutionary government and, against the will of count Ferdinando Cavalli (Ventura 1989, p. 53), a plebiscite on the annexation to the Sabaudian kingdom (Sect. 10.3 in Chap. 10). He also repeatedly (yet unsuccessfully) called upon all members of the associations to participate and make their voices heard during the assemblies (Anon. 1847k).

Given his emphasis on the spirit of collaboration, Andrea appears to be an heir of the Neapolitan enlightened scholar Francesco Mario Pagano (1748–1799), rather than indebted to Muratori's good sovereign or Zanotti's wise prince. For Pagano, who was deeply influenced by Vico, the sociability and perfectibility of man were the pillars of social order. Like Vico, he approached political philosophy through an evolutionary reconstruction of human societies, a conception of society as an interaction and balance of many powers, and by means of a basic analogy between physical forces, which hold together the universe, and a moral force, which holds together society (Cremaschi 2020). Andrea's distinctive understanding of associationism also became evident during the Venice Congress of 1847, when he announced a project to create a meta-organization among all the provincial agrarian associations. He called for a Veneto-wide network as a first step toward the 'fraternization' of all Italian associations (Anon. 1847c, p. 78, 111 and 123–124, 1847r). In his speech, he expressed the desire to see "one single family forming in the Venetian provinces, which will reach out its fraternal hand to the other Italian sisters" (Anon. 1847r, p. 337).[15] This declaration may seem naive, yet the names of the other four commissioners—Daniele Manin, the Milanese economist count Alessandro Porro (1814–1879)[16], the agronomist Francesco Luigi Botter, influential promoter of similar initiatives in Ferrara, and count Matteo II Thun (1812–1891) of Trento—attest to the seriousness (and political dimension) of the project, though it was soon upended by the political events of 1848. Together with Manin and count Gherardo Freschi (1805–1883) from Udine, Meneghini was charged to draft the statutes. The project foresaw a headquarter in Venice, but Meneghini and his companions did not plan to create a hierarchic organisation. Rather, they envisaged a consortium, that is, a pool of autonomous and equivalent parts unified by concrete and pragmatic common goals. This form of collaboration corresponded more closely to Andrea's idea of associationism than to Cittadella Vigodarzere's paternalism or the polycentric tradition of Veneto culture.

9.3 Andrea's *Il Tornaconto*

Andrea Meneghini had no difficulty forging a network of like-minded landowners for the planned Veneto-wide agrarian association, because many of them already were contributors to his journal *Il Tornaconto* (The

Profit). The dissemination of information was indeed of seminal importance for agrarian reform. For example, one reason for the failure of the first attempts to cultivate potatoes in Veneto (Sect. 9.4) was that such projects did not come with useful practical advice or support. The few farmers who were favorably disposed had to acquire the necessary experience themselves. The Society for Encouragement and *Il Tornaconto* complemented each other and worked together to ameliorate these shortcomings. Andrea Meneghini was the driving force in both institutions. Yet, the social composition of the journal's correspondents differed significantly from that of the inner circle of the agrarian associations. Therefore, the following analysis of *Il Tornaconto* thrusts a less prominent, yet very active group of agrarian reformers into the light. A glance at Andrea's journal also gives some insights into Padua's publishing landscape in the 1840s, into Andrea's social views, and into the role of science in the reformist programs.

The Republic of Venice had been quite active in the production of newspapers, especially after 1735. During the eighteenth century, nearly one third of all Italian-speaking periodicals were published in Veneto. Several of them, such as *Il Giornale d'Italia* (then *Nuovo Giornale d'Italia*), founded on 7 July 1764, dedicated some space to the natural sciences, in particular as applied to agriculture (Luzzatto 1928, p. 541–542). The complete name of the newspaper was in fact *Giornale d'Italia, spettante alla scienza naturale, e principalmente all'agricoltura, alle arti, ed al commercio.* It ceased to exist in 1797, together with the Republic of Venice. Until the 1840s, Padua did not have its own journal, even though it was one of Veneto's principal centers of printed media (Callegari 2012), a claim to fame that extended to the circulation of prohibited writings (Gloria 1927, p. xlvii). Matters changed during the reformist period. In 1841, six publishing houses were active in the city: Tipografia del Seminario (Fig. 2.1, B in Chap. 2), di Crescini (Fig. 2.1, F in Chap. 2), dei Fratelli Penada, di Angelo Sicca, della Minerva, and di Gaetano Longo (Salvagnini 1841, p. 52). Between November 1844 and January 1847, four newspapers started to circulate.

The *Giornale Euganeo* appeared for the first time in November 1844. During the first year, it sold 536 copies, about half of these in Padua and its province (Anon. 1844).[17] Under the strict surveillance of the Austrian authorities, it concealed its political messages in seemingly non-political articles on literature, art, and (occasionally) science, and it sought to involve all social classes in the diffusion of knowledge. Among the authors of scientific articles we find Giuseppe Meneghini, who contributed with his series *Lezioni di botanica popolare,* or Popular lessons on botany (G. Meneghini 1844–1846), Giuseppe Clementi who published his *Chimica popolare,* or Popular chemistry, and Carlo Conti (1802–1849), who was an enthusiastic botanist and creator of a herbarium, was author of a *Fisica popolare,* or Popular physics. Other contributors included the poet Jacopo Cabianca (1809–1878), who briefly became interested in agronomic experiments and gardening, Gustavo Bucchia (1810–1889), and Ferdinando Cavalli. Another weekly newspaper, *Il Caffé Pedrocchi,* was launched in January 1846. Like the coffee house of the same name, it mainly appealed to an aristocratic and upper class audience, but found itself, due to its polemic style, in constant wrangling with the Austrian censors.

Censorship was omnipresent (Berti 2007) yet, in these years of economic prosperity and of a relatively relaxed relationships between Austria and Veneto, not as strict as it had been. The Royal Imperial Censor for Books and Press (*I.R. Censore dei libri e delle stampe*) was the liberal abbot Giuseppe Onorio Marzuttini (1802–1849), professor of Theology and in 1845–1846 dean of the University of Padua. In addition, the fact that the police repeatedly attested to Andrea Meneghini's "unimpeachable moral, political, and social conduct" during the 1840s and issued a certificate of good conduct even during the period of his imprisonment in 1848 demonstrates that surveillance by Padua's police was not too strict (Miotto 1942, p. 13, 93, 94, and 100). Marzuttini owned and directed the first Catholic newspaper of Padua, the *Giornale dei Parrochi ed altri Sacerdoti,* published during the quadrennial 1846–1849. This newspaper enjoyed considerable success, in particular among the local priests. It disseminated the ideas of Padua's liberal clergy as well as Pope Pius IX's reforms throughout the province. Nevertheless, its insistence on traditional values, such as order, propriety, and authority, in the spiritual as well as the social and political dimensions of life, reflected the firm hierarchical conception of society of Veneto's clergy (Brunello 1975).

The first regular issue of Andrea Meneghini's weekly *Il Tornaconto, Giornale di agricoltura, orticoltura, industria, commercio ed economia comunale per le provincie venete,* advertised by means of a manifesto on 15 October 1846, appeared on 7 January 1847. The last of sixty issues was published on 24 February 1848, two weeks after Andrea's imprisonment. The editorial office was located in Meneghini's private house in Borgo Vignali number 3374. His co-director and co-editor was the chemist and botanist Giuseppe Clementi. Clementi, who in the meantime had become professor of physics at the lyceum of Bergamo, was Meneghini's host during his stays in Padua (Manganotti 1876, p. 396). He later also dedicated the clover *Trifolium meneghinianum* to Giuseppe (Clementi 1857, p. 267–268). The printing press was the ecclesiastic *Tipografia del Seminario*, formerly directed by Giuseppe Bernardi. The main audience were landowners, farmers, and industrialists across Veneto, and it sought to establish an enduring network among them. Meneghini's model was probably *L'Amico del Contadino*, or The Friend of the Farmer, which appeared in 1842–1848 in Venice. Its founder, Gherardo Freschi, had graduated from Padua in law and in natural sciences, and in 1846 created the *Associazione agraria friulana*. Another, similar journal, *L'Economista: Giornale di agricoltura teorico-pratica, ragioneria, amministrazione, tecnologia, commercio, ecc.* (*The Economist: Journal of theoretical-practical agriculture, accounting, administration, technology, commerce etc.*), appeared in Milan between 1842 and 1847.

Like in the registers of the associations, the names of future contributors and supporters of *Il Tornaconto*, presented in the manifesto, were listed in alphabetical order, thus disregarding social rank. Articles carried the bylines of about one hundred different authors. Some remained anonymous, others were signed with acronyms. Many authors contributed only one or few articles, but about ten were frequent contributors. Another ten regularly corresponded about the agronomical situation in their province. Most of latter were medium-sized landowners. Many of the frequent authors were professors at the University of Padua: Gustavo Bucchia, Gian Paolo Tolomei, and, obviously, Giuseppe Meneghini. The correspondents were active associates of the academies or societies of their provinces—Antonio Radice of the Agricultural Academy of Verona, Giovanni Battista Clementi of the *Accademia Olimpica* di Scienze Lettere ed Arti di Vicenza—or otherwise philanthropically active, like the Jewish banker Salvatore Anau (1810–1874) from Ferrara and the physician Jacopo Facen (1803–1886) from Feltre. Also noteworthy is the large number of contributing botanists and pharmacists—Francesco Fontana (1794–1867) from Treviso, Gaetano Grigolato from Rovigo, Antonio Manganotti (d. 1892) from Verona, Giuseppe Rocchetti (1799–1874) from Legnago, and Giovanni Battista Ronconi from Padua—and floriculturists, among them Angelo Giacomelli of Treviso (Sect. 8.2 in Chap. 8) and Cesare Schiavinotto, the gardener of the Treves garden. Andrea's early love for botany and herbal excursions may have translated into a special relationship with people who worked with plants. A final salient aspect concerns the European outlook of the journal. As most members of Padua's cultural elite were learned men with personal experience in and connections to European centers, *Il Tornaconto* was eager to carry news about the latest international developments.

Il Tornaconto did not promulgate explicit political opinions. We do not know whether this editorial policy was adopted to conceal anti-Austrian sentiments, a spirit that certainly animated Meneghini's later writings (e.g., A. Meneghini 1863; Meneghini 1865), or whether, at this time, his campaign against the Austrian authorities was principally grounded in economic rather than nationalistic concerns. In any event, Meneghini certainly feared a backlash by the censors against his journal. As his legal campaign against the Austrian authorities in late 1847 demonstrates, he was not a man who was afraid of personal consequences (Sect. 10.2 in Chap. 10). Whatever his reasons, in the introductory manifesto, Meneghini declared that the principal aim of his journal was to derive benefits from knowledge exchange. Some farms and manufactures worked marvelously, he wrote, but many farmers and industrialists, though open to improving themselves, did not have sufficient access to the necessary knowhow (Anon. 1846, p. 1). Therefore, *Il Tornaconto*'s aim was to provide technical-scientific information, reflecting an assumption that this type of information was immediately useful. "We want our farmers to harvest abundant wheat, corn, rice, grapes, cocoons, hemp; not that they get lost in useless experiments" (Anon. 1846, p. 2).[18] This was not to disregard science. On the contrary, the journal explicitly sought to counter the influence of those who considered science superfluous (Anon. 1847e, p. 9). Hence, the paper carried news about the latest scientific advancements, especially in botany, chemistry, and physics. However, Andrea promised to exercise utmost caution that these notions "will not stink of school" (ibid. p. 9). His idea of 'rational agriculture' recalled the teachings of Albrecht Thaer's (1752–1828) successful treatise *Grundsätze der rationellen Landwirthschaft,* or *Thoughtful principles of agriculture* (1809–1812), translated almost immediately into Italian (Thaer 1816–1818, 1818–1819). Like Thaer, Meneghini aimed to raise the productivity and profitability of farms with the help of the natural sciences. Meneghini and Clementi, however, advocated a more up-to-date version of this school of thought, adopting as their ideals the two chemists and agronomists Justus von Liebig (1803–1873) and Jean Baptiste Boussingault (1802–1887), who both personified the highest level of science and its profitable use in agriculture (Clementi 1845).

Andrea's epistemological philosophy was neither romantic nor positivistic. It can be characterized as a Baconian mixture of empiricism, inductive skepticism, and faith in the power of knowledge and progress. This did not amount to a blind belief in the infallibility of science. He admitted that the 'investigations of truth' had made errors, failed some experiences, did not succeed in dispelling a 'thousand uncertainties', and that much was left to be done. Nonetheless, science had made important contributions to progress and nobody ought to ignore its achievements (Anon. 1847e, p. 9). Again, we notice a subtle but important difference in the mindsets of Andrea Meneghini and of Cittadella Vigodarzere. They converged in privileging applied science in agricultural reform, yet Cittadella Vigodarzere's view was much more authoritarian. In the speech he delivered as general president of the Congress of the Italian scientists in Padua in 1842, the Count compared science to a fruit tree and labelled scientists as "society's principal benefactors" (Cittadella Vigodarzere 1843, p. xxvi). He complained about the scarcely effective collaboration, whereby "the land is in the hands of idiots, automatic repeaters of old habits", whereas the theoreticians "live afar from the fields". He envisioned an economy based on science and technology and a "major perfection of the social state when tightly connected to the progress of the physical and mathematical sciences" (ibid. p. xxvi).[19] Yet, revealingly, concurrent with the paternalistic management style he applied to his estates and his concept of a hierarchical and nature-given society, he hailed the natural sciences as guarantors of law and certainty. For him, they yielded not only economic and technical progress, but also a precise and secure moral science, and lasting social peace (ibid. p. xxvii-xxviii and xlvi-xlvii).

In contrast to Cittadella Vigodarzere, Meneghini's vision entailed an interplay between empirical skepticism and faith in the truth of science. He therefore held that he provided the instructions not *ex cathedra* but rather by way of trial (and error), and explicitly encouraged his audience to participate in this process by sharing their opinions, experiences, and objections. Hence, Meneghini's style of directing *Il Tornaconto* is further proof of his trust in the power of collaboration. As a director, he was quite humble. He saw his personal role—in striking similarity to Giuseppe's conception of the role of the brain in living organisms (Sect. 7.2 in Chap. 7)—as that of a coordinator between the parts, not as a principal or *pater familias*. In all his activities, Andrea did not stand out as an aloof visionary or a boss, but as someone who plans and lays the requisite foundations. He was not a republican but declared, "certainly we will have special love for those who rank higher in social importance. Yet this will not make us forget nor will it hinder us to treat the minor" (Anon. 1847d)[20] He wished to serve the *possidente* of ten thousand fields as well as the humble owner of few spans. In his mind, the journal was to create a network of landowners and farmers of all of Veneto and regardless of social class.

The motto and highest ambition of *Il Tornaconto* was usefulness in practice. Some decades later, Andrea Meneghini also put forward similar ideas of progress in agriculture, science, and education for the sake of a better society in his booklet on political economy (A. Meneghini 1851). He was not alone with this view of science and its role in society. On the contrary, the attitude was widespread in Padua and in Veneto. Utilitarian ends had been behind the foundation of the Botanical Garden as well as of the Agrobotanical Garden. A letter of 1847 from Niccolò Tommaseo to his friend Roberto de Visiani illustrates the ongoing influence of this practical attitude in Veneto. After applauding de Visiani's far-flung horticultural network for helping "the mind to flourish more than the gardens", he gestured toward the importance of botany for future pharmaceutical remedies and other useful products. "There will be a time (I do not doubt)," he wrote, "a time remote for centuries, when not only flowers and all herbs that earth brings forth will in some way serve life, and every single one of those hidden by the sea will pay their tribute to humanity" (Tommaseo 1847, p. 119).[21] The last part of this quote refers to Giuseppe Meneghini and Giovanni Zanardini's volume on the Dalmatian algae.

The same utilitarian conception of science also pervaded the Congresses of Italian scientists and was espoused by the group that elaborated the statutes for the future agrarian federation of Veneto. From the outset, the science congresses had been organized to promote the 'useful applications' of science.[22] Indeed, in the course of the 1840s, one detects a steady increase in practical projects and in accounts of applied empirical research, at the expense of pure research (Soppelsa 2001, p. 250). The founding commissioners of the agrarian association, too, invited science "to pass from speculation to practice, from the cabinet and the chair to the field and the workshop: in this way, busy industriousness benefits from knowledge and both conspire for the general welfare" (Anon. 1847r, p. 338).[23] This project provided for libraries, laboratories, exhibitions, prizes, schools, and model farms and factories for experimental cultivation and breeding.

All this points to the fact that, by mid-century, a significant shift in views about the role of science in Padua's intellectual debate had taken place. As we have seen, at the beginning of the nineteenth century, ideas about nature and its laws and thoughts about politics, economy, and social life developed in close proximity, often even coincided in the same person. In Padua, this was particularly true for the plant sciences. Around mid-century, however, we observe an increasing divergence, on the conceptual as well as on the biographical level. The less politicians, economists, and lawyers nurtured naturalistic passions, and the less they learned about scientific debates, the more their attitude toward science became utilitarian. In fact, the Baconian concept of useful science prevailed in most of the writings of the Lombardo-Veneto school of political economy (Chap. 6). The predilection for the productive 'facts' of applied science left little space for sustaining and understanding basic research or emerging scientific concepts like cell theory. To some extent, Giuseppe Meneghini acquiesced in this trend and contributed several articles to his brother's journal, writing on grafts, fossil fuel, and the potato disease, whereas his writings that were in spirit of pure science did not find the recognition they deserved (Sect. 7.4 in Chap. 7).

9.4 "Potatoes!"

The potato plays a triple role in the story of the interplay between politics and botany in Padua. The first is economic-political. Crops failures and famine, resulting in hunger and infection, were among the principal causes of popular unrests and revolutions. Political and agricultural disorder were always feared as two interconnected threats by contemporary politicians and scholars, who discussed the consequences of technical and methodological innovation, the perfectibility of nature, and the introduction of new species. A long-term study suggests that the introduction of potatoes into European agriculture led to a significant increase in agricultural productivity and general prosperity, and to a sizable reduction in conflict (Nunn and Qian 2011). On the other hand, severe potato failures had the opposite effect. The famine crisis of 1847 shortly after the vegetable had been introduced into Veneto agriculture is a case in point. Karl Marx interpreted the political crisis of 1847–1848 as, among other factors, a delayed reaction to the failure of the potato (Paping et al. 2006). Even if in Italy the effects were not as devastating as in Northern Europe, the issue was prominently debated by agronomists in Veneto. The second aspect is social and political, and regards the potato as a political metaphor. In the 1840s, the tuber was a symbol for 'the German', meaning the Austrian, which complicated a neutral debate. The third aspect is scientific. Like other European cryptogamists, Giuseppe Meneghini actively participated in trying to solve the problem as soon as the hypothesis of fungi as carriers of the disease emerged. For him, the quandary hence also became an occasion to demonstrate the indispensability of science and to display his expertise. The political, agronomical, social, and scientific dimensions were thus closely entangled in the potato debate of the 1840s.

In 1740, more than half of Veneto, and especially the most fertile areas in the lowlands, was in the hands of the nobility. Most of the owners did not belong to the rural nobility but resided in the cities, especially Venice. The simple farmers lived principally on the mountains, where farming was more difficult. Marino Berengo (1963, p. 3–4) regards the fact that the rich landowners increased their wealth by expanding what they owned rather than by increasing the productivity of their land as the principal cause of the backwardness of agriculture in Veneto. Another factor was the propensity, common among urban landowners, to concentrate on city life and to assume political and administrative positions in the major cities, while renting out their land and transferring the management to *fittanzieri* (renters) or local administrators. The consequences were often devastating. Farming became less and less profitable, unscrupulous administrators preyed upon the absence of the owners, judges were overwhelmed with litigation and lawsuits, and peasants became proletarianized and increasingly alienated from the landowners and from the land they cultivated (Berengo 1963, p. 5–8).

Veneto's agriculture was largely restricted to three products: wheat, corn, and wine (with very few parcels of land dedicated to olives, rice, and hemp). Few new products succeeded to penetrate the bulkwark to innovation consisting of a focus on immediate necessities, tradition, and mistrust. Enlightened scholars had recommended the cultivation of potatoes during subsistence crises from the eighteenth century. The first scholarly promotion of the tuber in that period in Italy was an article by Francesco Griselini (1717–1787), published in 1765 in Venice in the previously mentioned *Giornale d'Italia* (Biadene 1996, p. 30–33; Gentilcore 2012, p. 43). Griselini soon withdrew, disappointed by the indifference of the Venetian authorities to his reform projects. Similarly, academic discussions and writings by people like him, as well as the few practical forays of parsons like Abbot Agostino dal Pozzo (1732–1798) on the plateau of Asiago, and of

Abbot Antonio Marchesini in Recoaro near Vicenza, did not bear fruit. With the coming and going of Austrian troops, the main consumers of potatoes in this period, the tuber appeared and then disappeared again in the fields around the garrisons (Berengo 1963, p. 267–268). Veneto's people did not warm to that tuber that continued to be object of derision and pillory.

Initially, however, the potato was not necessarily an anti-Austrian metaphor. Michel Ragon describes in his romance *Les Mouchoirs Rouges de Cholet* how French peasants dismissed potatoes as symbols of the revolutionary regime (Ragon 1984). A similar politico-botanical link was made in Northern Italy under French occupation. In the spring of 1814, during the blockade of Venice, the people ridiculed the appeal judge Giovanni Battista Sanfermo, later denounced as a corrupt Napoleonic collaborator, for sowing potatoes on a public square in the lagoon of Venice (Dal Cin 2019, p. 325–328). 'Count Potato' he was nicknamed and became the symbol of French oppression. On 19 May, a straw puppet appeared on the Santa Caterina Bridge, with a potato in its mouth, two in its ears, a crown of potatoes on its head, and a basket of potatoes by its feet. At dawn, the puppet was sentenced to death, shot, burnt, and dragged through the streets, while people shouted "Death to the potato cropper!" The real Sanfermo, however, continued to stroll unmolested through the streets of Venice.

The great famine of 1816 changed this situation. A delegation of landowners from towns south of Padua—Este, Montagnana, Piove, and Battaglia—desperately petitioned Padua's mayor to supply them with seed potatoes for the starving population, but to no avail (Zadoks 2008). Encouraged by the Austrian authorities and guided by a number of scientific publications and practical advice, farmers began to sow sugar beet and to appreciate the potato for its resistance against climatic adversity and in alternation with maize growing (Monteleone 1969, p. 37). However, potatoes remained a food of the poor and for pig forage, whereas the urban markets, excluding the Austrian garrisons, continued to prefer wheat.

In 1846, after about a decade of prosperity, the most abundant years under Austrian domination, another famine descended upon Veneto. This time it concerned the potato. In 1845 and 1846, the potato harvest had been particularly bad throughout Europe, resulting in the great potato Irish famine. The crisis quickly spread to other north and central European lands, causing violent riots in the following years. In Veneto, this and several other factors spoiled the relationship between the population and the Austrian occupying forces, damaging the image of the tuber that still had no established place in Veneto's agricultural economy. In the spring of 1848, a popular jingle in Veronese dialect went:

> *Ferdinand, Ferdinand / Your reign will not stand / And Pius IX is to spread / The potatoes will go bad.*
>
> *(Ferdinando, Ferdinando / El to regno el va calando / E Pio Nono se ingrandisse / Le patate se smarsisse.)*
>
> (Solinas 2008, p. 121)[24]

Alongside popular occasions, the Congresses of Italian scientists, too, provided audiences for the anti-Austrian potato metaphor. A clandestine observer who attended the Venice congress in 1847 reported to the Austrian police that the attendees took advantage of the occasion to indulge veiled ironic anti-Austrian remarks. He also reported a rather explicit private comment by poet Giovanni Prati (1814–1884) to a friend: "among us, only the Germans are passionate about potatoes; let them eat these in holy peace in their countries, they should not dirty our soils with such a worthless fruit; I hope they'll go away soon" (quoted from Miotto 1942, p. 23; Manin 1852, 3, p. 356–357).

Curiously, at the Venice Congress in 1847, it was Baron von Hügel, so an Austrian attendee, who came out against the potato, whereas the Italian discussants spoke in its favor. Von Hügel advocated the replacement of the potato with *Apios tuberosa* (today *Apios americana*) from the Americas, because this crop variety was, according to him, less vulnerable to maladies (Anon. 1847c, p. 61). He reported that *Apios* had been introduced into Europe as a botanical curiosity in 1540 and that it was already cultivated in Bohemia. The botanist Bartolomeo Biasoletto (1793–1859) from Trieste affirmed this view. He added that experiments with *Apios* were being undertaken in the Botanical Garden of Mantua and that farmers in the vicinity knew and appreciated it as *castagnuola* or *trogne* (Anon. 1847c, p. 61). However, the fact that the potato bean or American groundnut takes two years to deliver acceptable yields did not work in favor of its diffusion (Reynolds et al. 1990). The Italian scholars, instead, tried to find ways to save the cultivation of potato in Veneto and discussed ways to avoid or attenuate the failures. In spring 1847, Padua's Agricultural Society, although well aware of the disease, still considered the promotion of potato cultivation worth a prize of 300 Lire (Anon. 1847h, p. 65–67).

At the Italian scientific congress of 1845 in Naples, the potato disease still did not gain attention. But only one year later, in Genoa, on three out of the thirteen conference days, twenty-four attendees participated in a discussion of the potato (Anon. 1847b, p. 115–118, 137–138, and 371–375). None of them favored a temporaneous or definite substitution.

Another year later, in Venice, the interest appears to have faded. Only seven landowners from several parts of Lombardy-Veneto—Father Innocenzo Ratti (1806–1883) from Milan, Jacopo Facen from Feltre, the agronomist and botanist Giuseppe Comolli (1780–1849) from Como, Francesco Gera (1803–1867), author of a study on the potato disease, Giovanni Battista Clementi from Vicenza, Count Matteo Thun from Trento, and Count Faustino Sanseverino (1801–1878) from Cremona—discussed personal experiences and proposals, and agreed to collect all studies and data in one center, the Physio-Medical-Statistical Academy of Milan, founded in 1843 (1847c, p. 54). All seven speakers were actively involved in agricultural reform and improvement and several of them were also regular authors in Andrea Meneghini's *Il Tornaconto*. Almost all reported only a few failures in 1846 and even less losses in 1847.

In spring 1847, around sowing time, Giuseppe Meneghini entered the debate about the 'unjustly neglected' potato by publishing a long article in the opening pages of his brother's journal, which then also appeared in Milan in the journal *L'Economista* (G. Meneghini 1847b, 1847c). In his general overview, he summarized the communications from each of the Veneto provinces and confirmed that the disease had been observed in many places, but seemed to have seriously afflicted only the lands around Belluno, a province close to the Austrian border. The most interesting part of Giuseppe's account is a quotation from a letter from Hugo von Mohl, one of the leading European botanists of the 1840s. Mohl had been in contact with Meneghini since 1836 and attended several Italian congresses. Meneghini reported that Mohl, engaged in an analogous commission by the Society of German Naturalists in Nürnberg (Anon. 1845), had observed an unusual quantity of the fungus *Botrytis infestans* on affected plants.[25] In summer 1845, two French and Belgian mycologists, Camille Montagne (1784–1866) and Marie-Anne Libert (1782–1865), first identified *Botrytis* as the potato blight (Montagne 1845; see also Bourke 1991). The question thus became a research focus for cryptogamists like Mohl and Meneghini. However, by the time of Mohl's letter to Giuseppe Meneghini in January 1846, most European agricultural societies and journals had almost unanimously dismissed the fungus theory and come out in favor of meteorological causes, pest invasions, or theories of soil deterioration. Mohl himself was not convinced that he had sufficient proof to draw a definite conclusion. Meneghini shared this view. His chief suspect was the fungus, though he did not exclude contributory meteorological causes. It was not until 1861 that Anton de Bary's (1831–1888) description of the complete life cycle of the fungus settled the controversy in favor of *Botrytis*, now called *Phytophthora*.

By virtue of their long-practiced tradition in microscopic anatomy and cryptogamology, North Italian and Veneto botanists probably had less difficulty accepting the idea of a small, invisible *contagium vivum* than traditional botanists and agronomists. Of the Italian cryptogamists, the Lombard official and naturalist Agostino Bassi (1773–1856) stands out in particular. He had published several studies on potatoes in the 1810s. In 1846, he defended the contagious nature of the disease (Bassi 1846, p. 26–31). Matters like this helped Meneghini to convert his expertise about microscopic life forms into tangible political authority. Indeed, Giuseppe concluded his article by reflecting on the practical consequences of his findings (G. Meneghini 1847b, p. 76–77). In September, he published a follow-up essay, *Popular instruction about the potato disease* providing concrete practical advice how to start and cultivate a robust potato crop (G. Meneghini 1847d). About two weeks later, on 22 September 1847, the Austrian government of the Veneto distributed almost identical guidelines (Biadene 1996, p. 128).

Revealing of Giuseppe and Andrea Meneghini's attitude is the way they criticized the work of Cesare Cantù (1804–1895), published in the following issue of *Il Tornaconto*. A man of many interests and a famous historian, novelist, and politician, Cantù also wanted to contribute to the potato debate. Yet, his 'intrusion' was not welcome. In a short note, an anonymous author in *Il Tornaconto*—maybe Giuseppe Meneghini himself, or his brother—severely censured not Cantù's results, but his methodology:

> The authority of important names cannot dazzle anybody in 1847. As for us, we declare frankly that the manner of proposed and narrated experiments by the illustrious writer cannot lead to any decisive result. This because practicing many methods on many varieties of potatoes and other plants of different families like *Helianthus tuberosus*, on twenty areas of land cannot but produce inextricable confusion.

The note continued even more explicitly:

> Excellent is the advice to contribute with all means, and all forces, to stave off the terrible disease from a cultivation, which merits being animated and extended. Yet one of these means is, according to us, to let speak only those who are at a minimum initiated to the sciences that are directly related to the treated argument. (Anon. 1847i, p. 85)[26]

In a debate still dominated by landowners, who tried to cope with potato blight by collecting a series of experiences in their fields, Meneghini sought to stand out as a man of science and as an expert in *real* experimental science. And this approach was vindicated. The mid-nineteenth century saw an increasing level not only of experimentalization, but also of institutionalization and professionalization of the life sciences (e.g., Barton 2003; Morrell 1990; Rheinberger and Hagner 1993). Meneghini's message was clear. Severe agricultural difficulties were no longer the playground for enlightened amateurs, but required the attention of professional scientists. As we have already seen in the introduction, professionalism played an essential role in Giuseppe Meneghini's self-understanding and construction of power relationships. The fact that several amateur mycologists entertained the fungus theory as early as July 1845, but were silenced by professional botanists (Bourke 1991), shows that the process of professionalization and scientific progress was not as smooth and simple as some narratives may claim *a posteriori*.

For Giuseppe Meneghini it was of utmost importance to present himself as an expert and holder of valuable knowledge. He was still professor of physics, chemistry, and botany for students of surgery. He always remembered this period as a great sacrifice because it had meant a heavy workload, low pay, little didactic satisfaction, and even less social recognition (Canavari 1889, p. 15). His personal ambitions, fueled by the international praise he had received for his works of 1836 and 1838, led him to not content himself with this position. De Visiani, who held the professorship of botany, was still young, and the possibility of making his way, like De Zigno, privately as a rich amateur disappeared when his father was ruined through unfortunate financial speculations.

Giuseppe actually enjoyed the fact that he was not forced to spend most of his energy on taxonomic questions, the main task of a botanist in those years. He could address general anatomical, physiological, and organographic questions, or, as he called it, "the philosophical part [...] those synthetic deductions that we rightfully regard as laws of organization and of life" (G. Meneghini 1844–1846, 1, p. 319). However, professionally his career risked reaching a dead end. Particularly in those years, the University of Padua produced an exceptionally high number of graduates in law and medicine who could not find adequate positions. A peak was reached in 1843, with 677 graduates in law and 600 in medicine, whose families had made enormous financial sacrifices to allow their sons to graduate in hope for social advancement. These unemployed graduates formed an ill at ease group of young men (Meriggi 1987, p. 153–172). A possible solution for Meneghini could have been the establishment of a chair for plant physiology. Such a position was created in Vienna for Franz Unger, a scholar with similar scientific ideas and interests, during the university reform after the 1848 revolution (Klemun 2016). Meneghini's subsequent endeavors in applied and agricultural botany may have been an attempt to convince reformist Paduan policy-makers of the economic usefulness of plant science and to persuade them to create new chairs and career opportunities. A similar institutional strategy had been adopted at the time by the physiologist Emil Du Bois–Reymond in Berlin (Lenoir 1992, p. 18–52) and François Raspail in Paris (Vienne 2017, p. 638). Unfortunately for Meneghini and for Padua, events took a different course (Sect. 10.3 in Chap. 10).

The potato was prominent, but it did not dominate discussions of agriculture in Veneto in the late 1840s, not even the agronomical sections of the Italian science congresses. On the other hand, despite all the popular satire and anti-Austrian propaganda jingles, the potato had found a place in agricultural practice and associate debates in Veneto. As Meneghini complained, many of the contributions by landowners were simple observations and personal assessments. The discussion nevertheless reflected an increasing level of experimentalization and transregional, and even international, collaboration. At the Venice congress, the German-Russian diplomat Baron Wilhelm von Freygang (1782–1849), who died in 1849 during the siege of Venice, offered himself as intermediary for the import of a more resistant potato variety from Russia (Anon. 1847c, p. 61). Likewise, Giuseppe Meneghini's correspondence with von Mohl, Andrea Meneghini's promotion of information exchange in *Il Tornaconto*, and the efforts to centralize all regional data collections in Milan point to the growing importance of networks that went beyond local circles. The potato crisis actually prompted a concerted international research effort.

After 1848, new varieties were developed. Some researchers even went back to the origins and investigated the primordial potato in the Americas. These activities were stunningly successful. Before 1848, there were 221 varieties in Europe, in the late 1870s they were more than 600 (Gentilcore 2012, p. 70–71). Several of these new lines were more disease resistant, others had better taste. Both improvements were fundamental for the potato to become an essential, even a 'traditional' part of Veneto's cooking tradition.

The potato failure was but one event that confirmed, in the eyes of Giuseppe Meneghini, the urgent need for the profound modernization of agriculture. Similar tales can be told for muscardine, a disease that afflicts silkworms, powdery mildew, which affects grape leaves, and other plagues. The days of amateur agronomists were over, as were the days of priest agronomists like Luigi Configliachi. The contribution of experimental botanists was fundamental but could not serve all the functions of proper professional agricultural science. On an organizational level, efficient ways had to be found to quickly and efficiently disseminate information among the landowners and farmers and beyond the local level. The *Società d'incoraggiamento* and specialized journals like *Il Tornaconto* illustrate how Padua started to merge with this international trend.

A final point that the history of the potato illustrates concerns the status of plants in Veneto in general. The debates and innovations of the 1840s, which revolved around the potato cultivation, to some extent paralleled the transformation of floriculture, described in Chap. 8. In both domains, a tacit transition took place, namely, from holistic romantic conceptions to more sober and pragmatic attitudes. This indirectly influenced the ideological status of plants in Veneto in general. They no longer embodied romantic metaphors or symbols of elitism and began to be viewed as simple objects of mass consumption.

Notes

1. When, in 1844–1846, the chairs of mathematics were separated from philosophy and conjoined in an autonomous faculty, the natural history lessons were held in the Philosophical faculty, and Rural Economy lessons (for future surveyors) in the Mathematical faculty (Fumian 1996, p. 85–87).
2. Selvatico (1842, p. 231): "municipali per l'applicazione, ma italiane per 'azione morale."

3. Selvatico (1842, p. 230): "menti di alto pensare vanno preparando questo sacro pane del popolare insegnamento."
4. Cavalli (1853, p. 6): "quella bella vicenda di reciproci soccorsi e scambievoli servigi, la quale costituisce la vita civile."
5. In 1830, the enrolment fee of the Reading cabinet had been 30 Lire, and the user charge was 6 Lire every two months.
6. Cittadella Vigodarzere (1851, 1, p. vii–viii): "Imperciocchè noi stimiamo falsa, ingiusta e dannosa l'opinione di quelli, che vogliono affliggere l'ordine e la quiete sociale all'ilotismo delle infime classi, e vogliono condannati perpetuamente I rustici ad essere poco più che un materiale istrumento agrario."
7. Cittadella Vigodarzere (1856, 1, p. 21): "Un proprietario di terre, che intende ad ammigliorare campi ed agricoltori; che suscita la fertilità in quelli, mentre spande su questi benevolenza, istruzione, soccorsi [...] oh! cotesto proprietario è una qualche imagine della Providenza divina."
8. Cittadella Vigodarzere (1851, p. xiii–xiv): "Noi, popolo essenzialmente agrario, non distolga dalla cura dei campi l'avidità di altri più larghi e più rapidi, ma meno sicuri, profitti; perché la prestigiosa ricchezza che viene dall'ardimento di macchinose manifatture e dal rischio di ampli commerci, patisce la frequente vicenda dei sùbiti e miserandi soqquadri."
9. Cittadella Vigodarzere (1851, p. xi–xii): "la più utile delle arti [...] una lunga esperienza incotta dai raggi del sole, macerata dalle piogge, invecchiata fra le nebbie, le brine ed i venti."
10. Cavalli (1853, p. 13): "L'agricoltura è scuola di filosofia morale, mercecchè essa faccia tuttogiorno toccare con mano la varietà delle sollecitudini necessarie al ben vivere".
11. Cavalli (1851, p. 173): "La terra è il laboratorio massimo delle ricchezze; ma la terra abbandonata a sé medesima o è del tutto sterile, o è poco fruttuosa. Affinché essa dispieghi l'immenso apparato de' suoi tesori è necessaria la mano fecondatrice dell'uomo, che ne svolga ed anzi ne accresca la forza produttiva."
12. Tommaseo (1870, p. 24): "più che il terreno, egli intendeva migliorare la condizione de' suoi contadini: e, facendosi del diritto un dovere, tutti li conosceva a uno a uno, e li trattava con quella semplicità dignitosa, con quelle maniere affettuosamente signorili, per cui si fece meritevole d'esser chiamato *decoro della nobiltà, perla della nobiltà*, [...] e giustamente notato come la ricchezza, usata così, possa invero assomigliarsi a *fonte* viva, senza seccarsi, distribuita in *rigagnoli*, i quali se tutti, non avendo un capo, fossero fili d'acqua tenui, si perderebbero inutile nella sabbia" (his emphases).
13. Selvatico (1842, p. 233): "da povera pianticella crescerà presto in albero robustissimo".

14. Cittadella Vigodarzere (1843, p. xl): "Un giardino botanico [...] simiglia un impero composto da molti e diversi popoli [...] impero lieto e leggiadro, che se teme le ire del cielo, non è però mai sovverso da ribellioni, turbato da guerre; e cresce ogni dì sotto lo sguardo amorevole del naturalista, che n'è il monarca."
15. Anon. (1847r, p. 337): "si formasse anche nelle Venete Provincie una sola famiglia, la quale poi avesse a tendere fraterna mano alle altre sorelle italiane."
16. Alessandro Porro's brother was the botanist and patriot Carlo Porro (1813–1848), who was accidentally shot as a hostage of general Radetzky.
17. In Padua 286 copies were sold and in the Paduan province 21. Other major locations with associates were Venice (55), Vicenza (54), Verona (24), Treviso (23), and Rovigo (21). Lombard readers bought 20 copies. Another 40 copies were exchanged with other journals.
18. Anon. (1846, p. 2): "Noi vogliamo che i nostri Agricoltori raccolgano abbondantemente del frumento, del grano-turco, del riso, delle uve, dei bozzoli, del canape; non che si perdano in futili esperimenti."
19. Cittadella Vigodarzere (1843, p. xxvi): "la maggior perfezione dello stato sociale collegarsi strettamente al progredimento delle scienze fisiche e matematiche."
20. Anon. (1847d): "certamente avremo speciale amore per quelli che più montano nella sociale importanza. Ma non per questo oblieremo o ci faremo ostacolo a trattare i minori."
21. Tommaseo (1847, p. 119): "Verrà tempo (io non dubito), tempo lontano di secoli quando non i fiori soltanto e l'erbe tutte che la terra porta, serviranno per qualche maniera alla vita, ma e ciascuna di quelle che il mare asconde, daranno all'umanità il suo tributo."
22. The first article of the statues declared, "The aim of the Congresses of the lovers of natural sciences is the benefit of progress, the diffusion of these sciences, and their useful application" (Anon. 1840, p. lii): "Il fine delle Riunioni dei cultori delle scienze naturali si è di giovare ai progressi, ed alla diffusione di tali scienze, e delle loro utili applicazioni."
23. Anon. (1847r, p. 338): "Per tal modo la scienza viene chiamata dalle speculazioni alla pratica, dal gabinetto, dalla cattedra al campo, all'officina: così l'industre operosità si giova del sapere, che cospira con quella al benessere generale."
24. I thank Luca Ciancio for his help in translating the jingle.
25. See also the letter from Hugo von Mohl to Giuseppe Meneghini of January, 1846. Biblioteca di Scienze naturali e ambientali di Pisa (BSNAP) Fondo Meneghini.
26. Anon. (1847i, p. 85): "L'autorità dei grandi nomi non può abbagliare alcuno nel 1847. In quanto a noi francamente dichiariamo, che la maniera

di esperimenti proposti e narrati dall'illustre scrittore non può condurre ad alcun decisivo risultamento. E ciò perché tanti metodi praticati con tante varietà di pomi di terra e di altre piante di diversa famiglia, come l'Helianthus tuberosus, in venti tavole di terreno non possono che produrre inestricabile confusione. Ottimo è il consiglio di contribuire con tutti i mezzi, e quanto più si possa, ad allontanare il terribile flagello da una coltivazione che merita di essere animata ed estesa. Ma uno di questi mezzi è, a nostro parere, il lasciar parlare quelli soltanto che siano almeno iniziati nelle scienze, le quali hanno diretta relazione coll'argomento trattato."

References

Ambrosoli, Mauro. 1998. Alberate imperiali per le strade d'Italia: La politica dei vegetali di Napoleone. *Quaderni storici* 33 (3): 707–738.

Anon. 1840. Regolamento generale per le annuali riunioni italiane dei cultori delle scienze naturali. In *Atti della Prima Riunione degli scienziati tenuta in Pisa nell'ottobre del 1839*, LII–LIV. Pisa: Tipografia Nistri.

———. 1842a. *Guida di Padova e della sua provincia*. Padova: Coi Tipi del Seminario.

———. 1842b. Progetto preliminare al regolamento d'una Società d'incoraggiamento agricola che la Commissione nominata dalla Sezione d'agronomia presenta ai membri della Riunione per le sottoscrizioni. *Giornale agrario Lombardo-Veneto e continuazione degli Annali di Agricoltura* 18 (2): 234–235.

———. 1844. Elenco alfabetico degli associati al Giornale Euganeo nel primo anno 1844. *Giornale Euganeo* 1: appendix, 1–8.

———. 1845. Protokoll aus der Commissionssitzung zur Berathung über die dermalige Kartoffelkrankheit. *Allgemeine Medicinische Central-Zeitung* 14: 641–644.

———. 1846. Manifesto. *Il Tornaconto* s.d. [October 1846].

———. 1847a. *Seconda esposizione di piante e fiori tenuta dalla Società promotrice del giardinaggio nell'I. R. Orto botanico di Padova*. Padova: Crescini.

———. 1847b. *Atti della Ottava Riunione degli scienziati tenuta in Genova dal XIV al XXIX settembre 1845*. Genova: Tipografia Ferrando.

———. 1847c. *Diario del Nono Congresso degli scienziati convocati in Venezia nel settembre MDCCCXLVII*. Venezia: Co' Tipi di Giovanni Cecchini.

———. 1847d. Ai signori agronomi, industri, commercianti, economi, ingegneri e publici amministratori delle provincie venete. *Il Tornaconto* 1: 8.

———. 1847e. Prologo ai nostri articoli di agricoltura. *Il Tornaconto* 2: 9–11.

———. 1847f. Gestione della Società d'incoraggiamento nella provincia di Padova a tutto 1846. *Il Tornaconto* 3: 22.

———. 1847h. Aspiri ai premj che la Società d'incoraggiamento nella provincia di Padova metteva a public concorso co' suoi programmi 7 marzo 1846. *Il Tornaconto* 9: 65–67.

———. 1847i. Appendice all'articolo sulla malattia delle patate del numero precedente. *Il Tornaconto* 11: 85.

———. 1847j. Piano di una Società di mutuo soccorso tra i possidenti. *Il Tornaconto* 12: 97–98.

———. 1847k. Sull'intervento alle adunanze comunali e sociali. *Il Tornaconto* 17: 136–137.

———. 1847r. Sulle associazioni agrarie. *Il Tornaconto* 41: 337–340.

———. 1848. Istituto medico-chirurgico-farmaceutico di mutuo soccorso in Padova. *Il Tornaconto* 2: 13–14.

Arduino, Pietro. 1766. *Memorie di osservazioni e di sperienze sopra la coltura, e gli usi di varie piante che servono, o che servir possono utilmente alla tintura, all'economia ecc.* Padova: Stamperia del Seminario.

Arduino, Luigi. 1807. *Catalogo primo delle piante che si coltivano nel R. Orto di agricoltura di Padova nonché di quelle che vi crescono spontanee, a cui si aggiunge l'elenco delle opere si stampate che inedite di Pietro Arduino [...] e del dottore Luigi di lui figlio [...].* Padova: Fratelli Penada.

Autiero, Carlo. 2006. *Un percorso semiotico nel parco romantico jappelliano.* Roma: Meltemi editore.

Barton, Ruth. 2003. 'Men of science': language, identity and professionalization in the Mid-Victorian scientific community. *History of Science* 41: 73–119.

Bassi, Agostino. 1846. *Discorsi sulla natura e cura della pellagra, sulla malattia contagiosa che attaccò l'anno scorso ed attacca tuttora in diversi stati d'Europa i pomi di terra, e come si possa arrestarla.* Milano: Giuseppe Chiusi.

Berengo, Marino. 1963. *L'agricoltura veneta dalla caduta della Repubblica all'Unità.* Milano: Banca Commerciale Italiana.

Bergamasca, Evelina. 2004. L'associazionismo operaio a Monselice: il dualismo tra i Figli del Lavoro e la Società operaia. In *Spazi laici: Strutture e reti associative tra Ottocento e Novecento*, eds. Renato Camurri and Marco Fincardi, 159–180. Venetica: Rivista di storia contemporanea 17: 1–221.

Bernardello, Adolfo. 1997. *Veneti sotto l'Austria: ceti popolari e tensioni sociali (1840–1866).* Verona: Cierre.

Berti, Giampietro. 2007. Censura e cultura nel Veneto austriaco. In *Potere e circolazione delle idee: Stampa, accademie e censura nel Risorgimento italiano*, ed. Domenico Maria Bruni, 237–247. Milano: Franco Angeli.

Bevilacqua, Piero ed. 1989–1991. *Storia dell'agricoltura italiana in età contemporanea.* 3 volumes. Venezia: Marsilio.

Biadene, Giovanni. 1996. *Storia della patata in Italia dagli scritti dei Georgici (1625–1900).* Bologna: Avenue media.

Biagioli, Giuliana. 2004. "Agricoltura come manifattura": Le condizioni per lo sviluppo agricolo. In *Agricoltura come manifattura: Istruzione agraria, professionalizzazione e sviluppo agricolo nell'Ottocento*, ed. Giuliana Biagioli and Rossano Pazzagli, 63–80. Firenze: Olschki.

Biagioli, Giuliana, and Rossano Pazzagli, eds. 2004. *Agricoltura come manifattura: Istruzione agraria, professionalizzazione e sviluppo agricolo nell'Ottocento*. Firenze: Olschki.

Bourke, Austin. 1991. Potato blight in Europe in 1845: the scientific controversy. In *Phytophtora*, ed. J.A. Lucas, R.C. Shattock, D.S. Shaw, and Louise R. Cooke, 12–24. Cambridge: Cambridge University Press.

Brunello, Piero. 1975. Mediazione culturale e orientamenti politici nel clero veneto intorno al 1848: il 'Giornale dei parrochi ed altri sacerdoti'. *Archivio veneto* 106: 139–186.

Busatto, Maria Teresa. 2015. *Luigi Configliachi. L'uomo, le opere*. Padova: Il Torchio.

Callegari, Marco. 2012. *Stampatori e librai a Padova nella prima metà dell'Ottocento*. Saonara: Il Prato.

Canavari, Mario. 1889. *Commemorazione di Giuseppe Meneghini fatta nell'aula magna dell'Università Pisana ai XXIV Marzo MDCCCLXXXIX*. Pisa: Tipografia T. Nistri.

Casellato, Sandra. 1996. Luigi Configliachi. In *Professori di materie scientifiche all'Università di Padova nell'Ottocento*, ed. Sandra Casellato and Luisa Pigatto, 73–75. Trieste: Edizioni Lint.

Casetta, Pietro ed. 2012. *Cenni scientifici del Viaggio fatto l'autunno 1834 nell'Ungheria inferiore, nella Gallizia, nella Slesia e nella Moravia: manoscritto XVII B 30 della Biblioteca del Museo Nazionale di Praga*. Padova: s.e.

Cavalli, Ferdinando. 1851. Studi economici sulle condizioni naturali e civili della provincia di Padova. In *Scritti raccolti e pubblicati dalla Società d'Incoraggiamento per la provincia di Padova*, vol. 1, 1–189. Padova: Tip. Sicca.

———. 1853. Allocuzione. In *Atti della Società d'incoraggiamento per l'agricoltura e l'industria in Padova: Atti della seconda distribuzione de' premj eseguita dalla Società d'incoraggiamento per l'agricoltura e l'industria in Padova, il giorno 12 giugno 1853 nella Gran sala della Ragione*, vol. 2, 5–23. Padova: Co' tipi di A. Sicca.

Celetti, David. 2008. *Il bosco nelle provincie venete dall'Unità ad oggi: Strutture e dinamiche economiche in età contemporanea*. Padova: CLEUP.

Cittadella Vigodarzere, Andrea. 1843. Discorso letto nell'adunanza del 15 settembre 1842. In *Atti della Quarta Riunione degli scienziati tenuta in Padova nel settembre del 1842*, xxv–xlix. Padova: Co' Tipi del Seminario.

———. 1846. Discorso. *Atti delle adunanze dell'I. R. Istituto veneto di Scienze, Lettere ed Arti* 5: 456–457.

———. 1851. Ragione del libro. In *Scritti raccolti e pubblicati dalla Società d'Incoraggiamento per la provincia di Padova*, vol. 1, v–xiv. Padova: Tip. Sicca.

———. 1856. Antonio Cittadella Vigodarzere. In *Memorie funebri antiche e recenti raccolte dall'ab. Gaetano Sorgato*, vol. 1, 18–22. Padova: Coi Tipi del Seminario.

Clementi, Giuseppe. 1845. La chimica applicata all'agricoltura ed alla fisiologia di Giusto Liebig, tradotto dal sig. dott. G. Netwald sulla quinta edizione tedesca, dall'autore diffusa ed ampliata (Vienna 1844). *Giornale euganeo* 2 (1): 413–433.

———. 1857. Sertulum orientale seu recensio plantarum in Olympo bithynico, in agro byzantino et hellenico nonnullisque aliis orientis regionibus, annis 1849–1850. *Memorie della R. Accademia delle Scienze di Torino s* 2 (16): 239–338.

Configliachi, Luigi. 1831. Dissertazione inedita sull'agricoltura del territorio padovano. In *Per le faustissime nozze Zara-Piazza*, 5–38. Padova: Tipografia del Seminario.

Cremaschi, Sergio. 2020. The Italian Enlightenment and the rehabilitation of moral and political philosophy. *The European Legacy* 25: 743–759. https://doi.org/10.1080/10848770.2020.1758411.

Dal Cin, Valentina. 2019. *L'élite veneta fra rivoluzione e restaurazione (1797–1815)*. Venezia: Edizioni Ca' Foscari.

Dandolo, Vincenzo. 1817. *La coltivazione dei pomi di terra considerata nei suoi rapporti colla nostra agricoltura, col benessere delle famiglie coloniche, dei possidenti e dello Stato*. Milano: Sonzogno.

Ferrari, Maria Luisa. 2012. *"Quies inquieta": Agricoltura e industria in una piazzaforte dell'Impero asburgico*. Milano: Franco Angeli.

Fumian, Carlo. 1984. Proprietari, imprenditori, agronomi. In *Storia d'Italia: Le regioni dall'Unità a oggi. Il Veneto*, ed. Silvio Lanaro, 99–162. Torino: Giulio Einaudi editore.

———. 1996. *Possidenti: Le élites agrarie tra Otto e Novecento*. Catanzaro: Meridiana Libri.

———. 2004. Il 'senno della nazione': I congressi degli scienziati e la cultura agronomica. In *Agricoltura come manifattura: Istruzione agraria, professionalizzazione e sviluppo agricolo nell'Ottocento*, ed. Giuliana Biagioli and Rossano Pazzagli, 203–251. Firenze: Olschki.

Gentilcore, David. 2012. *Italy and the potato: a history, 1550–2000*. London: Continuum.

George, Samantha. 2005. The cultivation of the female mind: enlightened growth, luxuriant decay and botanical analogy in eighteenth-century texts. *History of European Ideas* 31 (2): 209–223.

Gloria, Andrea. 1927. *Il Comitato Provvisorio Dipartimentale di Padova dal 25 marzo al 13 giugno 1848*. Padova: Tip. del Messaggiero.

Goeth, Sarah. 2017. Fortpflanzung—Verpflanzung: Vegetabile Metaphern in den Wissenskulturen des 18. und 19. Jahrhunderts. In *Metaphorologien der Exploration und Dynamik (1800/1900)*, ed. Gunhild Berg, Martina King, and Reto Rössler, 61–82. Hamburg: Felix Meiner Verlag.

Gullino, Giuseppe. 1996. *L'Istituto Veneto di Scienze, Lettere ed Arti: Dalla rifondazione alla seconda guerra mondiale (1838–1946)*. Venezia: Istituto Veneto di Scienze, Lettere ed Arti.

Klemun, Marianne, ed. 2016. *Einheit und Vielfalt: Franz Ungers (1800–1870) Konzepte der Naturforschung im internationalen Kontext*. Wien: Vienna University Press.

Lanaro, Silvio. 1984. Genealogia di un modello. In *Storia d'Italia: Le regioni dall'Unità a oggi. Il Veneto*, ed. Silvio Lanaro, 5–96. Torino: UTET.

Lazzarini, Antonio. 2013. I boschi del Veneto prima dell'Unità. *Archivio veneto* VI/5: 7–18.

———. 2014–2018. Boschi, legnami, costruzioni navali: L'Arsenale di Venezia fra XVI e XVIII secolo. *Archivio veneto* VI/7: 111–175, and VI/15: 85–153.

Lenoir, Timothy. 1992. *Politik im Tempel der Wissenschaft: Forschung und Machtausübung im deutschen Kaiserreich*. Frankfurt a.M.: Campus Verlag.

Luzzatto, Fabio. 1928. Prime linee di politica agraria negli scrittori veneti d'agricoltura del secolo XVIII. *Giornale degli Economisti e Rivista di Statistica* 68 (7): 537–578.

Lyttleton, Adrian. 1979. Landlords, peasants and the limits of liberalism. In *Gramsci and Italy's passive revolution*, ed. John A. Davis, 104–135. London: Routledge.

Maccabelli, Terenzio. 2000. La Società d'incoraggiamento di Padova e l'associazione per il progresso degli studi economici (1846–1878). In *Associazionismo economico e diffusione dell'economia politica nell'Italia dell'Ottocento: Dalle società economico-agrarie alle associazioni di economisti*, ed. Massimo M. Augello and Marco E.L. Guidi, vol. 2, 299–328. Milano: Franco Angeli.

Manganotti, Antonio. 1876. Della vita e degli studi del prof. Giuseppe cav. Clementi socio corrispondente dell'Accademia d'Agricoltura, Arti e Commercio di Verona. Memoria letta nella Seduta pubblica del giorno 19 di Agosto 1876. *Memorie dell'Accademia d'Agricoltura Arti e Commercio di Verona* 54: 379–410.

Manin, Daniele, ed. 1852. *Carte segrete e Atti ufficiali della polizia austriaca in Italia dal 4 giugno 1814 al 22 marzo 1848. 3 volumes*. Torino: Libreria Patria.

Marino, Giuseppe Carlo. 1974. *La formazione dello spirito borghese in Italia*. Firenze: La Nuova Italia.

Meneghini, Giuseppe. 1844–1846. Lezioni di Botanica Popolare. *Giornale Euganeo* 1: 314–320 and 394–402; 2(1): 197–207 and 407–412; 2(2): 31–39, 266–273, and 301–311; 3(1): 251–259; 3(2): 194–208, 432–450, and 497–519.

———. 1847b. Sulla malattia delle patate. *Il Tornaconto* 10: 73–77.
———. 1847c. Sulla malattia delle patate. *L'Economista* 5 (1): 129–135.
———. 1847d. Istruzioni popolari intorno alla malattia delle patate. *Il Tornaconto* 36: 293–294.
Meneghini, Andrea. 1851. *Elementi di economia sociale ad uso del popolo*. Torino: Cugini Pompa e comp.
———. 1863. *Le imposte nella Venezia e nella Lombardia*. Torino: Unione tipografico-editrice.
———. 1865. *Sulla condizione finanziaria delle province italiane tuttora soggette all'Austria*. Torino: Stamperia dell'Unione tipografica editrice.
Meriggi, Marco. 1987. Il Regno Lombardo-Veneto. In *Storia d'Italia*, ed. Giuseppe Galasso, 18(2). Torino: UTET.
Miotto, Onelio. 1942. *Andrea Meneghini: patriota inedito del nostro Risorgimento (1806–1870)*. Padova: Cedam.
Mogavero, Valeria. 2017. La Guerra del dopoguerra: Il Veneto consegnato ai notabili. *Venetica: Rivista di storia contemporanea* 52 (1): 39–67.
Molesti, Romano. 2006. *Economisti e accademici nel Settecento veneto: Una visione organica dell'economia*. Milano: Franco Angeli.
Montagne, Camille. 1845. Observations sur la maladie des pommes de terre. *Bulletin de la Société Philomatique de Paris* 13: 312–314.
Monteleone, Giulio. 1969. La carestia del 1816–1817 nelle province venete. *Archivio Veneto* 86: 23–86.
Morrell, J.B. 1990. Professionalisation. In *Companion to the history of modern science*, ed. Robert C. Olby, Geoffrey N. Cantor, John R.R. Christie, and Jonathan S. Hodge, 980–989. London: Routledge.
Nunn, Nathan, and Nancy Qian. 2011. The potato's contribution to population and urbanization: evidence from a historical experiment. *Quarterly Journal of Economics* 126 (2): 593–650.
Paping, Richard, Eric Vanhaute, and Cormac Ó. Gráda, eds. 2006. *When the potato failed: causes and effects of the last European subsistence crisis, 1845–1850*. Turnhout: Brepols.
Preto, Paolo. 1982. Andrea Cittadella Vigodarzere. In *Dizionario Biografico degli Italiani*, vol. 26, 62–63. Roma: Treccani.
Ragon, Michel. 1984. *Les Mouchoirs Rouges de Cholet*. Paris: Albin Michel.
Reynolds, Berthal D., William J. Blackmon, E. Wickremesinhe, M.H. Wells, and R.J. Constantin. 1990. Domestication of Apios americana. In *Advances in new crops*, ed. J. Janick and J.E. Simon, 436–442. Portland: Timber Press.
Rheinberger, Hans-Jörg, and Michael Hagner, eds. 1993. *Die Experimentalisierung des Lebens: Experimentalsysteme in den biologischen Wissenschaften 1859/1950*. Berlin: Akademie Verlag.
Salvagnini, Antonio Augusto. 1841. *Statistica della città e provincia di Padova*. Padova: Tipi di Angelo Sicca.

Schnitter, Joachim. 2011. *Anguis in herba: Gartenpädagogik und Weltveredelung im Leben des schwedischen Agitators Olof Eneroth*. Hamburg: Disserta Verlag.
Scolari, G.L. 1846. Biografie dei presidenti generali dei seguiti congressi. *Il Mondo Illustrato: Giornale illustrato* specimen copy (19 settembre 1846): 3–5.
Selvatico, Pietro. 1842. L'opportunità delle società d'incoraggiamento per l'agricoltura. *Giornale agrario Lombardo-Veneto e continuazione degli Annali di Agricoltura* 18 (2): 230–233.
Solinas, Giovanni. 2008. *Verona e il Veneto nel Risorgimento*. Verona: West Press Editrice.
Solitro, Giuseppe. 1930. *La 'Società di cultura e di incoraggiamento in Padova nel suo primo centenario: Un secolo di vita padovana*. Padova: Società di cultura e di incoraggiamento.
Soppelsa, Maria Laura. 2001. Immagini della cultura scientifica veneta nei Congressi degli scienziati italiani di Padova (1842) e Venezia (1847). In *Dopo la Serenissima: Società, amministrazione e cultura nell'Ottocento veneto*, ed. Donatella Calabi, 233–270. Venezia: Istituto Veneto di Scienze, Lettere ed Arti.
Tenzer, Heike. 2017. "Die wilde harbkesche Baumzucht": Zum Zusammenspiel von Botanik und Ästhetik am Beispiel der Parkanlagen von Harbke und Ostrau. *Annals of the History and Philosophy of Biology* 22: 123–136.
Thaer, Albrecht. 1816–1818. *Principij ragionati d'agricoltura*. Napoli: Tipografia di G. B. Seguin.
———. 1818–1819. *Principij ragionati d'agricoltura*. Florence: Stamperia Piatti.
Tommaseo, Niccolò. 1847. *Intorno a cose dalmatiche e triestine: Scritti*. Trieste: L. Papsch & C.
———. 1870. *Il conte Andrea Cittadella Vigodarzere: Memoria*. Padova: P. Prosperini.
Trautmann, Leopold. 1810–1811. *Versuch einer wissenschaftlichen Anleitung zum Studium der Landwirthschaftslehre*. Vol. 2 vols. Wien: Beck.
Trautmann, Leopoldo. 1820–1821. *Elementi di economia rurale. Prima traduzione italiana dall'originale tedesco con annotazioni dei signori professori ab. Luigi Configliachi e Giuseppe Moretti*. 3 vols. Pavia: tipografia di Pietro Bizzoni.
Treviranus, Ludolf Christian. 1806. *Vom inwendigen Bau der Gewächse und von der Saftbewegung in denselben*. Göttingen: Heinrich Dieterich.
———. 1822. *Memoria sopra dei vasi e del sugo organizzatore dei vegetabili, traduzione del Professore Luigi Configliachi*. Padova: Tipografia del Seminario.
Tusset, Gianfranco. 2000. Paternalismo ed economia all'Accademia Galileiana di scienze, lettere ed arti di Padova (1866–1876). In *Associazionismo economico e diffusione dell'economia politica nell'Italia dell'Ottocento: Dalle società economico-agrarie alle associazioni di economisti*, ed. Massimo M. Augello and Marco E.L. Guidi, vol. 2, 95–116. Milano: Franco Angeli.
Ventura, Angelo. 1989. *Padova*. Bari: Edizioni Laterza.

Vienne, Florence. 2017. Worlds conflicting: the cell theories of François-Vincent Raspail and Theodor Schwann. *Historical Studies in the Natural Sciences* 47 (5): 629–652.

Visconti, Agnese. 2013. Il trasferimento delle piante nella Lombardia austriaca negli ultimi decenni della dominazione asburgica. *Altre modernità. Rivista di studi letterari e culturali* 10: 39–51.

Zadoks, J.C. 2008. The potato murrain on the European continent and the revolutions of 1848. *Potato Research* 51 (1): 5–45.

Zanetti, Pier Giovanni. 1992. L'orto agrario e la cattedra di agricoltura dell'Università di Padova. *Padova e il suo territorio* 35: 16–20.

———. 1996. L'orto agrario di Padova e l'agricoltura nuova. *Rivista di Storia dell'agricoltura* 36 (1): 5–68.

Zangrando, Mario. 2017. Rivoluzioni, costellazioni, cospirazioni: Enrico Nestore Legnazzi tra Risorgimento e post Risorgimento. *Venetica: Rivista di storia contemporanea* 52 (1): 13–38.

CHAPTER 10

Revolutions and Their Failures

What happened in Padua during the spring of 1848 cannot properly be called a revolution, not even an upheaval. Except during its short and unexpected initial and final phases, the short-lived Paduan revolution did not involve physical violence or noteworthy military conflicts. Nevertheless, the withdrawal of the Austrian troops and the investiture of Andrea Meneghini and his peers as the city's governing body represented a profound break in the city's history. Moreover, the unrest in February can been seen as a prologue to the revolutions in Lombardy-Veneto (Ventura 1989, p. 48). Padua's place in the European landscape of 1848 revolutions and on Italy's path to national unity is far too complex to be satisfactorily discussed in a chapter of this book. My aim is to concentrate on how these events and their aftermath spelled an end to the peculiar interplay between plants and politics in Padua as well as to the dream of the Meneghini brothers to actively shape the city's civil society. The various actors we encountered in the foregoing chapters played rather different roles in this process. For moderate progressivists like Andrea and Giuseppe Meneghini, the advent of a participatory form of government probably appeared as a logical consequence of a historical development, a conception that may have also guided Andrea Cittadella Vigodarzere's involvement, which was, however, also moved by the desire to curb potentially

A. Dröscher, *Plants and Politics in Padua During the Age of Revolution, 1820–1848*, Palgrave Studies in the History of Science and Technology, https://doi.org/10.1007/978-3-030-85343-3_10

dangerous developments and to preserve, as much as possible, the existing political and social setting. Giuseppe Bernardi represents the collaboration between 'revolutionists' and the liberal clergy, and Achille De Zigno's role was that of a mediator between the Paduan and the Austrian actors, before, during, and after the revolt.

10.1 Padua and the European Appeal

Earlier surveys of the events of 1848 have stressed the diversity and the local dimensions of the uprisings that broke out in many parts of the continent. In the last two decades, historians have instead again focused on the European character and on the importance of these movements for the European identity (Haupt and Langwiesche 2008; Sperber 2005). Padua's revolution has been somewhat neglected by international as well as by Italian scholarship. Some even refuse to call it an insurrection (Laven 2001; Ventura 2001). Indeed, it did not see military battle or heroic sieges like in Milan, Venice, Brescia, or Vicenza. On the other hand, it represented an important pawn in the broader scene of insurrections in Veneto. Moreover, Padua was certainly a minor political event, yet it displayed a considerable level of organization and involvement by different parts of society. In contrast to the creed of the Paduan Giacobins, a small number of mainly aristocrats who transformed the ideals of the French Revolution into a municipalist and aristocratic-conservative seclusiveness (Ventura 1989, p. 38–39), the 1848 revolution showed democratic and participatory tendencies. Padua exhibits a series of peculiarities with respect to the European, Italian, and even Veneto landscape, but likewise has many features in common: temporal simultaneity, the concert of many basic aspirations, like democratization, liberalization, and nationalism, and the sense of striving for this together with many of the protagonists across Europe, whether they were republicans, socialists, or moderates (Demantowsky 2007, p. 70–75). Moreover, like in other European countries, the revolution brought along a lasting transformation of the means and modes of communication (Francia 2013, p. 238), and of the venues, persons, and modes of political activity.

The two main peculiarities in the North Italian context were the role of religion and anti-Austrian patriotism. Many Paduan clerics, like Giuseppe Bernardi and Padua's Bishops Francesco Scipione Dondi dall'Orologio and his successor Modesto Farina (1771–1856), cherished liberal ideas. On 16 June 1846, Giovanni Maria Mastai Ferretti (1792–1878) was elected Pope Pius IX. Applauded by Europe's liberals, he instituted an amnesty for political prisoners, carried out a series of reforms, permitted the creation of journals, improved education and scientific institutions, and promoted a customs union among the Italian states. These events were met with enthusiasm and raised the hopes of many reformers and nationalists. In the eyes of many, the Pope became a symbol and a possible head of a unified Italian state. Moreover, it convinced many rather conservative figures to participate in the Risorgimento movement (Francia 2013, p. 12). Whereas most European revolutionaries were anticlerical, in Italy the Pope became the 'glue' that held together, at least temporarily, a politically and socially fragmented society. He united nationalists and internationalists, federalists and centrists, monarchists, constitutional monarchists, republicans and liberals, and fueled insurrectional fervor. At the same time, the prospect of a Pope-King placated the great fear of new Parisian *terreur* among the more conservatives.

The second peculiarity was anti-Austrian patriotism. As we have seen in the preceding chapters, the relationship between Padua's elite and the Habsburg regime was complex. David Laven is probably the key author who admonishes against a one-sided perspective of Italian heroic nationalism against an ever-hated Austrian tyranny (Laven and Parker 2014). In fact, the years between 1837 and 1846 had been the most prosperous of Austria's rule in Veneto (Berengo 1963, p. 82–83). A series of plentiful harvests led to a good general satisfaction and demographic growth. In about half a century, the population of the city of Padua had grown from nearly 42,000 to over 50,000, the number of university students from about 500 to over 2000, and the military barracks from one to five, with a total of about 4000 soldiers, 1000 veterans, and 100 officers in 1848 (Del Negro 2000). Under the Austrian domination, Veneto's road network had been considerably improved and expanded. The first section of the railroad line (Padua—Marghera) was inaugurated on 13 December 1842, and soon connected to the Milan-Venice line. Roads were repaired and new ones constructed. Public gas lighting was introduced in July 1847.

Regardless of these improvements, those who wanted change in the economic and political situation soon made out that the main enemy was in Vienna. The idea of overcoming the 'Habsburg rule' became a goal that unified a heterogeneous group of activists, and it hid general discord among them about the aims of the revolution as well as the means to achieve them (Soldani 2008). Ultimately, the growing influence of organicist thinking also contributed to the increasing desire for autonomy. According to the state-as-organism analogy, the state was a natural entity, whose parts, like those of a living organism, had common genealogical origins (Chap. 6). The state thus ought to be founded on a historically consolidated group of persons, a *Volk*. Supporters of this view tended to regard the Austrian government as 'other', that is, as an alien body that exerted an artificial, unnatural power. Paduan naturalists, on the other hand, as seen in the previous chapters, had cultivated much stronger bonds with Austrian colleagues than the artists, poets, and scholars of the humanities. Whereas the latter often extolled the imaginary future nation as a natural community founded on remote claims of kinship and territory (Banti 2000, p. 73),[1] the vision of the former was more moderate. Flowers and gardens had performed a bridging function, too, and, finally, the natural sciences were not the exclusive domain of the revolutionaries. Even Metternich himself practiced them (Kadletz-Schöffel 1992). The fact that Padua's elite was more drawn to the naturalistic than the artistic currents may be one reason why the relationship between Padua and Vienna remained ambivalent throughout the period treated in this book.

10.2 "Meneghini for President!"

The year 1846 was a turning point in the Habsburg Veneto relationship. In 1845, after many relatively good years, the harvest was poor. The next year, it failed almost completely and this re-exposed all the weaknesses of Veneto's economy (Chap. 9). Some reforms were happened, yet the situation soon became critical. Even if the harvest of 1847 was abundant, the price of bread and polenta remained high and hit the peasant masses who still lived in great poverty. Diseases of deficiency like pellagra spread over the countryside. The situation was further exacerbated, when the Austrian government increased taxes. Due to internal quarrels and enormous debt, the Austrian government was unable to respond adequately to the crisis, leading to discontentment among the peasants, who joined—although not across the board—the 1848 revolts.[2] In addition, the trading and banking crises of late 1847 provoked resentment among the Venetian merchants and businessmen and disaffection with the Habsburg rule (Ginsborg 1974, p. 195–197). Those who hoped for political change promptly politicized the ensuing uprisings. Andrea Meneghini was one of them. He preferred legal battles and the written word to military confrontation. On 4 January 1848, in the wake of Daniele Manin and Giovanni Battista Nazzari's (1791–1871) complaints issued in the two Central Congregations in Venice and Milan to urge the Austrian authorities to respect previous agreements and to alleviate the situation of the peasants, Andrea started a similar appeal in the Provincial Congregation of Padua. Without the endorsement of other Paduan *notabili* and even defying all their attempts to bypass his motion, he took the occasion to openly declare the desire of the Lombardy-Veneto people to become an autonomous nation (Ferrari 1958, p. 195–197; Tomasoni 1870, p. 7). In line with his collaborative approach, he availed himself of all his networks and tried to initiate actions in all of the provinces.

Meanwhile, antagonism also grew among the urban population, leading to sporadic acts of outright hostility and culminating in the dramatic February events. The epicenter was the University and the *Caffè Pedrocchi* (Fig. 2.1, I in Chap. 2; Fig. 10.1), where notables, students, and people of lower rank met. The University accounted for more than 2000 of the city's 50,500 inhabitants. Although it had lost its status as an international center of science, it still attracted many students from outside, above all from the provinces of Veneto and from Lombardy. Paduan students were

Fig. 10.1 The Caffè Pedrocchi in Padua in 1842. Lithograph by G. B. Cecchini. From Anon. (1842, p. 262). By courtesy of Biblioteca Capitolare di Padova, Coll. 700.AA5.7

notorious for their unruly behavior. Fights and violence were frequent (Del Negro 2002; Del Negro 2019; Laven 2001). Therefore, the Austrian authorities were slow to recognize the nationalist dimension of the uprising. Subsequent accounts, however, emphasize the increasing invoke of nationalist and sometimes socialist mottos and growing confrontation with Austrian soldiers, with whom, some months earlier, they had still drunk together in the taverns. On 8 February 1848, the most violent unrest erupted. It was initiated by students, but backed by a good part of the urban population (Del Negro 2003; Legnazzi 1892; Zangrando 2017). Five Austrian soldiers and two students died and about one hundred were injured. Yet, the fight was uneven and fast, and the Austrians reestablished order that same evening. Padua's moderates helped to calm the mood, yet Andrea Meneghini openly accused the Austrian regime of being responsible for the deaths.

As a consequence of the unrest, seventy three students and four professors were expelled from the university. The fights, however, created a sense of communality between students, professors, and the urban population, even if this was short-lived. The day after, Andrea Meneghini was arrested together with Guglielmo Stefani, editor of the journals *Giornale Euganeo* and *Il Caffè Pedrocchi*, and they were imprisoned, together with the leaders of Venice's insurrection, Niccolò Tommaseo and Daniele Manin, in the lagoon city. Andrea Cittadella Vigodarzere and Carlo Leoni travelled to Venice and implored the authorities not to order an execution (Meneghini 1848, p. 5; Preto 1982, p. 62–63). But Austria had more serious problems at home. Upon the revolution in Vienna and Prince Metternich's resignation (13 March), all four were freed (17 March). Andrea's return to Padua was a triumph that was celebrated by the whole city (Miotto 1942, p. 46).

Padua's liberation came about non-violently a few days later. On 18 March, the Austrian Emperor proclaimed a constitutional government for Lombardy and Veneto. On 22 March, the Austrian troops left Venice, triggering a domino effect in the other cities. Two days later, the Austrian commander, Field-Marshal-Lieutenant Konstantin Freiherr d'Aspre von Hoobreuck, left his personal belongings in the custody of Achille De Zigno and negotiated a peaceful retreat of his troops from Padua with the counts Andrea Cittadella Vigodarzere, Giovanni Cittadella, and Francesco Papafava (d. 1848). Padua's new municipal administration, however, did not resist the popular desire for a more radical renewal. On the 25th, De Zigno stepped back from the office of *podestà* and, in this moment of political vacuum, called for the creation of a Civil Guard, to protect the public order, and for a new consultative body. Eminent figures of the local civil society were appointed to this body, among them Andrea Meneghini, Giovanni Cittadella, Giuseppe Jappelli, Andrea Cittadella Vigodarzere, Carlo Conti, and Abbot Giuseppe Bernardi (Leoni 1976, p. 195n). The creation of the Civil Guard amounted to a similar break with the past. It not only prevented serious harassment of the Imperial troops during their withdrawal but also became an institutional symbol of the new times (Francia 2008). It soon consisted of about 960 men, also admitted non-Paduans and volunteers from the lower classes, and soon took a clear anti-Austrian stance, adopting the name 'National Guard'.

The city now became part of the Republic of San Marco. On 24 April, Padua acceded to the Provisional Government in Venice, and Andrea Meneghini was nominated temporary regent. On the 25th, in an act of freedom, self-determination, and collaborative spirit, he called for elections. Every voter could elect seven candidates. Meneghini won 933 out of about 2000 votes (Magarotto 1848). The other six elected representatives were the lawyer Barnaba Zambelli (1799–1862) (682 votes), Carlo Leoni (638 votes), Carlo Cotta (532 votes), Alessandro Gritti (521 votes), Giovanni Battista Gradenigo (249 votes), and Ferdinando Cavalli (224 votes), the only *possidente*. Andrea was proclaimed president of the *Comitato provvisorio dipartimentale* and thus assumed command of the city. The Bishop of Padua, Modesto Farina, backed the revolution by performing the *Ambrosiano Hymn* in San Marco's Cathedral and blessing the Italian flag. He ordered prayers in the churches and called the peasants to arms. Parish priests indeed were at the head of many volunteer troops (Ginsborg 1974, p. 519–520).

Yet, the joy did not last. The Austrians had not been definitively defeated. Rather, d'Aspre had cleverly assembled the Austrian troops in Verona in order to rescue General Radetzky, who had survived the defeat in Milan, and was thus able to prepare the reconquest of Northern Italy. Veneto's volunteer troops were far too few in number and too poorly armed to resist. The Paduan troops were composed in good part of university students and, in the higher ranks, of university professors from the technical and scientific disciplines (Del Negro 2000). Promises of help were sent by the Pope, the South Italian monarch Ferdinand II and, in particular, the Savoyan king Carlo Alberto (1798–1849), who had 60,000 men and was eager to present himself as Italy's future leader. The ensuing disillusionment was great when these troops were defeated at Cornuda, near Treviso (8th–9th May). The Pope had already withdrawn from the national movement before the battle began. Padua had sent many volunteer fighters to Venice, but the lagoon city refused to come to Padua's defense. In addition, signs of disunity emerged among Padua's political leaders. On 5th May, for instance, new norms were enacted that excluded non-Paduan students and representatives of the popular classes from the National Guard. A few days later, the aristocrats Achille De Zigno, Francesco Gaudio, and Pietro Sinigaglia were named captains of the National Guard, and on 11 June Andrea Cittadella Vigodarzere became its commander. We do not know what role Andrea Meneghini played in these developments. At any rate, the new configuration of the Guard is clear evidence of the end of the, if limited, egalitarian and democratic project of communality between Padua's different social groups.

Andrea's brother Giuseppe Meneghini, too, joined in the action. On 13th May, he was elected lieutenant of the 1st company of the National Guard, and traveled, as a last desperate attempt, together with Giuseppe Clementi in an official mission to the Savoyan king to request his support (Leoni 1976, p. 74).[3] After the Pope had abandoned the national cause, Carlo Alberto di Savoia was Veneto's last hope. In a plebiscite (18 May), 62,259 Paduans voted for accession to the Savoyan kingdom, but in an official letter the provisional government of the Venetian Republic refused to acknowledge the result. The nearby city of Vicenza fell on 12th June. Padua knew it would be the next. When two small battalions of volunteers from Milan and Naples arrived, Andrea Meneghini and his Comitato acclaimed the "union of brothers from all over Italy" (Legrenzi 1848, p. 31), and the people of Padua prepared to resist recapture. But a few hours later, the two battalions decamped toward Venice, and Padua was alone. The defensive walls of the city had not been significantly upgraded since the Renaissance and were not up to withstand an attack by 30,000 Austrian soldiers and 120 cannons. Moreover, Padua's National Guard, commanded by Cittadella Vigodarzere, made it very clearly that it had no intention to participate in a military conflict. Even the local police withdrew.

Several eyewitness accounts and a letter of justification of Andrea Meneghini provide a vivid and multilayered picture of the dramatic events that followed (Dondi dall'Orologio 2011 [1848]; Legrenzi 1848, p. 43–46; Tolomei 2011 [1848]). That same evening the Comitato reinstated the Municipal Congregation and Achille de Zigno as Podestà and passed the powers of the Comitato to the National Guard. When the population learned that Andrea and the Comitato had decided to capitulate, an upheaval broke out, with looting of the rooms of the Comitato. The plundering spread over the whole city and a bullet nearly struck Andrea.[4] The National Guard did not intervene, but succeeded in restoring order the following morning. During that same night, the Meneghini family fled from Padua. On 14th June, d'Aspre and his soldiers re-took the city without shooting a single bullet.

10.3 Broken Dreams

In 1860, Carlo Leoni estimated for Padua more than 4000 emigrants from the city and another 2000 from the province (Leoni 1976, p. 545 and note 14). For the Meneghini brothers, the short revolutionary dream ended in almost the worst possible scenario. They were both still alive, but on the run. They had lost their homes and all their belongings and they had jeopardized their families. They had been abandoned by their 'Italian brothers', by the Pope, by Venice and its revolutionaries, by many of Padua's moderates, and even by Padua's people who accused them of betrayal, cowardice, and even of having looted the Comitato's money (Dondi dall'Orologio 2011, p. 47). They fled together with the Gritti family, Clementi, and the politician and amateur ornithologist count Giovanni Battista Camozzi (1818–1906) first to Ferrara, which was part of the Papal State. When Ferrara was occupied by the Austrians (17th July) they went further south to Bologna, then to Florence and Collina Tonti, near Pistoia. In Florence, they parted company. Andrea and his family went to Ancona, Giuseppe to Pisa, Clementi decided to go to Greece, Camozzi to Milan and then Piedmont, and the Gritti family had obtained the permission to return to Padua, where their palazzo had been seized by d'Aspre, however.

Andrea's odyssey was arduous and tragic (Miotto 1942). Upon the restoration of the Grand Duchy of Tuscany, he, his wife, and his two sons could not remain in Florence. They went to Ancona, but when the town was bombed, they boarded, with great difficulty, an English ship and travelled to the island of Corfu (until 1797 under Venetian rule) and then to Athens. In July 1848, Andrea accepted an invitation from Sebastiano Tecchio (1807–1886), a refugee from Vicenza, to join a politically active group of exiles in Turin. On 12th August 1849, General Radetzky's proclaimed an amnesty for the Lombardo-Veneto emigrates, yet Andrea's was the first name of six Paduans who were expressly excluded (Leoni 1976,

p. 291; Girardi 2017–2018, p. 286). However, he obtained permission to leave his family in Padua, and settled by himself in Turin. In 1857, he was granted permission to return to Padua (Leoni 1976, p. 515), but after the battle and the armistice of Villafranca (July 1859), in which Austria lost Lombardy to the Piedmont, he once again had to flee. He returned to Turin, where he lived absorbed by political and editorial activities, but also in great financial difficulties. This time, he was accompanied by his now twenty-six years old son Agostino.

It was not before the end of 1853 that Giuseppe resumed correspondence with Andrea, and another two years before they once again became intimate. When Andrea's son Agostino died of tuberculosis in May 1862 (Galletto 1992, p. 53), Andrea was inconsolable. This painful period seems to have re-kindled a close relationship between both brothers. Giuseppe sent him 200 francs[5] and wrote a heartfelt letter to alleviate his grief.[6] Professionally, Andrea's situation slowly improved. Together with Tecchio, Alberto Cavalletto (1813–1897), Giovanni Battista Giustinian (1816–1888), later mayor of Venice, and several others, he was a member of the *Comitato politico centrale Veneto* (Central political Venetian committee), a moderate movement that represented the Veneto emigrées and supported the Savoyan king Victor Emanuel II. He also published a series of politico-economic treatises, among these his *Elementi di economia sociale* (Sect. 6.3 in Chap. 6). It took another six years and an alliance with Bismarck in the Austro–Prussian war before Veneto became part of the Kingdom of Italy. In July 1866, Andrea, who had for a short time become deputy in the Italian parliament, returned to Padua for good. In December he was appointed, not without difficulty and polemics (Zangrando 2017, p. 31–32; Mogavero 2017), its first mayor, a position he held until his death on 17th November 1870. His funeral roused the whole city (Leoni 1976, p. 670–671; Tomasoni 1870, p. 17–23).

Giuseppe found a new home in Pisa where he stayed for the rest of his life. Throughout the entire time he was fleeing, he could rely on a network of botanists, colleagues, friends, and even Austrian military personnel who were anything but the friends of Italian revolutionaries. Among them was Baron von Hügel, now the plenipotentiary Austrian minister in Tuscany. Giuseppe had acquainted him during his stay at Vienna, and then met him again as secretary of the Gardening Society and during the Congresses of Italian scientists (Sects. 5.3, 6.2, and 9.4 in Chaps. 5, 6, and 9, respectively). Together with Field-Marshal-Lieutenant Konstantin d'Aspre, who had invaded Tuscany in May 1849 and responded to any resistance with utmost force, he protected Giuseppe from his enemies, who wanted to expel him from Tuscany for being a subversive. In September 1849, Feldmarshall-Lieutenant Franz von Wimpffen, Giuseppe's successor as owner of villa Meneghini, signed his passport and Hügel personally guaranteed for his safety as he travelled to Padua to visit his family during the summer months (Corsi 2001, p. 920). In 1854, general Radetzky rescinded the confiscation of the properties of Giuseppe and 160 other exiles, yet there was nothing left to restore to the Meneghinis (Leoni 1976, p. 493; Girardi 2017–2018, p. 177 and 289). However, Giuseppe recovered means when, in 1861, he married the rich Paduan widow Anna Lipparini (Sect. 3.2 in Chap. 3). In 1886, he was named senator of the Italian kingdom.

Other botanists lent further support to Giuseppe during the post-1848 years. The brothers Paolo and Pietro Savi supported, with the help of De Visiani, his nomination for the chair of Geology at the University of Pisa, an appointment that was made on 20 January 1849.[7] The chair had become vacant when Leopoldo Pilla (1805–1848), a volunteer in a battalion of Tuscan and Neapolitan university students and professors, was killed during the battle of Curtatone and Montanara in Lombardy. Hitherto Meneghini had not distinguished himself as a geologist, apart from some studies of the Venetian soil. But from now on, he focused his scientific activities on paleontology, re-established international connections, mainly with British and French museums, and founded an important school (Ciancio 2013, p. 341–342; Corsi 2008).

Despite enormous disenchantment with the political and their personal situation, Andrea's as well as Giuseppe's fundamental beliefs seem to have endured over time. Andrea's approach became markedly anti-Austrian (e.g., A. Meneghini 1859; A. Meneghini 1863), but continued to confide in the lawful development of history. After returning to Padua, he declared that "even if the reaction has triumphed for some time, by virtue of the natural law [...] this could not have lasted a long time, in particular in Italy" (quoted from Tomasoni 1870, p. 11). Giuseppe, too, continued to develop his evolutionary ideas—no longer with algae and cells, but with fossils. Indeed, he was one of the first in Italy to read and defend Charles Darwin, yet his concept of evolution was still that of a progression actuated by divine providence (Corsi 2001, p. 924). Similar ideological conservatism holds for the other figures we have met in this book. Whilst the revolutionary events profoundly impacted on the personal and professional lives of all members of Padua's networks, it appears that the influence on the development of their political thoughts remained minor.

These beliefs continued to revolve around the same pillars as during the pre-revolutionary period: moderatism, progressivism, social solidarity, paternalism, and Catholicism (Chap. 6).

Nevertheless, the failed revolution and its aftermath left an imprint on Padua and plant studies in the city. The defeat of the Piedmontese troops at Custoza (27th July 1848), the surrender of Venice (22nd August 1849), and the Pope's dissolution of the Roman parliament (26th August 1849) marked the definitive end of the 'Spring of the people' and the beginning of a long period of isolation. Journals were closed, the national science congresses stopped, and communication among the single states prevented. In few other places did the post-1848 repression have such a profound negative impact on the development of science as in Italy. This was particularly consequential for disciplines that were not yet institutionally entrenched, such as plant physiology or cytology. Whereas the University of Vienna was radically reformed and modernized, creating for instance the chair of plant physiology for Franz Unger, who thus was able to develop and teach his cytological knowledge (Dröscher 2016; Klemun 2016, p. 35–36), no such professorship was created in Padua or at any other Italian university.[8] Giuseppe Meneghini would have been best suited for such a position, yet on 13 August 1848, he was officially deposed from all offices he held in Padua. In addition, the many Italian intellectuals forced into exile (Fournier-Finocchiaro 2014) included a considerable number of scientists, in particular those most receptive toward the new theories circulating in Europe.

Giuseppe Meneghini did not resume his botanical and microscopic studies. Nor did Achille De Zigno. In 1848, De Zigno resumed his position as *Podestà*, whereupon he tried to re-establish public order and to placate Austrian anger. He occupied several high positions in the Austro-Venetian administration and participated in Imperial geological projects (Klemun 2012, p. 88). Yet, after Veneto became Italian, his strategy to mitigate the Austrian repressions and taxes through 'friendly cooperation' was harshly attacked as collaborationism. This fate he shared with Cittadella Vigodarzere, who had pursued a similar strategy and received high decorations from the Austrians while his requests for regional autonomy, agrarian reform, and technical and scientific progress went mostly unheard. Both men were named senators of the Italian kingdom, but increasingly withdrew to private study. De Zigno devoted himself to paleontology and geology and created a noteworthy private collection and library that Giuseppe Meneghini consulted with great profit when he stayed in Padua during the summer months.

The other members of Padua's botanical network, except for De Visiani, did not embark on an academic career. Giuseppe Clementi lived in Greece and Turkey for several years. He found a mentor in Giovanni Battista Camozzi, with whom he had fled from Padua. Camozzi financed his herborizing excursions in Montenegro, Western Turkey, and the Near East. Clementi bestowed on him his private herbarium of almost 16,000 species (Andreis 1988). In 1851, he became a schoolteacher first in Genoa and San Remo and finally in Turin. Giovanni Zanardini practiced as a physician in Venice and cultivated his naturalistic studies privately. The more he lost touch with Meneghini, the more he abandoned the cellular approach and focused exclusively on the taxonomy of algae. Zanardini's sole post-1848 essay on cytology even seemed to revert to seventeenth-century fibre theory (Zanardini 1855). Vittore Trevisan assumed several administrative posts in Padua and taught at the local lyceum. In 1853, he retired to his estate and devoted himself to intense naturalist and lichenological studies. After the 1848 revolution, he made no further contribution to cytology but concentrated on a quarrel with Abramo Massalongo about novel classificatory systems for lichenized fungi (Nimis and Hawksworth 1994). Massalongo worked as a teacher of natural history at the lyceum of Verona. In the 1850s he actively participated in the secret circle *La Compagnia dell'Ibis*, or *The Ibis Company*, a kind of democratic counterinstitution to the official Veronese *Accademia di Agricoltura*, which met at the *Caffè Coraini* and dedicated itself, until its closure by the Austrian authorities in 1858, to the divulgation of agricultural and natural science (Curi and Delaini 1999–2000). Among the corresponding members of the Ibis Company were also Giuseppe Meneghini, Roberto De Visiani, and Alberto Parolini's daughter Elisa Parolini Ball (1830–1867), wife of the Irish politician and naturalist John Ball (1818–1889). Interestingly, Massalongo's pseudonym as member of this society was *Reivas*, inspired by the Middle-Iranian legendary ancestor plant of mankind. In the European Middle Ages, reivas was evocated as the tree of life and for Massalongo, as a product from an illusory land, an imagery of freedom and independence (Delaini 2010). Massalongo died in 1860 at the early age of thirty-six (De Visiani 1860–1861). Thus, De Visiani and Trevisan in Padua, Massalongo in Verona, and Zanardini in Venice continued their studies of taxonomy, paleobotany, and cryptogamology at a high scientific level, but the ideas about biological constitution and the promising lines of cytological thought of the 1830s and 1840s, initiated by Meneghini, were not developed further. Therefore, notwithstanding the promising beginnings, plant studies did not see a revolution in Padua, not institutionally, not scientifically.

The situation in the rest of Italy was not much better. At the University of Naples, the largest in Italy, cell theory had become part of some rather original physiological and pathological concepts in the late 1840s. Yet, the violent repression truncated further development in Naples as well, leading to the closure of institutions and removal of many scientists or forcing them into exile. By 1849, cell research had not sufficiently penetrated the Italian institutions to withstand this assault. Schleiden and Schwann's *Zellenlehre* remained almost completely unknown. Thereafter, cell research slowly resumed as late as the 1860s as an import of Virchow's *Cellular pathology,* and then developed mainly in the medical faculties (Dröscher 1996, 2012).

Notes

1. Banti's view has been strongly contested by Laven (2009).
2. Marco Meriggi (2008) maintains that after the 1848 revolution the Austrians succeeded in regaining the goodwill of most peasants by selectively punishing their 'bosses'. Yet poor harvests during the following years compromised this goodwill again.
3. In the editor's note, Clementi is erroneously called Carlo.
4. Antonio Brusoni reported this incident in a letter to his father (Miotto 1942, p. 95).
5. Letter from Andrea to Giuseppe Meneghini of 20 Nov 1860 and 2 Dec 1860—BSNAP.
6. Letter from Andrea to Giuseppe Meneghini of 28 May 1862—BSNAP.
7. Letter from Giuseppe Meneghini to Roberto De Visiani of 1 Nov 1849—BOBP.
8. For a short period, between 1857 and 1862, such a professorship was created in Pavia for the Neapolitan mycologist and expatriate Guglielmo Gasparrini (1803–1866). After that, it took sixty-three years until another university chair for plant physiology was created in Italy (in Rome).

References

Andreis, Carlo. 1988. L'erbario generale (ex Clementi) del Museo di scienze naturali 'E. Caffi'. *Rivista del Museo Civico di Scienze Naturali "E. Caffi"* 13: 227–235.

Anon. 1842. *Guida di Padova e della sua provincia*. Padova: Coi Tipi del Seminario.

Banti, Alberto Mario. 2000. *La nazione del Risorgimento: Parentela, santità e onore alle origini dell'Italia unita*. Torino: Einaudi.

Berengo, Marino. 1963. *L'agricoltura veneta dalla caduta della Repubblica all'Unità*. Milano: Banca Commerciale Italiana.

Ciancio, Luca. 2013. I segni del tempo: Teorie e storie della Terra. In *Il contributo italiano alla storia del pensiero*, ed. Antonio Clericuzio and Saverio Ricci, 332–343. Roma: Istituto dell'Enciclopedia Italiana.

Corsi, Pietro. 2001. La geologia. In *Storia dell'Università di Pisa*, vol. 2 (1737–1861), 889–927. Pisa: Pisa University Press.

Curi, Ettore, and Paolo Delaini. 1999–2000. L'Ibis (1856–1858). *Atti e Memorie della Accademia di Agricoltura Scienze e Lettere di Verona* 176: 37–50.

De Visiani, Roberto. 1860–1861. Commemorazione di Abramo Massalongo (1804–1878). *Atti del R. Istituto Veneto di Scienze, Lettere ed Arti* 19: 91–118.

Del Negro, Piero. 2000. Padova 1848: gli aspetti militari. In *La "primavera liberale" nella terraferma veneta 1848–1849*, ed. Alba Lazzaretto Zanolo, 69–185. Venezia: Marsilio.

———. 2002. La partecipazione degli studenti dell'Università di Padova alla rivoluzione e alla guerra del 1848–1849. In *Universitari italiani nel Risorgimento*, ed. Luigi Pepe, 109–137. Clueb: Bologna.

———. 2003. L'8 febbraio 1848 a Padova: un moto studentesco? *Archivio Veneto* a. 134 v. 160: 63–96.

———. 2019. Il volontariato studentesco padovano del 1848–49. In *L'ateneo di Padova nell'Ottocento: Dall'Impero asburgico al Regno d'Italia*, ed. Filiberto Agostini, 12–34. Milano: Franco Angeli.

Delaini, Paolo. 2010. The strange case of the plant *rivas*. How the Middle-Persian tree of life became protagonist of an episode of Austrian history. In *Ancient and Middle Iranian Studie*, ed. Maria Macuch, Dieter Weber, and Desmond Durkin-Meisterernst, 53–60. Wiesbaden: Harrassowitz Verlag.

Demantowsky, Marko. 2007. Die 1848er Revolution als europäische Veranstaltung. Forschungsstand und didaktische Konsequenzen. In *Europa in historisch–didaktischen Perspektiven*, ed. Bernd Schönemann and Hartmut Voit, 61–79. Idstein: Schulz-Kircher Verlag.

Dondi dall'Orologio, Michele. 2011. Cronaca padovana, 27 marzo—7 ottobre 1848. In … *a notte avanzata si scorgeva il fuoco dei canoni …: Avvenimenti padovani del Quarantotto in una cronaca di Michele Dondi Dall'Orologio*, ed. Pietro Gnam, 37–66. Padova: Biblioteca Universitaria di Padova.

Dröscher, Ariane. 1996. *Die Zellbiologie in Italien im 19. Jahrhundert*. Halle: Deutsche Akademie der Naturforscher Leopoldina.

———. 2012. 'Fallaci sistemi forestieri': i docenti italiani di fronte alla riforma della medicina. In *Le Università e l'Unità d'Italia (1848–1870)*, ed. Alessandra Ferraresi and Elisa Signori, 217–231. Bologna: Clueb.

———. 2016. 'Lassen Sie mich die Pflanzenzelle als geschäftigen Spagiriker betrachten': Franz Ungers Beiträge zur Zellbiologie seiner Zeit. In *Einheit und Vielfalt: Franz Ungers (1800–1870) Konzepte der Naturforschung im internationalen Kontext*, ed. Marianne Klemun, 169–194. Wien: Vienna University Press.

Ferrari, Giorgio E. 1958. L'attitudine di Padova verso Venezia nella crisi veneta del Quarantotto (esordio ad un bilancio bibliografico). In *Miscellanea in onore di Roberto Cessi*, vol. III, 183–234. Roma: Edizioni di Storia e Letteratura.

Fournier-Finocchiaro, Laura. 2014. La nazione degli esuli del Risorgimento. In *"Già troppe volte esuli": Letteratura di frontiera e di esilio*, ed. Novella Di Nunzio and Francesco Ragni, vol. 1, 163–179. Perugia: Università degli Studi di Perugia.

Francia, Enrico. 2008. Difendere la rivoluzione, conservare l'ordine: La Guardia nazionale nell'Ottocento. In *Gli italiani in guerra: Conflitti, identità, memorie dal Risorgimento ai nostri giorni. Vol. 1 (Fare l'Italia: Unità e disunità nel Risorgimento)*, ed. Mario Isnenghi and Eva Cecchinato, 156–163. Turin: UTET.

———. 2013. *1848: La rivoluzione del Risorgimento*. Bologna: Il Mulino.

Galletto, Pietro. 1992. *Galantuomini padovani dell'Ottocento: cenni biografici ed istantanee*. Padova: Libreria Draghi Randi.

Ginsborg, Paul. 1974. Peasants and revolutionaries in Venice and the Veneto, 1848. *The Historical Journal* 17 (3): 503–550.

Girardi, Giacomo. 2017–2018. *I beni degli esuli: I sequestri austriaci in Veneto tra controllo politico e prassi burocratica (1848–1861)* (Dissertation thesis). Milano: Università degli Studi di Milano. Dipartimento di Studi storici.

Haupt, Heinz-Gerhard, and Dieter Langwiesche. 2008. The European revolution of 1848: its political and social reforms, its politics of nationalism, and its short- and long-term consequences. In *Europe in 1848: revolution and reform*, ed. Dieter Dowe, Heinz-Gerhard Haupt, Dieter Langwiesche, and Jonathan Sperber, 1–24. New York: Berghahn.

Kadletz-Schöffel, Hedwig. 1992. *Metternich und die Wissenschaften*, 2 vol. Wien: VWGÖ.

Klemun, Marianne. 2012. National 'consensus' as culture and practice: the geological survey in Vienna and the Habsburg empire (1849–1867). In *The nationalisation of scientific knowledge in the Habsburg empire (1848–1918)*, ed. Mitchell G. Ash and Jan Surman, 83–101. London: Palgrave Macmillan.

———. 2016. Franz Unger (1800–1870): multiperspektivische wissenschaftshistorische Annäherungen. In *Einheit und Vielfalt: Franz Ungers (1800–1870) Konzepte der Naturforschung im internationalen Kontext*, ed. Marianne Klemun, 15–92. Wien: Vienna University Press.

Laven, David. 2001. Disordini studenteschi all'Università di Padova, 1815–1848. In *Studenti, Università, città nella storia padovana*, ed. Francesco Piovan and Luciana Sitran Rea, 489–504. Trieste: Edizioni Lint.

———. 2009. Why patriots wrote and what reactionaries read: reflections on Alberto Banti's *La nazione del Risorgimento. Nations and Nationalism* 15 (3): 419–426.

Laven, David, and Paula Parker. 2014. Foreign rule? Transnational, national, and local perspectives on Venice and Venetia within the multinational empire. *Modern Italy* 19 (1): 5–19.

Legnazzi, Enrico Nestore. 1892. *L'8 febbraio 1848 in Padova: Commemorazione letta nell'Aula Magna della R.* In *Università di Padova l'8 febbraio 1892.* Padova: Fratelli Drucker.

Legrenzi, Domenico. 1848. *Cinque giorni del mese di giugno 1848 in Padova: Cronachetta. Padua: tipi di Angelo Sicca.*

Leoni, Carlo. 1976. *Cronaca segreta de' miei tempi, 1845–1874; con prefazione e note di Giuseppe Toffanin jr.* Quarto d'Altino: Rebellato.

Magarotto, Cesare. 1848. Fatti di Padova. *Il Caffè Pedrocchi* 1 (6 April 1848): 2–3; and 2 (10 April 1848): 7–8.

Meneghini, Andrea. 1848. Lettera a G. Stefani. *Il Caffè Pedrocchi* 1 (6 April 1848): 5.

Meriggi, Marco. 2008. Austriaci ed austriacanti. In *Gli italiani in guerra: Conflitti, identità, memorie dal Risorgimento ai nostri giorni*, ed. Mario Isnenghi and Eva Cecchinato, vol. 1, 226–232. Turin: UTET.

Miotto, Onelio. 1942. *Andrea Meneghini: patriota inedito del nostro Risorgimento (1806–1870).* Padova: Cedam.

Mogavero, Valeria. 2017. La Guerra del dopoguerra: Il Veneto consegnato ai notabili. *Venetica: Rivista di storia contemporanea* 52 (1): 39–67.

Nimis, Pier Luigi, and David L. Hawksworth. 1994. The lichenological activity of Vittore Trevisan earl of San Leon (1818–1897). *Opera Naturalistica Classica* 2: 13–27.

Preto, Paolo. 1982. Andrea Cittadella Vigodarzere. In *Dizionario Biografico degli Italiani*, vol. 26, 62–63. Roma: Treccani.

Soldani, Simonetta. 2008. Approaching Europe in the name of the nation: The Italian revolution, 1846/49. In *Europe in 1848: revolution and reform*, ed. Dieter Dowe, Heinz-Gerhard Haupt, Dieter Langwiesche, and Jonathan Sperber, 59–72. New York: Berghahn.

Sperber, Jonathan. 2005. *The European revolutions, 1848–1851.* 2nd ed. Cambridge: Cambridge University Press.

Tolomei, Giampaolo. 2011 [1848]. *Vera storia dei fatti di Padova nei giorni XII e XIII del giugno 1848 comprovata con documenti.* Padova: Tip. di Antonio Bianchi, 1848. In ... *a notte avanzata si scorgeva il fuoco dei canoni ...: Avvenimenti padovani del Quarantotto in una cronaca di Michele Dondi Dall'Orologio*, ed. Pietro Gnam, 83–109. Padova: Biblioteca Universitaria di Padova.

Tomasoni, Giovanni. 1870. *Andrea Meneghini sindaco di Padova: Commemorazione*. Padova: Tipografia Luigi Penada.
Ventura, Angelo. 1989. *Padova*. Bari: Edizioni Laterza.
———. 2001. L'8 febbraio 1848 nella storia dell'Università di Padova. In *Studenti, Università, città nella storia padovana*, ed. Francesco Piovan and Luciana Sitran Rea, 707–720. Trieste: Edizioni Lint.
Zanardini, Giovanni. 1855. Sulla struttura elementare della cellula organica. *Atti del Reale Istituto Veneto di Scienze, Lettere ed Arti* 2/VI: 303–306.
Zangrando, Mario. 2017. Rivoluzioni, costellazioni, cospirazioni: Enrico Nestore Legnazzi tra Risorgimento e post Risorgimento. *Venetica: Rivista di storia contemporanea* 52 (1): 13–38.
Corsi, Pietro. 2008. *Fossils and Reputations: A Scientific Correspondence: Pisa, Paris, London, 1853–1857*. Pisa: Edizioni Plus/Pisa University Press.

CHAPTER 11

Conclusion

This inquiry into the economic, social, and intellectual world of Padua during the years preceding the 1848 revolution has revealed the existence of a small, but extremely close-knit group of figures that aspired to cultural and then, during the brief revolutionary period, political leadership. Plants played an important role in these circles, also on historical grounds. Ties between the socio-political elite and the realm of botany had always been strong in Veneto; for several reasons, such as a longstanding tradition in the commerce and processing of plants and herbs, the role of vegetal imagery in pastoral dramas and other genres of North Italian literature and art, the embrace of flowers and trees within the passion for collecting, and the domination of crop production in agriculture. Padua, thanks to its university and Botanical Garden, was an intellectual center where all these currents met. In addition, during the first half of the nineteenth century, romantic movements cultivated a particular and strong emotional and symbolic as well as intellectual relationship with plants. These currents found expression in landscape parks and in the cross-fertilization between the underlying conceptions of botany and political philosophy.

By virtue of their magnitude, the events of 1848 left little scope for intellectuals not to become involved in politics in some way. Likewise, the broad uptake of a naturalistic philosophy by the intellectual elites of Veneto as well as their insistence on a role for knowledge and education in

A. Dröscher, *Plants and Politics in Padua During the Age of Revolution, 1820–1848*, Palgrave Studies in the History of Science and Technology, https://doi.org/10.1007/978-3-030-85343-3_11

progress left little scope for politicians to ignore science. The foregoing chapters have thrown into relief three kinds of intersection in the multilayered interplay of botany and politics during the pre-revolutionary period. At the personal level—at first glance, the most striking—numerous scientists were politically active, and many influential figures in politics were involved in scientific and naturalistic enterprises. At the conceptual level, many views emerged contemporaneously within the spheres of naturalistic and of political discourse. Finally, at the policy level, political decisions taken during the reformist period created new opportunities, whereas those associated with the restoration had markedly negative repercussions on the development of botany and science in general in Padua.

A comparison with other European countries shows that many of the socio-cultural transformations and political aspirations we encountered in Padua paralleled those of the age of revolutions in general. The features recognizable in many European settings include socio-economic transformations, the role of journals, a growing desire for democratization and political liberties, attempts to create a new civil society, associationism, and increasing public visibility and political activism by scientists and naturalists. Other features such as the Risorgimento and the influence of the Catholic Church were peculiar to Italy. The prominent role of plants and botanical knowledge in the political events in Padua was not unique. Similar cases of overlap can be found in Tubingen, Berlin, Jena, Paris, Milan, and other cities. Nevertheless, Padua stands out in this regard. In the form of parks, precious plant collections, crops, social symbols, and sources of inspiration, the vegetal world provided a fundamental link between the various personalities considered in this book.

Unsurprisingly, several similar botanical and socio-political concepts, though they were never directly related, emerged contemporaneously at this intellectual interface. Thoughts about the dignity of lower plants followed similar lines as those about the dignity of lower social classes, unitary (cellular) concepts of living beings paralleled those of organicist theories of the state, and ideas about cooperation between bodily parts those about the active cooperation of citizens in the associations. Notions of historicism, evolution, transformism, and progress pervaded political as well as biological debates. Finally, the view that the Divine Providence guided bodily processes had a counterpart in the hope of many that the Pope would guide the process of national unification.

Despite much intellectual cross-fertilization and notable discursive overlap, neither Giuseppe Meneghini nor any of his like-minded peers showed an inclination to explicitly use biological concepts to illustrate or legitimize political visions. Whereas in Germany analogies between the biological and the hoped-for societal constitution and interaction of parts were common among liberals, no such analogies were to be found among their Paduan counterparts. Organic in particular vegetal imagery was frequently used, especially from the mid-1820s, but remained rather vague. The innovative biological concepts of the 1830s and 1840s did not penetrate general political debate. Most of the protagonists of this book entertained similar ideas of sociality and constitution, yet it appears that they were not *au fait* with the latest arguments in plant anatomy and physiology. Even those politicians and political theorists, who, like Andrea Meneghini, had cultivated an interest in botany in their youth and probably had conversed with Giuseppe during the excursions or at the Caffé Pedrocchi about his anatomical-physiological findings, did not allude to the more complex possible cross-references regarding constitution, because, I suggest, they knew that their audience was unlikely to understand them. The fact that Giuseppe Meneghini did not succeed in creating a school of plant physiology in Padua held back the education of members of the socio-political elite during the second half of the century who, like Georges Clemenceau (1841–1929) in France for example, could avail themselves of cutting-edge scientific knowledge.

The use of bio-political metaphors during the period under consideration was even rarer. The popular equation between 'the potatoes' and the Austrians shows that the Italians were not averse to flowery language, especially if it provided a loophole to escape censorship. Regarding the use of bio-political metaphors in the 1830s and 1840s, however, Padua's debate is in line with what happened elsewhere in Europe. Cellular conceptions of living bodies, initially developed mostly in the botanical sciences, and civil conceptions of the state were both still not mature and widespread enough to be fused into a common discourse. After 1848, we have seen a divergence, with many famous metaphors coming up in Britain, France, and Germany, but not in Italy. Censorship may have been one reason. Even if the censors in Padua were not very strict during the years from 1840 to 1847, the fear of being monitored by the Austrian police was ever-present and probably convinced many not to air their political ideas too openly. Another reason was the increasingly positivist understanding of science that demanded a separation of the empirical and philosophical spheres. Finally, the repression during the post-revolutionary period was particularly severe in Italy. It isolated Italy from the rest of

Europe and silenced many young scientists who, like Giuseppe Meneghini, had been in the vanguard of the latest scientific developments.

If not through explicit analogies and metaphors, the foregoing chapters demonstrate, however, that the world of plants did influence intellectual debates and social life in Padua in more subtle ways. The emergence of organicist views was less prominent in Veneto than in other places in Europe, especially in the German lands, but played an important, hitherto neglected part in conceptions of social order. Often, these views went hand in hand with the use of botanical and agricultural tropes. Similarly, historicist and developmental views were commonly assimilated and often combined with images of life history and of seeds, flowers, and fruits. Scholars like Cittadella Vigodarzere, whose lives revolved around their estates and the annual production of crops, tended to favor cyclic visions, whereas scholars of the natural and earth sciences like Giuseppe Meneghini and Achille De Zigno thought in terms of longer stretches of time and tended toward concepts of continuous perfection. It is almost impossible to determine whether naturalistic ideas influenced political ones, or vice versa. Giuseppe Meneghini espoused his biological ideas long before the appearance of his brother's treatise on social economy. On the other hand, in the 1830s, Giuseppe was still too young to have a major intellectual impact on other members of Padua's social elite. Most probably, both the botanical and the political sets of ideas sprang from the same soil contemporaneously.

On the institutional level, the relationship between plants and politics was even closer and produced some peculiar forms of sociality in Padua. Participation in the landscape garden movement was much more than a fashionable trend. For the moderate circles of Veneto, it was a vehicle of group identification and a political manifesto. By creating a landscape garden, the park owners outed themselves, among other things, as promoters of innovation, yet displayed awareness and respect for tradition; as open to European currents, but also defending the roots and cultural importance of (Northern) Italy; and as promoters of liberalism and opponents of absolutism. The Gardening Society and the Festival of Flowers elevated the status of Padua's Botanical Garden from a place of decline to a key site of urban and even regional sociality. Moreover, they fostered a sense of community among the members of moderate circles and accommodated their need of public self-display. At the same time, the society and the festival reflected the complex relationships between Padua's social elite and the Austrian rulers and the social and economic transformations of the time, in particular women and members of the lower bourgeosie entering the public stage. Finally, agrarian associations and networks brought together most of the influential figures of Padua and were the main venues for political discussion and action. Andrea Cittadella Vigodarzere's peculiar agro-botanical social theory evinces the close coupling between all of the intellectual currents and institutional arenas mentioned.

Andrea and Giuseppe Meneghini illustrate all three of these intersections. The focus on their lives and thoughts has given us a deeper understanding of the connection between the two apparently distant areas of political economy and political action, on the one hand, and of botanical science, on the other. It also allowed us to present two individual viewpoints and hence provided a more nuanced picture of the age of revolution in Padua. The biographical approach has helped us to compare the views of both brothers to those of other European debates, and to emphasize the commonalities but also differences in the standpoints of several members of Padua's moderate circle. The concept of cooperation, for instance, set the scientific and the political philosophies of the two Meneghinis apart from those of other members of Padua's elite, in particular from Andrea Cittadella Vigodarzere's paternalistic social theory. Finally, we also found some points of disagreement between Andrea and Giuseppe, for example in the major role attributed to the agency of the individual parts by Andrea.

While the first decades of the nineteenth century were a period of exceptional interaction and cross-fertilization between botany and politics, from the late 1840s, we observe that political interest in plants became increasingly instrumental and utilitarian. The few but distinctive streaks of romanticism in North Italian philosophy gave way to positivistic views. Plants and politics grew apart, resulting in diminished interest in and influence on politics by botanists and, on the other hand, growing scientific and naturalistic illiteracy among politicians (although many of them continued to treat in vague and often outdated scientific ideas in their political declarations).

I had promised to tell a loser's story. Indeed, Andrea Meneghini's plans to reform Padua on a socio-political level as well as Giuseppe Meneghini's ambition to modernize Padua's scientific-institutional settings can be described as failures. In the short run, this harsh judgment holds true, and the personal dramas of the two brothers are exemplary of the shortcomings and miscarriages of the age of revolution in general. Ultimately, however, many of the projects came to fruition, and both brothers contributed to these outcomes. Both achieved their professional ambitions, even if under different circumstances: Andrea as mayor of Padua and Giuseppe as university professor and head of an important school in Pisa. Likewise, Italian independence and unification were attained, and many civil reforms were enacted. What had definitely ended was the particular interplay between plants and politics in Padua.

Index[1]

[1] Note: Page numbers followed by 'n' refer to notes.

A. Dröscher, *Plants and Politics in Padua During the Age of Revolution, 1820–1848*, Palgrave Studies in the History of Science and Technology, https://doi.org/10.1007/978-3-030-85343-3

H

I

J

K

L

The manufacturer's authorised representative in the EU is Springer Nature Customer Service Centre GmbH, Europaplatz 3, 69115 Heidelberg, Germany. If you have any concerns regarding our products, please contact ProductSafety@springernature.com

Printed and bound by CPI Group (UK) Ltd, Croydon, CR0 4YY

30/09/2026

02225492-0005